Groundwater Hydraulics

WATER RESOURCES MONOGRAPH SERIES **9**

Groundwater Hydraulics

Joseph S. Rosenshein and
Gordon D. Bennett, Editors

American Geophysical Union
Washington, D.C.
1984

Water Resources Monograph Series

Groundwater Hydraulics

Joseph S. Rosenshein and Gordon D. Bennett, Editors

2000 Florida Avenue, N.W.
Washington, D.C. 20009

First Printing 1984
Second Printing 1986

Library of Congress Cataloging in Publication Data

Main entry under title:

Groundwater hydraulics.

(American Geophysical Union water resources monograpl
Based on papers from the John Ferris Symposium on Groundwater Hydraulics held at the American Geophysical Union's 1981 spring annual meeting in Baltimore, Md,. sponsored by the Groundwater Committee, American Geophysical Union and the U.S. Committee, International Association of Hydrogeologists.
Includes bibliographies.
1. Hydrogeology--Congresses. 2. Hydraulics--Congresses. I. Rosenshein, J. S. (Joseph S.), 1929- II. Bennett, G. D. (Gordon D.) III. John Ferris Symposium on Groundwater Hydraulics (1981 : Baltimore, Md.) IV. American Geophysical Union. Groundwater Committee. V. International Association of Hydrogeologists. U.S. Committee. VI. Series: Water resou monograph ; 9.
GB1003.2.G77 1983 628.1'14 83-15844
ISBN 0-87590-310-X

Printed in the United States of America

CONTENTS

PREFACE

John Ferris's research has covered a broad range of hydrologic problems in groundwater, including pollution, storage, recharge, and pumping of groundwater; drainage design; hydraulics of aquifer systems; saltwater encroachment; and application of geophysics to groundwater development. These contributions have been made as a researcher and teacher of national and international renown during a career that has lasted more than four decades. During these four decades, the methodologies for application of hydraulics to solving groundwater problems have continued to evolve and improve. The development of aquifer test techniques and analytical solutions been commonplace in the 1950's, 1960's, and early 1970's have increasingly been supplemented by the use of numerical methods and automated parameter estimation techniques.

In the last 5 to 10 years, emphasis in the literature on groundwater hydraulics has been placed chiefly on advancing the hydrologist's ability to simulate complex flow systems and to address the problems of mass and heat transport. Although a strong underlying interest still exists in the groundwater community in aquifer hydraulics as applied to analysis of field data, a relatively small amount of literature has been published on this subject during the period.

This monograph is an outgrowth of the John Ferris Symposium. The symposium addressed the principal areas of major interests and concerns of the theoretician, academician, and applied hydrologist in the field of groundwater hydraulics. The principle subject areas covered by the symposium contributions were aquifer hydraulics, heat and moisture transport, and modeling. The monograph provides good insight into the state of the science of groundwater hydraulics and the state of the art of application of hydraulics to solving groundwater problems. The general areas of interest covered by this monograph will continue to be those of concern to the ground-water hydrologist well into the future.

The John Ferris Symposium on Groundwater Hydraulics was held at the American Geophysical Union's 1981 Spring Annual Meeting in Baltimore, Maryland. The symposium was jointly sponsored by the Groundwater Committee, American Geophysical Union, and the U.S. Committee, International Association of Hydrogeologists. The symposium was held in honor of John's contributions to general application of principles of hydraulics to the solution of ground-water problems and his scientific contribution to hydrology.

The editors of this monograph wish to express their appreciation to P. E. LaMoreaux of P. E. LaMoreaux and Associates, who represented the cosponsoring society, the International Association of Hydrogeologists, and presided over part of the symposium sessions. Acknowledgment is also due the Groundwater Committee, AGU Section of Hydrology, who cosponsored the symposium and provided peer reviews for part of the papers included in the monograph. The editors also wish to express their appreciation to the many peer reviewers who contributed their time to help assure that the content of this monograph meets the high standards set by the American Geophysical Union for its publications.

JOSEPH S. ROSENSHEIN and
GORDON D. BENNETT, Editors

1 INTRODUCTION

The stage was set in 1935 for development of the science of ground water hydraulics as we know it today in the United States. This stage was set by the publication of the paper by C. V. Theis on the relation between the lowering of piezometric surface and the rates and duration of discharge of a well using groundwater storage. This paper was appropriately published in the Transactions of the American Geophysical Union. Publications of the American Geophysical Union continued to serve in the late 1930's through the mid 1950's as a principal outlet for publications on aquifer hydraulics.

In the 25 years that followed the noteworthy publication by Theis, marked progress was made in the theory and application of groundwater hydraulics to addressing a wide range of groundwater problems in the field. Notable contributions were made by C. E. Jacob, S. W. Lohman, J. G. Ferris, R. W. Stallman, M. S. Hantush, M. I. Rorabaugh, and J. F. Poland as well as many other groundwater scientists.

The need to solve groundwater problems on a larger scale while at the same time more effectively taking into consideration the complexities of aquifer systems led to application of resistance-capacitance networks and numerical methods to problems of groundwater hydraulics principally through use of groundwater models. Applications were made at first through use of electric analog models and computers and are currently being made through use of digital computer models. In the last decade, marked advances have been made in the theoretical aspects and the application of theoretical aspects of modeling to problems of groundwater hydraulics.

2 AQUIFER HYDRAULICS

The nine papers composing this chapter address a wide range of hydraulic problems of concern to both the practitioner and the theoretician. The papers reflect the current state of the art of groundwater hydraulics. Theoretical solutions are provided for field problems concerned with optimum location of a well near a stream, prediction of transient movement of solid particles in response to pumping confined aquifer systems, average regional land subsidence equations for artesian aquifers, approximations of drawdown patterns of multiple well systems in layered soils, and unsteady drawdown in the presence of a linear discontinuity. New type curve solutions obtained by numerical inversion of Laplace transform solutions are presented for analysis of pumping test data. The current status of the hydrologist's ability to apply hydraulics to solve groundwater problems in fractured aquifer formations is reviewed and demonstrated by analysis of test data from several pumping tests. In addition, two papers present methods of obtaining aquifer parameters from analysis of field tests: one from tracer-injection tests and the other from slug tests.

Optimum Location of a Well Near a Stream

Edwin P. Weeks
U.S. Geological Survey, Lakewood, Colorado 80225

Charles A. Appel
U.S. Geological Survey, Reston, Virginia 22092

Introduction

Kernodle [1977] points out that the steady state drawdown in a production well completed in an aquifer near a stream with a semipervious bed may be greater than that in an equivalent well located some greater distance away from the stream and that an optimum distance exists at which such drawdown is minimized. This finding is counterintuitive, as one generally assumes that the drawdown in a well nearer a constant head boundary, such as a stream, should always be less than that in a more distance equivalent well. Consequently, this paper expands on Kernodle's [1977] work to prove the existence of the minimum in a mathematical sense and to explore the physical basis behind this nonintuitive result.

In addition to finding the optimum location of a well in such a system in nondimensional terms, this paper considers steady state drawdown in a well in the center of a circular island rimmed by a thin relatively low-permeability layer and examines the flow nets for a well near a straight stream. These analyses indicate that the larger drawdowns for wells near the stream result from the fact that flux through the semipervious bed is relatively large if the well is quite close to the stream, resulting in substantial hydraulic head loss at the semipervious bed-aquifer interface. This hydraulic head loss needs to be added to the hydraulic head loss associated with moving the water through the aquifer to produce total drawdown at the well. As the distance from the well to the stream increases, the hydraulic head loss in the semipervious bed decreases rapidly, while the additional hydraulic head loss needed to drive the water through the aquifer from the semipervious bed- aquifer interface to the well increases more slowly, resulting in a decreased aquifer interface to the well increases more slowly, resulting in a decreased total drawdown. Beyond the optimum distance, on the other hand, the aquifer hydraulic head loss increases more rapidly with increasing distance from the well than the semipervious bed hydraulic head loss decreases.

Semi-Infinite Aquifer Near a Stream With a Semipervious Bed

A solution for groundwater flow to a well in a horizontal unconfined aquifer near a stream separated from the aquifer by a bed of materials having a hydraulic conductivity appreciably less than that of the aquifer has been given by Hantush [1965]. That development uses the Dupuit-Forchheimer assumptions. However, it will suffice for this paper to consider the less general case of a confined aquifer in which the water level is nowhere drawn down below the top of the aquifer. The geometry considered here is depicted in Figure 1.

Letting Hantush's dependent variable Z for the water table case be equal to 2bs, his equation may be expressed as follows:

$$s(x,y,t) = \frac{Q}{4\pi T}\left\{\left[W\left(\frac{r^2S}{4Tt}\right) - W\left(\frac{r'^2S}{4Tt}\right)\right]\right.$$

$$\left. + 4\int_1^\infty \frac{\exp[-\alpha(v-1) - \beta(v^2+\delta^2)]}{v^2+\delta^2}\, v\,dv\right\} \qquad (1)$$

where

s(x,y,t) drawdown at point x, y and time t, L;

Q well discharge, L^3/T;

T transmissivity of the aquifer, equal to Kb L^2/T;

K mean hydraulic conductivity of aquifer materials, L/T;

b aquifer thickness (Figure 1), L;

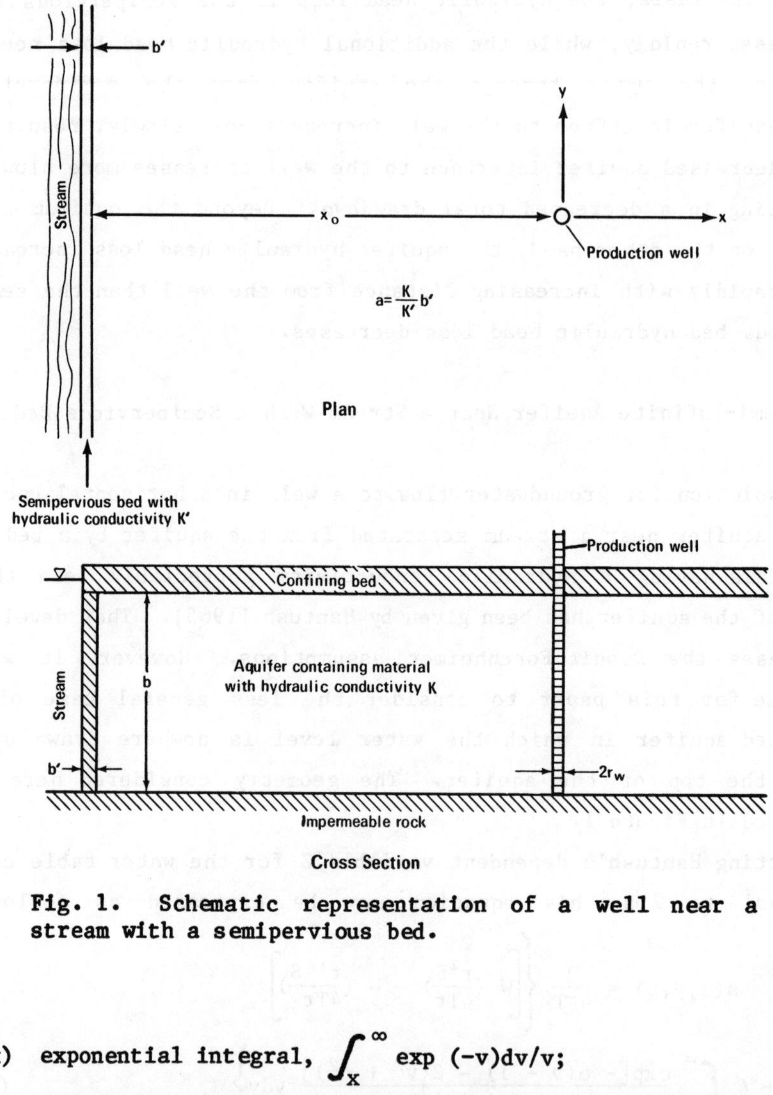

Fig. 1. Schematic representation of a well near a stream with a semipervious bed.

- $W(x)$ exponential integral, $\int_x^\infty \exp(-v)dv/v$;
- r radial distance from production well to observation well, L;
- S storage coefficient of the aquifer, dimensionless;
- t elapsed time since the start of pumping, T;
- $\alpha = (2x_0 + x)/a$, dimensionless;
- x_0 distance from production well to the semipervious bed-aquifer interface (Figure 1), L;
- x distance from production well to observation well in direction perpendicular to stream, as defined in Figure 1, L;

a Hantush's retardation factor Tb'/T', L;

T' transmissivity of semipervious bed, equal to $K'b$, L^2/T;

K' hydraulic conductivity of semipervious bed, L/T;

b' thickness of semipervious bed (Figure 1), L;

v dummy variable of integration;

$\beta = [(2x_0 + x)^2 S]/4Tt$;

$\delta = y/(2x_0 + x)$;

$r' = [(2x_0 + x)^2 + y^2]$, L; and

y distance from production well to observation well in direction of stream (Figure 1), L.

For equilibrium conditions, (1) reduce to

$$s(x,y) = \frac{Q}{2\pi T}\left\{\ln\frac{r'}{r} + 2\int_1^\infty \frac{\exp[-\alpha(v-1)]v\,dv}{v^2+\delta^2}\right\} \qquad (2)$$

Location and Magnitude of the Minimum Drawdown

Existence of a minimum steady state drawdown in a pumped well located at some distance from the stream can be determined by the following analysis. Consider only the drawdown at the pumping well, which is taken to be at ($x = r_w$, $y = 0$), where r_w is the radius of production well, L. If x_0 is very much greater than r_w, (2) is reduced to

$$s(r_w,0) = \frac{O}{2\pi T}\left\{\ln\left(2\frac{x_0}{r_w}\right) + 2\int_1^\infty \frac{\exp[-\alpha(v-1)]}{v}\,dv\right\} \qquad (3)$$

Letting $(v - 1) = p$, note that

$$\int_1^\infty \frac{\exp[-\alpha(v-1)]}{v}\,dv = \int_0^\infty \frac{\exp(-\alpha p)}{1+p}\,dp \qquad (4)$$

which is equal to $e^\alpha W(\alpha)$ [Gautschi and Cahill, 1965, section 5.1, equation 28]. Because $x = r_w$, (3) can be written as

$$s(r_w,0) = \frac{Q}{2\pi T}\left[\ln\left(\frac{2x_o}{r_w}\right) + 2\exp\left(\frac{2x_o + r_w}{a}\right) W\left(\frac{2x_o + r_w}{a}\right)\right] \quad (5)$$

Substituting selected values of (x_o/a) into (5) for fixed values of (r_w/a) indicates that $s(r_w,0)$ does, in fact, decrease away from the streambed to a minimum value of some optimum distance from the stream, and then increases beyond that distance, as shown in Figure 2.

Proceeding more formally, the optimum value of (x_o/a) may be determined for the value of (x_o/a) for which

$$\frac{\partial[s(r_w,0)]}{\partial(x_o/a)} = 0 \quad (6)$$

Taking the appropriate derivative of (5) and setting it to zero can be shown to give

$$\Theta e^{\Theta} W(\Theta) = \tfrac{1}{2}$$

where

$$\Theta = \frac{2x_o + r_w}{a} \quad (7)$$

The solution of (7), as described in the appendix, gives $\Theta = 0.610$. Thus the minimum drawdown in the pumping well will occur when

$$\frac{x_o}{a} = 0.305 - \frac{r_w}{2a} \quad (8)$$

Because $r_w/2a$ is small for most cases of interest, (8) virtually is equal to

$$x_o/a = 0.305 \quad (9)$$

as shown in Figure 2 and as found by Kernodle [1977] in dimensional terms.

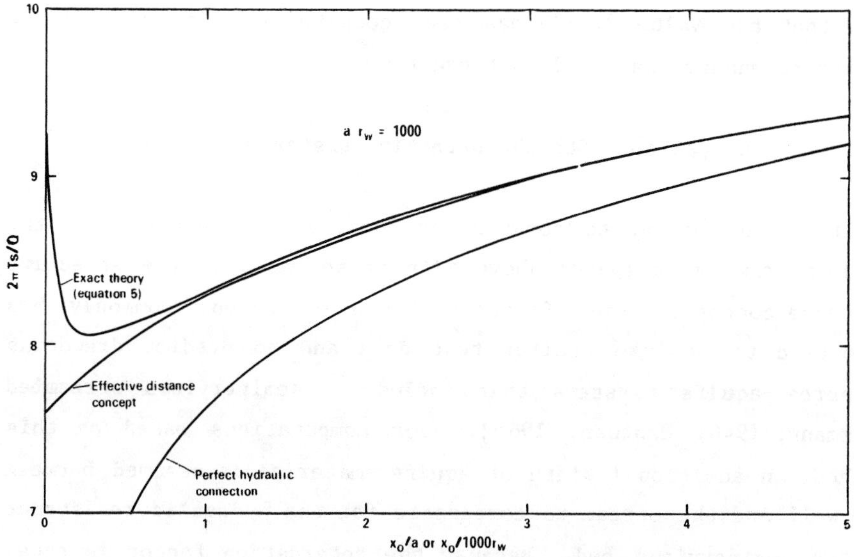

Fig. 2. Dimensionless steady state drawdown in a production well near a stream with a semipervious bed as a function of dimensionless distance from the stream, as computed both from the exact theory of Hantush [1965] and from the effective distance concept. Drawdown in a well near a stream with perfect hydraulic connection is drawn for comparison.

The result of (7) also may be used to evaluate the magnitude of the minimum drawdown as a function of a/r_w. The term $2e^{\Theta}\ W(\Theta)$ in (5) may be evaluated at the location of minimum drawdown by substituting $\Theta = 0.610$ to yield a value of 1.64. Also, at the optimum location, $2x_0 = 0.610a - r_w$. Substitution of this value into the logarithmic expression in (5) yields the expression

$$\frac{2\pi T s_{wm}}{Q} = \ln\left(\frac{0.610a}{r_w}\right) + 1.64 \tag{10}$$

where s_{wm} is the minimum drawdown in the production well (L), and other terms are as previously defined. Equation (10) may be simplified by rewriting it

$$\frac{2\pi T s_{wm}}{Q} = \ln\left[\frac{0.610a\ \exp(1.64)}{r_w}\right] = \ln\left(\frac{3.1425a}{r_w}\right) \tag{11}$$

Note that the value 3.1425 has been computed to five significant digits to ensure that it is not exactly π.

Comparison With the Effective Distance Concept

It is interesting to compare the location and magnitude of the minimum drawdown computed above with those based on the effective distance concept. The effective distance concept commonly has been used to analyze aquifer test data and to predict drawdowns in stream-aquifer systems that include a semipervious streambed [Kazmann, 1946; Hantush, 1965]. For computations based on this method, an additional width of aquifer material is assumed between the well and the stream to compensate for the hydraulic resistance of the semipervious bed. Because the retardation factor is equal to the width of an aquifer strip that has the same hydraulic resistance as the semipervious bed, effective distance may be computed by the equation

$$x_e = x_0 + a \tag{12}$$

where x_e is the effective distance, L. Based on this definition of effective distance, the steady state drawdown in the production well is given, assuming that r_w is small relative to x_0, by the equation

$$s_w = \frac{Q}{2\pi T} \ell n \left[\frac{2(a + x_0)}{r_w} \right] \tag{13}$$

Drawdowns computed by use of this equation are plotted in Figure 2 for comparison with those given by the more exact theory.

Equation (13) indicates that minimum drawdown in the production well would occur if the well were located at the semipervious bed-aquifer interface (x_0 = 0) and would be given by the equation

$$s_{wm} = \frac{Q}{2\pi T} \ell n \left(\frac{2a}{r_w}\right) \tag{14}$$

Thus, the effective distance concept fails to predict that the minimum drawdown occurs at some distance from the semipervious bed-aquifer interface and substantially underpredicts drawdown in the production well if the well is located less than the optimum distance from the stream.

These differences between results for the two methods occur because although it compensates for the effects of hydraulic resistance in the semipervious bed, the effective distance concept allows flow components in the y direction within the semipervious bed. Hantush's equation, on the other hand, assumes that flow is perpendicular to the stream in the semipervious bed, an assumption that has been shown by Neuman and Witherspoon [1969, p. 804] to be valid for an analogous case if the hydraulic conductivity of the aquifer is more than 100 times that of the semipervious bed.

Flow Nets

Flow nets graphically demonstrate the differences between the results of the exact theory and those derived from the effective distance concept and help to explain the minimum drawdown paradox. Flow nets for three different well locations relative to the retardation factor and for a well near a stream in perfect hydraulic connection with the aquifer are shown in Figure 3. The flow net equipotentials for the exact theory were determined from equation (2). The integral in (2) was initially evaluated by a series expansion for values of α and δ, both less than 0.5 and 0.8, respectively, and by Gauss-Laguerre integration, as described by Todd [1954] for larger values of the two terms. However, a subsequent comparison showed that the series expansion and Gauss-Laguerre integration gave identical results to five significant figures (the maximum printed out). Flow lines were computed by evaluation of the equation [Hantush, 1965, equation (13); Todd, 1954, p. 314]

$$\frac{2\pi b\psi}{Q} = \left\{ w - w' - 2\int_0^\infty \exp(-v) \left[\frac{y/a}{\left(\frac{2x_o + x}{a} + v\right)^2 + (y/a)^2} \right] dv \right\} \qquad (15)$$

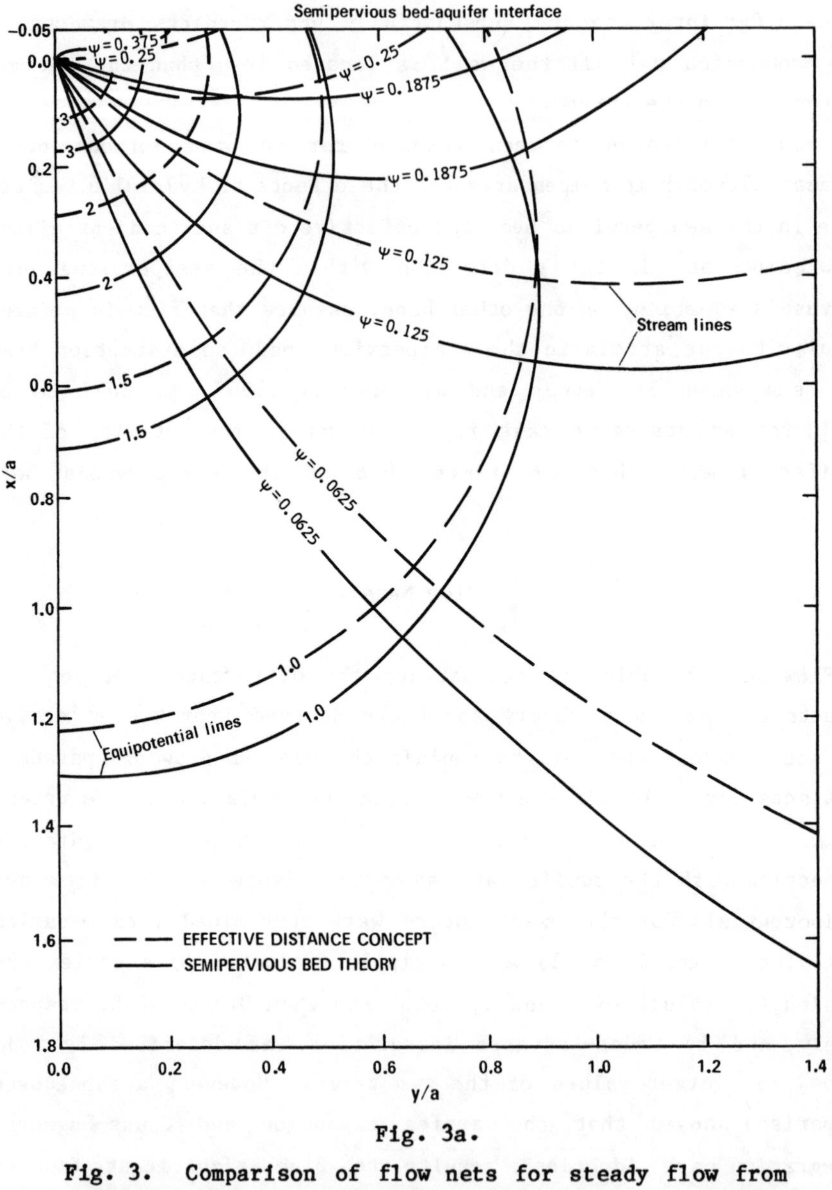

Fig. 3a.

Fig. 3. Comparison of flow nets for steady flow from a stream with a semipervious bed toward a production well, as derived by the exact theory of Hantush [1965] and by the effective distance concept. Production well located at a distance (a) $x_0/a = 0.05$, (b) $x_0/a = 0.305$ (optimal distance, (c) $x_0/a = 1.0$, and (d) $x_0/a = \infty$ (stream in perfect hydraulic connection with the aquifer).

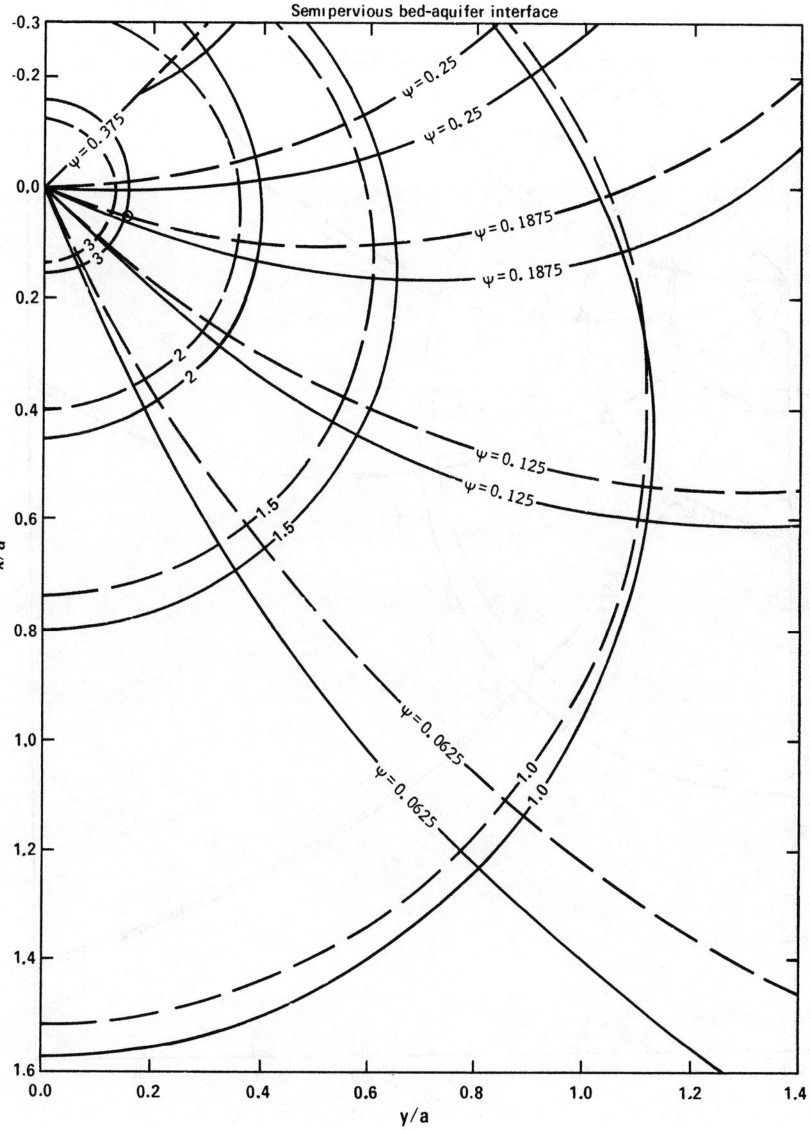
Semipervious bed-aquifer interface
-0.3
-0.2
0.0
0.2
0.4
0.6
0.8
1.0
1.2
1.4
1.6
x/a
0.0
0.2
0.4
0.6
0.8
1.0
1.2
1.4
y/a
ψ=0.375
ψ=0.25
ψ=0.25
ψ=0.1875
ψ=0.1875
ψ=0.125
ψ=0.125
ψ=0.0625
ψ=0.0625
3
3
2
2
1.5
1.5
1.0
1.0

Fig. 3b.

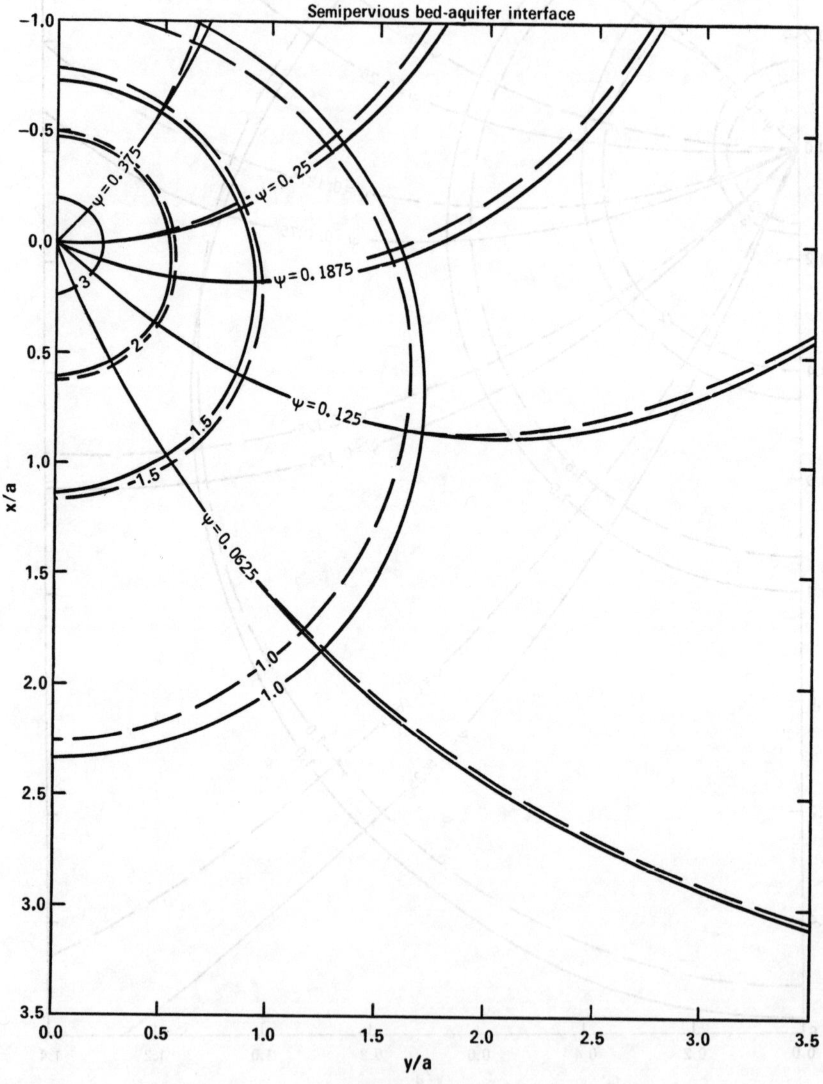

Semipervious bed-aquifer interface
x/a
y/a
ψ=0.375
ψ=0.25
ψ=0.1875
ψ=0.125
ψ=0.0625
3
2
1.5
1.0

Fig. 3c.

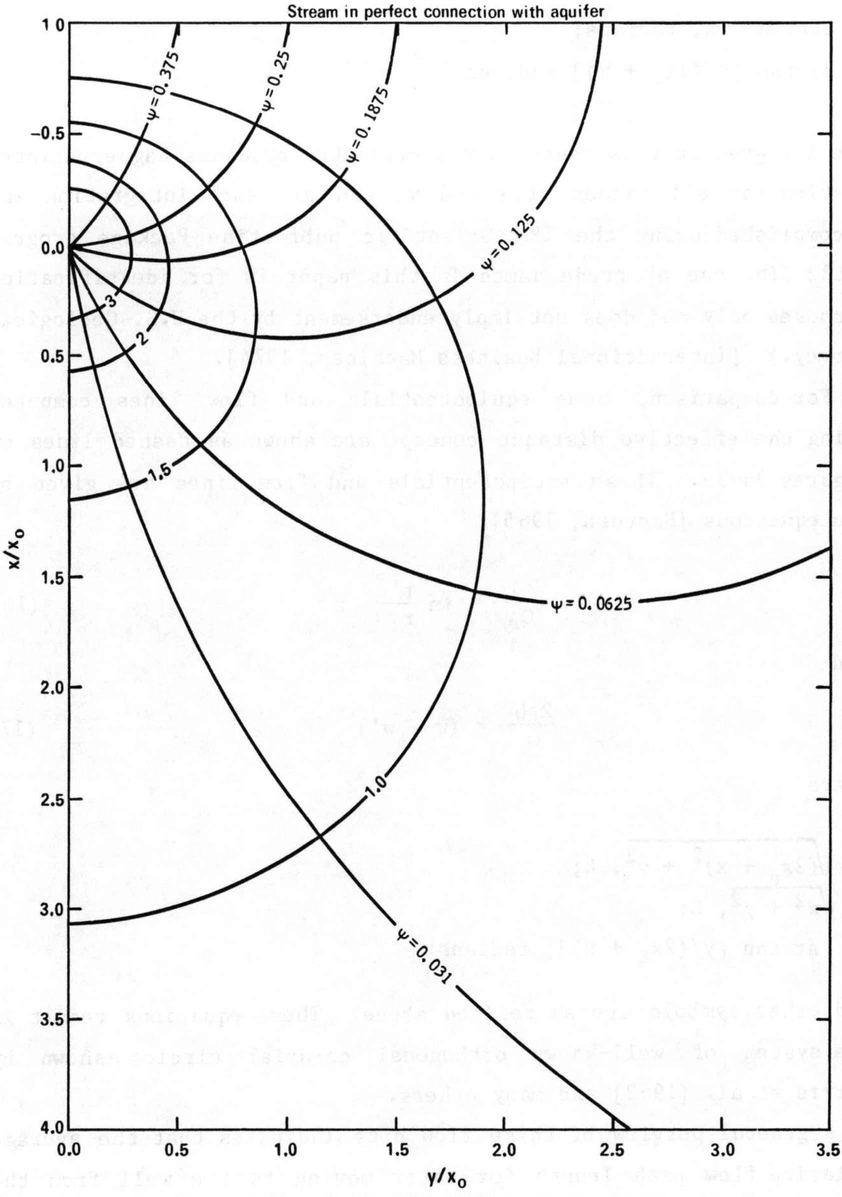
Stream in perfect connection with aquifer
ψ = 0.375
ψ = 0.25
ψ = 0.1875
ψ = 0.125
ψ = 0.0625
ψ = 0.031
3
2
1.5
1.0
x/x₀
y/x₀
1.0
−0.5
0.0
0.5
1.0
1.5
2.0
2.5
3.0
3.5
4.0
0.0
0.5
1.0
1.5
2.0
2.5
3.0
3.5

Fig. 3d.

where

ψ stream function, L^2/T;
w arctan y/x, radians;
w' arctan $[y/(2x_0 + x)]$ radians.

The integral in this equation was evaluated by Gauss-Laguerre integration for all values of x and y, and all such integration was accomplished using the IBM Scientific Subroutine Package program DGL32 (The use of trade names in this paper is for identification purposes only and does not imply endorsement by the U.S. Geological Survey.) [International Business Machines, 1974].

For comparison, some equipotentials and flow lines computed using the effective distance concept are shown as dashed lines on Figures 3a-3c. These equipotentials and flow lines are given by the equations [Hantush, 1965]

$$\frac{2\pi Ts}{Q} = \ell n \frac{r'}{r} \tag{16}$$

and

$$\frac{2\pi b\psi}{Q} = (w - w') \tag{17}$$

where

r' $\sqrt{(2x_e + x)^2 + y^2}$, L;
r $\sqrt{x^2 + y^2}$, L;
w' arctan $[y/(2x_e + x)]$, radians;

and other symbols are as defined above. These equations result in the system of well-known orthogonal co-axial circles shown by Ferris et al. [1962] and many others.

A general purview of these flow nets indicates that the average relative flow path length for water moving to the well from the stream is greatly increased by the presence of the semipervious bed if the well is near the stream. This increase in average flow path length increases the drawdown in the production well beyond that occurring when the stream and aquifer are in perfect hydraulic connection. Thus the drawdown in the production well near the stream is increased both by large hydraulic head losses through the semipervious bed and by the increased average flow path resulting from the fact that more of the flow is induced toward the well from more distance reaches of the stream.

For these flow nets the flow line $\psi = 0.25$ is of particular interest, as it defines the reach within which one fourth of the flow toward the well is induced. For a well completed in an aquifer that is a perfect hydraulic connection with the stream, this flow line (Figure 3d) is represented by a quarter circle that has its center at the stream and passes through the well. Hence, based on symmetry, one half the flow will be derived from a reach of length $\pm x_0$. However, from Figure 3a, it can be seen that for $x_0/a - 0.05$, the $\psi = 0.25$ line intersects this stream at about $14x_0$. Moreover, the flow line rather than entering the well on a tangent parallel to the stream sweeps considerably to the landward of the well before reaching it. Thus more than one half the flow will enter the well from the landward side under these conditions.

The $\psi = 0.25$ line (Figure 3b) in the flow net for a well located at the optimum distance from the stream ($x_0 = 0.305a$) indicates that one half the flow enters the aquifer from the stream within the reach $y = \pm 3.3x_0$. Also, this flow line enters the well along the y axis, to which it is nearly tangent for some distance, indicating that, as for the case where the stream has perfect hydraulic connection, one half the flow enters from the stream side of the well. For this case, flow lines and equipotentials computed using the effective distance concept are not greatly different from those computed using the exact theory, although the $\psi = 0.25$ line plots somewhat nearer the well than it does for the exact theory.

The third flow net (Figure 3c), prepared for a well located at $x_0/a = 1$, indicates that one half the flow enters the aquifer from the stream reach $y \lessapprox \pm 1.7\ x_0$ and that the flow net is quite similar to that for a well near a perfectly hydraulically connected stream, shown in Figure 3d. Also, for this case, the flow net computed using the effective distance concept is almost identical to that based on the more exact theory. Hence, for well locations $x_0 \geq a$, use of the effective distance concept adequately approximates the effects of a semipervious bed.

Percentages of Flow Per Unit Reach

The fact that the flow path length increases dramatically as the value x_0/a decreases to less than the optimum value is demonstrated by the flow nets shown in Figure 3. These effects and their implications also are shown by the percentage of flow derived in the reach $y = \pm\ x_0$, defined here as a unit reach, for various values of x_0/a and the length of reach from which one half the flow is derived (Figure 4). Data for Figure 4 were obtained by evaluating equation (15) for a number of values of x_0/a at $x = x_0$ (the semipervious bed-aquifer interface) and determining the y values at which $\psi = 0.25$. These computations show that for a production well located at $x_0/a = 0.05$, only about 9% of the flow is derived from a unit reach, and the reach from which one half the flow is derived is 14.0 unit reaches long. This compares with a value of 6.4 unit reaches derived by the effective distance method. At the optimum distance, about 23% of the flow is derived from a unit reach, and one half the flow is derived within about 3.3 unit reaches, as compared to 2.75 unit reaches derived from the effective distance method.

The reach length from which one half the flow is obtained also may be compared to the retardation factor a. In this case, x_0 is expressed as a decimal fraction of a, and that fraction is multiplied by N x_0. Thus at $x_0 = 0.05a$, the reach length is 0.7a; at $x_0 = 0.1a$, 0.77a; and at $x_0 = 0.305a$, 1.0a. Similar computations for larger values of x_0/a indicate that the reach length increases slowly but monotonically in terms of the retardation factor and does not have a minimum. Nonetheless, it is interesting that length for one half the diversion is equal to the retardation factor if the production well is located at the optimum distance.

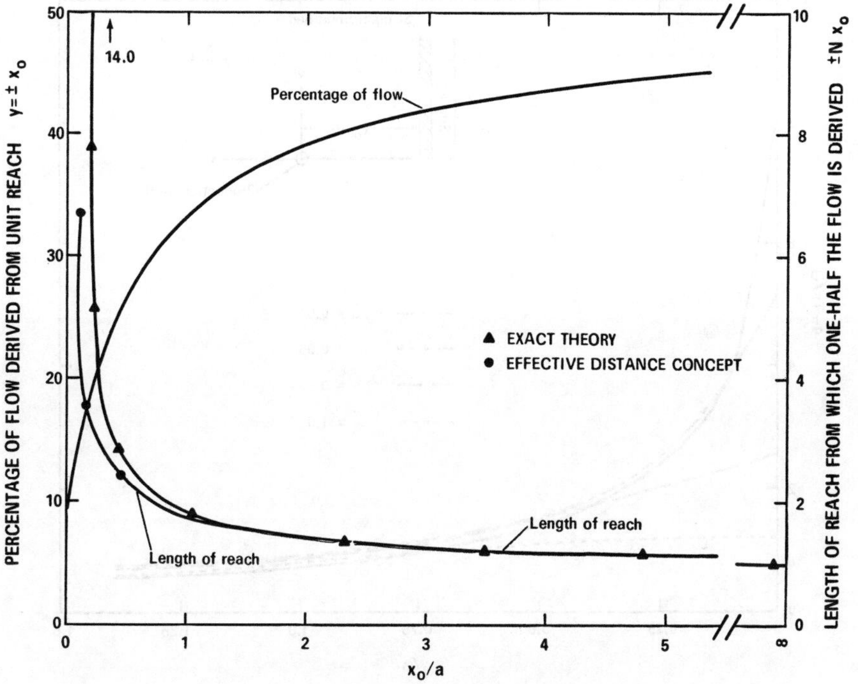

Fig. 4. Fraction of total flow derived from the unit reach ($y = \pm x_0/a$) centered about the production well and the number of unit reaches required to supply one half the flow to a well located near a stream with a semipervious bed.

Drawdown at the Interface

In addition to flow path length, the minimum drawdown paradox may result from large drawdowns at the semipervious bed-aquifer interface, particularly as indicated by the circular symmetry case described below. Consequently, drawdowns along the interface were computed from equation (2). These drawdowns, shown in Figure 5, are extremely large for values of x_0/a less than 0.3, the optimum location, indicating that hydraulic head loss through the semipervious bed is an important factor in accounting for increased drawdown when the production well is located closer to the stream than the optimum distance.

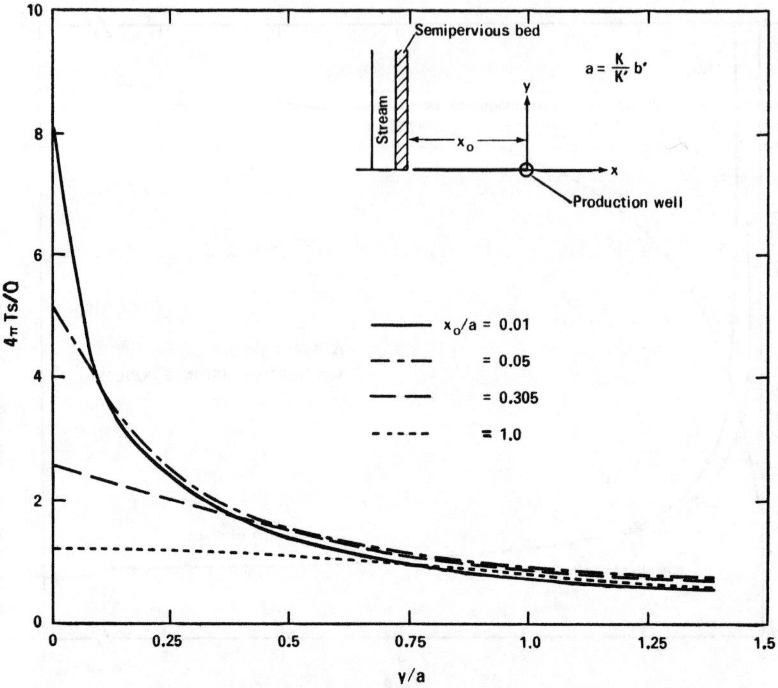

Fig. 5. Plot of dimensionless steady drawdown at the semipervious bed-aquifer interface due to production from a well located at a distance x_0/a from the interface.

Circular Island Surrounded by a Rim of Relatively Low-Permeability Materials

Although the mathematics for the above case clearly indicates that a minimum drawdown occurs if the well is located some distance from the stream, the physics of the problem is not obvious. Consequently, to gain further insight on the physical phenomena involved, the problem of radial flow to a well from a concentric circular, fixed hydraulic head boundary with a relatively low-permeability zone near that boundary (Figure 6) is considered.

The subject radial flow problem can be described by the following:

$$\frac{\partial}{\partial r}\left(\frac{\partial h}{\partial r}\right) = 0, \quad r_w < r < R + W \tag{18}$$

$$T \frac{\partial h}{\partial r}(r = R^-) = T_c \frac{\partial h}{\partial r}(r = R^+) \tag{19}$$

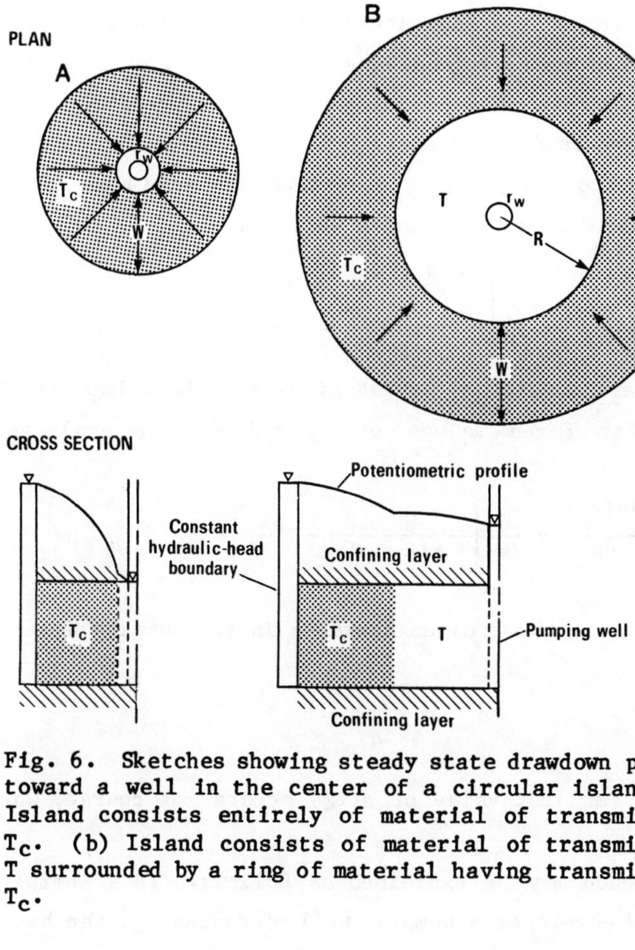

Fig. 6. Sketches showing steady state drawdown profiles toward a well in the center of a circular island. (a) Island consists entirely of material of transmissivity T_c. (b) Island consists of material of transmissivity T surrounded by a ring of material having transmissivity T_c.

where $T > T_c$

$$h(r = R^-) = h(r = R^+) \tag{20}$$

$$\lim_{r \to r_w} r \frac{\partial h}{\partial r} = \frac{-Q}{2\pi T} \tag{21}$$

$$h(r = R + W) = h_o \tag{22}$$

where

R distance from center of well to semipervious zone-aquifer interface, L;

T_c transmissivity of the concentric semiclogged zone, L^2/T;

W fixed width of semipervious zone, L.

Other symbols are as defined above.

The solution to the problem described by equations (18)-(22), in terms of drawdown in the pumping well, is

$$s(r_w) = \frac{Q}{2\pi}\left[\frac{\ln(\frac{R + W}{R})}{T_c} + \frac{\ln(\frac{R}{r_w})}{T}\right] \tag{23}$$

To determine the value of R that gives a minimum drawdown in the pumping well for fixed values of r_w and W, one evaluates the following:

$$\frac{ds(r_w)}{dR} = \frac{Q}{2\pi}\left[\frac{-W}{T_c R(R + W)} + \frac{1}{TR}\right] = 0 \tag{24}$$

that indicates that the minimum drawdown in the pumping well occurs when

$$R = W\left(\frac{T}{T_c} - 1\right) \tag{25}$$

If $T = T_c$, the smallest value of $s(r_w)$ occurs, of course, when $R = r_w$.

This phenomenon may be explained as follows. If a semipervious zone of width W envelopes a pumping well of radius r_w, the hydraulic head difference between the well and the constant hydraulic head boundary is affected by an average flow length W through an average perimeter of length $2\pi(W/2 + r_w)$, as shown by the example in Figure 6a. Next suppose that a thickness $(R - r_w)$ of more transmissive materials separates the well face from the semipervious zone. For small enough values of $(R - r_w)$, most of the hydraulic head loss between the constant hydraulic head boundary and the well occurs through the semipervious zone. Thus, although the flow length has increased to $(W + R - r_w)$, the semipervious zone now has a longer perimeter $[2\pi(W/2 + R)]$, and the result is as though the effective radius of the well had been increased. Thus the hydraulic head difference between the constant hydraulic head boundary and the pumped well is decreased. However, if the width of the intermediate zone ($R - r_w$) is large enough, the fraction of the total hydraulic head loss taking place through that zone would become significant relative to that through the fixed width of semipervious zone, and the drawdown increases with increasing values of ($R - r_w$) beyond the critical value $R = W(T/T_c - 1)$.

The above developments show that the production well drawdown is minimized if the well is located some distance from the semipervious zone-aquifer boundary, both for the circular island and for the semi-infinite aquifer cases. However, the relationship between the two cases is not clear. In particular, the entire flux to the well in the circular island must pass through a fixed length of semipervious material and is uniformly distributed throughout that length. Hence, the minimum drawdown phenomenon for this case may be explained solely by the intensity of flux and hydraulic head decline in the semipervious zone relative to those in the aquifer materials. For the case of the semi-infinite aquifer, however, the effect of locating a well near the stream results both in intensifying the flux through the semipervious bed and in inducing a greater part of the total flow from distances farther upstream and downstream from the well than would be the case if the stream were in perfect hydraulic connection with the aquifer. Thus the greater drawdown in a well near the stream results both from hydraulic head losses in the semipervious bed and from a relative increase in the mean flow path length.

Discussion

The fact that steady state drawdown in a well near a fully penetrating stream with a semipervious bed has a minimum for a location removed from the stream is mainly of academic interest because most the stream is mainly of academic interest because most streams do not fully penetrate the aquifer. However, these results are occasionally applicable to actual situations. Kernodle [1977] presents an actual situation that is approximately described by the theory given above. Other situations would occur when a stream with a semipervious bed flows along one side of an alluvial valley bounded on that side by bedrock of minimal permeability. Under these conditions, the optimum location of the production well would be at some distance from the stream. In addition, diversion of flow from the stream would be spread upstream and downstream substantially by the effects of the semipervious bed. These effects might result in contamination of the well from some point source of pollution located far upstream or downstream from the well. Hence the extent of clogging of the streambed under these circumstances presents implications both in regard to minimizing energy costs by decreasing pumping lift and in locating the production well to minimize pollution hazards.

Analytical expressions for estimating the drawdowns that result from steady state flow to a well near a stream that only partially penetrates the aquifer are developed by Boulton [1942] and Ernst [1979]. Both of them use an approximation to account for the effect of a streambed of relatively low hydraulic conductivity. The authors have not evaluated those analytical expressions to determine whether the drawdown in a pumping well near a partially penetrating stream lined with a bed of relatively low hydraulic conductivity has a minimum away from the river.

Conclusions

The steady state drawdown in a production well located near a fully penetrating stream with a semipervious bed has a minimum if the well is located at a distance equal to 0.305 times the retardation factor and the dimensionless drawdown at that distance depends only on the ratio of the retardation factor to the production well radius. Both these results differ from those derived using the effective distance concept, which predicts that the minimum drawdown would occur at the stream. In fact, the effective distance method inadequately predicts drawdown and the flow net if the production well is located nearer the stream than its optimal distance

(0.305a), but provides an adequate approximation if the well is located at greater distances.

The fact that drawdown in a production well is at a minimum if the well is located some distance from the water source is counter-intuitive and not readily explained in physical terms for the case of a semi-infinite aquifer. However, a similar phenomenon occurs for a production well located at the center of a circular island rimmed by relatively low-permeability materials. Under these conditions, most of the hydraulic head loss occurs through the low-permeability perimeter if the radius of the island is small, and the hydraulic head loss through the perimeter decreases as the length of perimeter increases. At distances less than an optimum radius, the decrease in hydraulic head loss through the perimeter exceeds that through the additional width of aquifer materials. The explanation for the semi-infinite case is less clear but appears to be accounted for by significant hydraulic head loss through the semipervious bed and increased flow path length if the production well is located less than the optimum distance from the stream.

Appendix: Solution of Equation (7)

From Gautschi and Cahill [1965] for $\Theta > 0$ and n = 1,2,3,. . .:

$$\frac{1}{\Theta + n} < e^{\Theta} W_n(\Theta) \leq \frac{1}{\Theta + n - 1} \tag{A1}$$

where $W_n(\Theta)$ are generalized exponential integrals.

From (A1),

$$\frac{\Theta}{\Theta + 1} < \Theta e^{\Theta} W(\Theta) \tag{A2}$$

Note that for $\Theta > 1$, $\Theta/(\Theta + 1) > ½$ and thus $\Theta e^{\Theta} W(\Theta) > ½$. Thus it is necessary only to consider values of $\Theta \leq 1$ in seeking the solution to (7).

Hurr [1966] found that the function $\Theta W(\Theta)$ has a maximum value of 0.28149. From that and the inequality that for $\Theta < 1$,

$$e^{\Theta} < \frac{1}{1 - \Theta} \tag{A3}$$

[Zucker, 1965] it follows that for $\Theta < 1$,

$$\Theta e^{\Theta} W(\Theta) < \frac{\Theta}{1 - \Theta} W(\Theta) \leq \frac{0.28149}{1 - \Theta} \tag{A4}$$

Denote by Θ' a solution of (7). From (A4) it follows that

$$\tfrac{1}{2} < \frac{0.28149}{(1 - \Theta)}$$

or

$$0.43702 < \Theta' \tag{A5}$$

It is necessary to seek solutions for (7) only for values of Θ between about 0.437 and 1.0. A systematic numerical evaluation of the left-hand side of (7) in this range indicates a single solution at about $\Theta = 0.610$.

Notation

a	retardation factor Tb'/T', L.
b	aquifer thickness, L.
b'	semipervious bed thickness, L.
h	hydraulic head in aquifer, L.
h_0	fixed hydraulic head bounding circular aquifer, L.
K	hydraulic conductivity of aquifer materials, L/T.
K'	hydraulic conductivity of semipervious bed materials, L/T.
p	$= v - 1$, dimensionless.
Q	well discharge, L^3/T.
R	radial thickness of aquifer materials for circular island case, L.
r	distance from production well, L.
r_w	radius of production well, L.
r'	$= [(2x_0^2 + x) + y^2]^{\frac{1}{2}}$, L.
S	storage coefficient of aquifer, dimensionless.
s	drawdown in aquifer.
s_w	drawdown in production well, L.

s_{wm} drawdown in production well at optimum location, L.

T transmissivity of aquifer, L^2/T.

T_c transmissivity of circular rim of semipervious materials, L^2/T.

T' transmissivity of semipervious bed, L^2/T.

t elapsed time since start of pumping, T.

v dummy variable of integration, dimensionless.

W fixed width of semipervious zone, L.

$W(x)$ exponential integral of x, dimensionless.

w = arctan y/x, radians.

w' = arctan $[y/(2x_0 + x)]$ or arctan $[y/2x_e + x)]$, radians.

x distance from production well to observation well in direction perpendicular to stream, L.

x_e effective distance from production well to stream, L.

x_0 distance from production well to semipervious bed-aquifer interface, L.

y distance from production well to observation well in direction parallel to stream, L.

α = $(2x_0 + x)/a$, dimensionless.

β = $[(2x_0 + x)^2 S]/4Tt$, dimensionless.

δ = $y/(2x_0 + x)$, dimensionless.

θ = $(2x_0 + r_w)/a$, dimensionless.

ψ stream function, L^2/T.

References

Boulton, N. S., The steady flow of groundwater to a pumped well in the vicinity of a river, Philos. Mag., 33, 34-50, 1942.

Ernst, L. F., Groundwater flow to a deep well near a rectilinear channel, J. Hydrol., 42, 129-146, 1979.

Ferris, J. G., D. B. Knowles, R. H. Brown, and R. W. Stallman, Theory of aquifer tests, U.S. Geol. Surv. Water Supply Pap., 1536-E, 69-174, 1962.

Gautschi, W., and W. F. Cahill, Exponential integral and related function, in Handbook of Mathematical Functions, edited by M. Abramowitz and I.A. Stegun, pp. 227-251, Dover, New York, 1965.

Hantush, M. S., Wells near streams with semipervious beds, J. Geophys. Res., 70(12), 2829-2838, 1965.

Hurr, R. T., A new approach for estimating transmissibility from specific capacity, Water Resour. Res., 2(4), 657-664, 1966.

International Business Machines, Scientific subroutine package, 307 pp., Poughkeepsie, N. Y., 1974.

Kazmann, R. G., Notes on determining the effective distance to a line of recharge, EOS Trans. AGU, 27(6), 854-859, 1946.

Kernodle, J. M., Theoretical drawdown due to simulated pumpage from the Ohio River alluvial aquifer near Silvam, Kentucky, U.S. Geol. Surv. Water Resour. Invest., 77-24, 37 pp., 1977.

Neuman, S. P., and P. A. Witherspoon, Theory of flow in a confined two-aquifer system, Water Resour. Res., 5(4),803-816, 1969.

Todd, J., Evaluation of the exponential integral for large complex arguments, J. Res. Natl. Bur. Stand., 52(6), 313-317, 1954.

Zucker, R., Elementary transcendental functions--Logarithmic, exponential, circular and hyperbolic functions, in Handbook of Mathematical Functions, edited by M. Abramowitz and I. A. Stegun, pp. 65-225, Dover, New York, 1965.

Analysis of Sedimentary Skeletal Deformation in a Confined Aquifer and the Resulting Drawdown

D. C. Helm
Lawrence Livermore National Laboratory
Livermore, California 94550

Introduction

The purpose of this paper is to develop a method for predicting transient three-dimensional movement of solid particles at depth in response to pumping a confined aquifer. Calculating particle movement allows us in turn to find the transient change in fluid pressure mathematically within deforming pore spaces whose change in volume is calculated as an intermediate step.

In order to accomplish this, equations for three-dimensional movement of solids will be derived with emphasis on what underlying assumptions are required. Using these equations, horizontal and vertical components of displacement will be calculated based on specified boundary and initial conditions. These conditions will be analogous to those used in the standard theory of leaky aquifers.

Analytic solutions for the axially symmetric displacement and the resulting strain and change in fluid pressure will be plotted non-dimensionally as type curves. These in turn will be compared to the commonly used theory of leaky aquifers.

The approach or conceptual sequence that will be followed is to start with an equation of motion, then to introduce equations of state, and, as a final step, to use the concept of mass balance. This simple procedure will yield governing equations in terms of the displacement field of solids. We shall then take the time derivative of each term in order to simplify the governing equations which will thereby be expressed in terms of the velocity field of solids.

If, instead of taking the time derivative of each term, we were to take the divergence of each term, the resulting equations would have volume strain as the dependent variable. Expressing the governing equations in terms of volume strain allows us to compare the present approach directly with more standard approaches [Biot, 1941; Mikasa, 1965; Verruijt, 1969; Bear and Pinder, 1978].

The standard approaches follow a different sequence in their mathematical development. Previously, the initial step rather than the final step has been to introduce the concept of mass balance. An equation of motion and equations of state are generally introduced subsequently rather than initially. As a result, one's governing equations are immediately expressed either in terms of volume strain or in terms of a corresponding transient change in fluid pressure.

At no place in the standard developments does the displacement field or velocity field of solids appear directly as an unknown. It is necessary, following the standard sequence, to add a subsequent step. This step is to integrate volume strain with respect to space and to assume in so doing that the displacement field has somehow been found. Transient displacement of reservoir material in a direction of interest, however, is actually the sum of appropriate directional components of volume strain. It is not the integral sum of volume strain itself. To find directional components of strain in three dimensions requires so many parameters and so many rheological assumptions that by using this approach a realistic solution for matrix movement becomes intractable under most field conditions.

The present paper will follow a simpler approach. Namely, we begin with an equation of motion. The significance of the present paper extends beyond developing a method to calculate the magnitude of the displacement of solids. Quantitatively, the cumulative displacement may be small or large. The dynamics of skeletal movement lies at the heart of the mechanics of aquifer systems during transient flow. It is a conceptual starting point for understanding complex drawdown patterns.

Equation of Motion

Our equation of motion is Darcy's law in a form developed for transient flow by Gersevanov [1937], Biot [1941], and many others. It includes the velocity field of solids $\bar{v}_s$ in the relation

$$\bar{q} \equiv \varphi(\bar{v}_w - \bar{v}_s) = -\bar{\bar{K}}\bar{\nabla}h \tag{1}$$

where

- φ average porosity of a specified elemental bulk volume V or surface S;
- $\bar{q}$ specific discharge as defined by the left-hand identity of (1);
- $\bar{v}_s$ average velocity of solids associated with V or S;
- $\bar{v}_w$ average velocity of water associated with V or S;
- $\bar{\bar{K}}$ hydraulic conductivity tensor associated with V or S;
- h hydraulic head.

The right-hand side of (1) is a relative-flow expression of Darcy's law. Figure 1a schematically illustrates the definition of $\bar{q}$ as expressed by the left-hand identity of (1). By simply adding $\bar{v}_s + \bar{\bar{K}}\bar{\nabla}h$ to both sides of the right-hand equality of (1), we find

$$\bar{v}_s = \varphi\bar{v}_w + (1-\varphi)\bar{v}_s + \bar{\bar{K}}\bar{\nabla}h \tag{2}$$

Defining a bulk flux $\bar{q}_b$ for saturated porous flow by

$$\bar{q}_b \equiv \varphi\,\bar{v}_w + (1-\varphi)\bar{v}_s \tag{3}$$

and substituting this definition into (2), we end up with

$$\bar{v}_s = \bar{q}_b + \bar{\bar{K}}\bar{\nabla}h \tag{4}$$

The definition of $\bar{q}_b$ in (3) is illustrated in Figure 1b. In turn, Figure 1c illustrates (4) in terms of Darcy's law (see equation (1)).

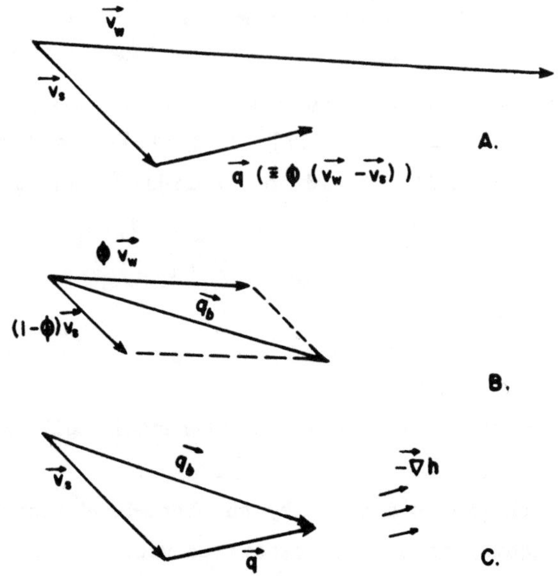

Fig. 1. Vector diagrams for transient flow with non-zero v_s.

Stress

Equation (4) is a fundamental equation of motion for solids based on Darcy's law. We now wish to express h in terms of effective stress $\bar{\bar{\sigma}}'$. First we express h in terms of fluid pressure p [Hubbert, 1940], namely,

$$h = z + p/\rho_w g \tag{5}$$

where

- z elevation of a point of interest in terms of a nondeforming vertical frame of reference (an example of a point of interest is the center of mass of a representative elemental volume V or surface S);
- ρ_w average density of interstitial water associated with V or S;
- p average fluid pressure within interstitial water associated with V or S (in excess of atmospheric pressure);
- g gravitational acceleration at z.

Note in (5) that we use the empirical definition of h. The appropriateness of this when analyzing $\bar{v}_s$ is discussed in detail by Helm [1982, Appendix II]. The principle of effective stress can be written

$$\bar{\bar{\sigma}}' = \bar{\bar{\sigma}} - \bar{\bar{I}}p \tag{6}$$

where

$\bar{\bar{\sigma}}'$ effective stress tensor;
$\bar{\bar{\sigma}}$ total stress tensor;
$\bar{\bar{I}}$ identity matrix.

Eliminating p from (5) and (6) and substituting the result into (4) yields

$$\bar{v}_s = \bar{q}_b + \bar{\bar{K}}\,[\hat{k} + \bar{\nabla}\,(\sigma_{tr} - \sigma_{tr}')/3\rho_w g] \tag{7}$$

where

$\hat{k}$ ($\equiv \bar{\nabla}\, z$) is a unit vertical vector, positive upward;
σ_{tr}' trace of $\bar{\bar{\sigma}}'$ (first invariant);
σ_{tr} trace of $\bar{\bar{\sigma}}$ (first invariant).

Physically, σ_{tr}' and σ_{tr} represent the sum of the orthogonal normal components of effective stress and total stress. The latter includes the influence of tectonic forces as well as overburden load.

Displacement

We wish now to express (7) in terms of the displacement field of solids. A cumulative displacement field of solids that slowly departs from an initial distribution pattern can be defined by

$$\bar{u}_{cum} \equiv \int_0^t \bar{v}_s dt \tag{8}$$

In the present paper we require $\bar{v}_s$ to represent the actual velocity (total derivative) of the center of mass of an arbitrarily specified group of solids (Lagrangian, Eulerian, or other). In differential form, (8) means that

$$d\bar{u}_{cum}/dt = \bar{v}_s \tag{9}$$

where the initial displacement field of solids is defined to be zero. The distinction between a total derivative as expressed in (9) and a material derivative (which is simply the total derivative of a Lagrangian group of solids) is discussed by Helm [1979b, 1982, Appendix I). All equations through (22') are valid for any specified volume element. Starting with (22"), the analysis is valid only for an element fixed in space.

Finite Volume Strain and Incremental Stress

We write

$$d\varepsilon_v/dt = \bar{\nabla} \cdot \bar{v}_s \tag{10}$$

where ε_v is finite cumulative volume strain of the skeletal frame, namely,

$$\varepsilon v = \int_0^t (d\varepsilon_v/dt)dt \tag{11}$$

Our choice of a reference frame based on the initial distribution of solids allows us to assume an initial unstrained state. Combining (8), (10), and (11) gives

$$\varepsilon_v = \bar{\nabla} \cdot \bar{u}_{cum} \tag{12}$$

If we were to take (12) as a definition of ε_v, then (10) follows from this definition. The term ε_v can also be interpreted as the trace (first invariant) of the skeletal strain tensor.

We now distinguish between an incremental change in effective stress σ' and the residual or initial unstrained effective stress σ_i' namely,

$$\sigma_{tr}' = \sigma_i' + \sigma' \tag{13a}$$

Equation (13a) merely states that the trace of the effective stress tensor σ_{tr}' can be separated into two parts, namely, a residual part σ_i' that is associated with an unstrained initial condition and an incremental part σ' that is associated with the induced skeletal volume strain ε_v of interest.

Similarly, we distinguish between an incremental change in total load σ and a residual or initial unstrained total load σ_i, namely,

$$\sigma_{tr} = \sigma_i + \sigma \tag{13b}$$

We now introduce a constitutive stress/strain relation, namely,

$$\sigma'/3\rho_w g = -\varepsilon_v/S_{sk} \tag{14}$$

which essentially defines our use of the skeletal component S_{sk} of specific storage S_s. The term $S_{sk}/3\rho_w g$ can be considered a three-dimensional field equivalent to the coefficient of volume strain m_v used in the laboratory by soil engineers [Lambe and Whitman, 1969]. More precisely, it is a scalar that relates skeletal volume strain to an incremental change in mean normal effective stress. Because these terms are invariant, no directional components of stress or strain need be known in order for S_{sk} to be used with no loss of generality. This is a distinct advantage over previous derivations. Directional components of skeletal strain will be seen to be controlled by the hydraulic conductivity tensor according to the present development.

Governing Equation in Terms of Displacement of Solids

Combining (7), (9), (12), and (14) gives

$$d\bar{u}_{cum}/dt - \bar{\bar{K}}\,\bar{\nabla}[(1/S_{sk})\,\bar{\nabla}\cdot\bar{u}_{cum}] = \bar{q}_b - \bar{q}_i + \bar{\bar{K}}\,\bar{\nabla}\,(\sigma/3\rho_w g) \tag{15}$$

where $\bar{q}_i$ represents the initial unstrained value of specific discharge $\bar{q}$, namely,

$$\bar{q}_i = -\bar{\bar{K}}\,\bar{\nabla}\,h_i = -\,\bar{\bar{K}}\,\bar{\nabla}\,[z + (\sigma_i - \sigma_i')/3\rho_w g] \tag{15a}$$

Equation (15) is a fundamental governing equation in terms of $\bar{u}_{cum}$. We shall now discuss various ways to modify (15) in order to eliminate terms in the right-hand side through appropriate simplifying assumptions.

Before continuing to develop (15) in a direction that will be used in the remainder of this paper, we shall digress briefly to show its relation to diffusion equations presently being used in soil mechanics, consolidation theory, petroleum reservoir engineering, and groundwater hydraulics.

Governing Equation in Terms of Volume Strain

Taking the divergence of each term in (15) gives

$$d\varepsilon_v/dt - \bar{\nabla}\cdot\bar{\bar{K}}\,\bar{\nabla}\,(\varepsilon_v/S_{sk}) = \bar{\nabla}\cdot\bar{q}_b - \bar{\nabla}\cdot\bar{q}_i + \bar{\nabla}\cdot(\bar{\bar{K}}\,\bar{\nabla}\,\sigma/3\,\rho_w g) \tag{16}$$

We shall now discuss under what conditions (16) simplifies to

$$d\varepsilon_v/dt - (\bar{\bar{K}}/S_{sk})\,\nabla^2\,\varepsilon_v = 0 \tag{16a}$$

Using the equations of state for incompressible constituents, namely,

$$\rho_s = \text{const} \tag{17a}$$

$$\rho_w = \text{const} \tag{17b}$$

where ρ_s is the density of individual solids, it has been shown [Helm, 1979a, 1982] from mass balance considerations alone that

$$\bar{\nabla}\cdot\bar{q}_b = 0 \tag{18a}$$

Equation (18a) is valid for transient flow of incompressible water past incompressible solids as well as for steady flow. Equation (18a) essentially states that if every constituent is incompressible, then from a bulk material point of view, $\bar{q}_b$ is distributed uniformly in space at any specified instant. However, (18a) says nothing of how solids and water redistribute themselves relative to each other. In other words, transient change in porosity is distinct from (18a).

If the compressibility of individual constituents were nonzero, this fact would enter the present analysis through the $\bar{\nabla} \cdot \bar{q}_b$ term which would no longer be zero valued. In fact, $\bar{\nabla} \cdot \bar{q}_b$ would become a function both of constituent compressibilities and of changing values of porosity. For the purposes of the present paper, however, we shall assume (17a) and (17b) from which (18a) follows directly.

If in addition we require steady relative flow under initial unstrained conditions, then

$$\bar{\nabla} \cdot \bar{q}_i = 0 \tag{18b}$$

If there is no incremental change in total load, the last term in the right-hand side of (16) reduces to zero. We need not make such a restrictive assumption if instead we merely require the second derivative of any nonzero incremental change in total load σ to be negligibly small. Combining this requirement with (18a) and (18b) reduces (16) to (16a) for homogeneous porous and permeable material.

Equation (16a) has been used in soil mechanics literature since Mikasa [1965] derived it using a different sequence of reasoning but using essentially identical assumptions. It is considered an improved modification of Terzaghis' classic theory of consolidation. It is important to note that (15) is more fundamental than either (16) or (16a). It is this fact that allows us to follow an entirely new direction of analysis later in this paper.

Governing Equation in Terms of Hydraulic Head

In order to show how equations from groundwater hydraulics and petroleum reservoir engineering relate to the present approach, we depart from the preceding line of reasoning at an earlier stage, namely, at (4). Taking the divergence of each term in (4) gives

$$d\varepsilon_v/dt = \bar{\nabla} \cdot \bar{q}_b + \bar{\nabla} \cdot (\bar{\bar{K}}\bar{\nabla}h) \tag{19}$$

where we used (10). Assuming incompressible constituents in accordance with (17a) and (17b), (19) becomes

$$d\varepsilon_v/dt - \bar{\nabla} \cdot (\bar{\bar{K}}\bar{\nabla}h) = 0 \tag{19a}$$

where we have used (18a). Equation (19a) is a straightforward continuity equation. Let us develop it a few steps further. Instead of using S_{sk} of (14), we define a skeletal component S_{sk}* of specific storage S_s by the relation

$$d\varepsilon_v/dt = -S_{sk}^* \, d(\sigma_{tr}'/3\rho_w g)dt \tag{20}$$

If S_{sk} and S_{sk}* were both assumed to be constant with respect to time, they would equal each other. Such an assumption, however, need not be made here. From (5) and (6), we write

$$\sigma_{tr}' = \sigma_{tr} - 3\rho_w g \, (h - z) \tag{21}$$

Combining (19a), (20), and (21) gives

$$dh/dt - (1/S_{sk}^*) \, \bar{\nabla} \cdot (\bar{\bar{K}}\bar{\nabla}h) = dz/dt + d(\sigma_{tr}/3\rho_w g)/dt \tag{19b}$$

For homogeneous porous and permeable material, (19b) simplifies to

$$dh/dt - (\bar{\bar{K}}/S_{sk}^*)\nabla^2 h = dz/dt + d(\sigma_{tr}/3\rho_w g)/dt \tag{19c}$$

If both the total load and the elevation z of the point of interest (for example, the center of mass of a specified elemental volume V or surface S) are constant with respect to time, (19c) reduces to

$$\partial h/\partial t - (\bar{\bar{K}}/S_{sk}^*)\ \nabla^2 h = 0 \qquad (19d)$$

which is frequently used in geohydrology [Jacob, 1940, 1950].

Petroleum reservoir engineers [Muscat, 1937] use (19d) with fluid pressure p rather than hydraulic head h as the unknown. They frequently consider the compressibility of interstitial fluids. Soil engineers [Taylor, 1948; Lambe and Whitman, 1969] follow Terzaghis' lead and use fluid pressure in excess of an ultimate equilibrium pressure as the unknown. Similar to the present analysis, soil engineers generally assume water and solids to be incompressible. Geohydrologists [Jacob, 1940, 1950] use hydraulic head h as it appears in (19d). However, in place of S_{sk}^* in (19d) they use a specific storage term S_s, which equals S_{sk}^* plus $\varphi \rho_w g \beta_w$ and thereby includes a component that accounts for the expansion of interstitial water. In accordance with (17b), we have assumed water to be significantly less compressible than the porous structure, namely, $S_{sk}^* \gg \varphi \rho_w g \beta_w$, where β_w is the compressibility of water. For practical purposes this translates to requiring S_{sk}^* to be greater than roughly $1 \times 10^{-5}\ m^{-1}$, which is a reasonable assumption for most compressible sedimentary deposits. Under field conditions when this requirement is not satisfied, the expansion of water should be included in the specific storage term. Note that use of (19d) implicitly requires a nondeforming frame of reference (to justify use of (5)) and a nonmoving representative elemental volume V or surface S (to justify assuming zero-valued dz/dt). A somewhat less restrictive assumption is simply to require that any change in mean normal total load with respect to time is offset by the rate of change in elevation of the center of mass of an appropriately selected elemental volume. It is not obvious whether such an elemental volume can be found that is not physically self-contradictory. The conceptual search for such an element is beyond the scope of the present paper.

Governing Equation in Terms of Velocity of Solids

We now return to our main line of reasoning, namely, to (15). Note that it is essentially an equation for slow motion of solids modified by a constitutive volumetric strain/stress relation in the form of (14) and a volumetric strain/displacement relation in the form of (12). For grain slipping slowly past grain the material is assumed to behave like a fluid, whereas for dilatation and compression of interconnected pores the material is modeled to behave like a solid.

Rather than take the divergence of each term, we now follow another conceptual path. Taking the time derivative of each term in (15) yields

$$d\bar{v}_s/dt - d[\bar{\bar{K}}\bar{\nabla}(\varepsilon_v/S_{sk})]/dt = d\bar{q}_b/dt - d\bar{q}_i/dt + d[\bar{\bar{K}}\bar{\nabla}(\sigma/3\rho_w g)]/dt \tag{22}$$

where we have used (9) and (12). For $\bar{\bar{K}}$, S_{sk}, and $\rho_w g$ constant in time and uniform in space, (22) simplifies to

$$\frac{d\bar{v}_s}{dt} - \frac{\bar{\bar{K}}}{S_{sk}}\bar{\nabla}(\bar{\nabla}\cdot\bar{v}_s) = \frac{\bar{\bar{K}}}{3\rho_w g}\bar{\nabla}\frac{d\sigma}{dt} + \frac{d\bar{q}_b}{dt} - \frac{d\bar{q}_i}{dt} \tag{22a}$$

At a point fixed in space, (22a) can be expressed [Helm, 1982] by local derivatives

$$\frac{\partial\bar{v}_s}{\partial t} - \frac{\bar{\bar{K}}}{S_{sk}}\bar{\nabla}(\bar{\nabla}\cdot\bar{v}_s) = \frac{\bar{\bar{K}}}{3\rho_w g}\bar{\nabla}\frac{\partial\sigma}{\partial t} + \frac{\partial\bar{q}_b}{\partial t} + \frac{\partial\bar{q}_i}{\partial t} \tag{22b}$$

It is generally assumed in finding an analytic solution for transient flow within an aquifer [Ferris et al., 1962] that the rate of withdrawal Q is a step function increasing from zero to a constant at $t = 0^+$. For a constant rate of withdrawal Q, we find

$$\partial\bar{q}_b/\partial t = 0 \tag{23a}$$

where we have assumed conditions (17a) and (17b).

For steady specific discharge under initial unstrained conditions ($t=0^-$), we write

$$\partial \bar{q}_i / \partial t = 0 \tag{23b}$$

Finally, we assume that after the initial instant $t = 0^+$ the rate of change in the sum of the normal components of total load is uniform in space, namely,

$$\bar{\nabla}(\partial \sigma / \partial t) = \partial(\bar{\nabla}\sigma)/\partial t = 0 \tag{23c}$$

Assumption (23c) ignores the gradient of change in the submerged weight of solids which is included in Helm's [1982] more complete theoretical analysis. In order to compare computational results of the present approach later in this paper with more standard hydrogeologic analysis, (23c) must be assumed. This is because the self weight of constituent material and changes in tectonic forces are uniformly ignored in the traditional theories of aquifer tests. Conditions (23a), (23b), and (23c) reduce (22b) to

$$\partial \bar{v}_s / \partial t - (\bar{\bar{K}}/S_{sk}) \; \bar{\nabla}(\bar{\nabla} \cdot \bar{v}_s) = 0 \tag{24}$$

Equation (24) is a partial differential equation in which $\bar{v}_s$ is the unknown.

Discussion of Assumptions

For the sake of emphasis, we list the assumptions contained in (24):

1. Darcy's law as expressed in (1).
2. Hydraulic head within a nondeforming frame of reference as expressed in (5).
3. The principle of effective stress for permeable porous material as expressed in (6).
4. A cumulative displacement field as expressed in (8) and (9).
5. Skeletal volume strain as expressed in (12).
6. A skeletal stress/strain relation for interconnected porosity as expressed in (14).

7. Incompressible constituents for saturated porous material as expressed in (17a) and (17b).

8. Mass balance as expressed in (18a) (see (23)).

9. Parameters $\bar{\bar{K}}$, S_{sk}, and $\rho_w g$ to be constant in time and uniform in space (see (22a)).

10. The representative elemental volume to be fixed in space (see (22b)).

11. A constant volume rate of fluid withdrawal Q (see (23a)).

12. An initial ($t = 0^-$) unstrained condition of zero or steady specific discharge as expressed in (23b).

13. The gradient of the sum of the normal components of any induced change in total load is constant with respect to time as expressed in (23c).

The assumptions listed above are essentially no more restrictive than those required for the more standard equations of comparable simplicity, namely (16a) and (19d). The exception is that Q is required to be constant. This assumption of constant Q is generally used in finding an analytic solution to (19d) but not in the derivation of (19d). In short, constant Q is not necessary for (18a) to be valid, but is necessary for (23a) to be valid. For time-varying Q one would require a nonzero $\partial \bar{q}_b / \partial t$ to appear in the right-hand side of (24).

Let us examine (23a) a bit further. This will help later in the discussion of initial conditions. From mass balance considerations for incompressible constituents, the bulk volume flux through any regional bounding surface Γ fixed in space (Figure 2) equals zero. This can be seen from substituting (18a) into the divergence theorem, namely,

$$\int_{\Gamma} \bar{q}_b\,(r, \theta, z, t) \cdot \hat{n} dS = \int_{V} (\bar{\nabla} \cdot \bar{q}_b) dV = 0 \qquad (25)$$

where $\hat{n}$ is a unit vector at (r, θ, z) normal to an arbitrarily specified regional bounding surface Γ. We separate Γ into two parts (namely, Γ_1 and Γ_2) such that (25) becomes

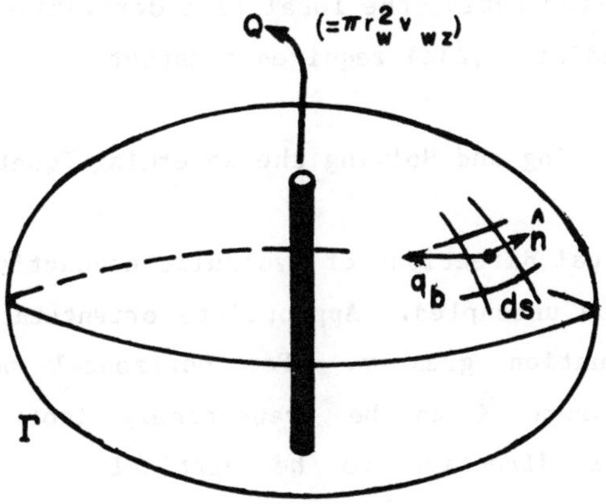

Fig. 2. The total bulk volume flux Q is being withdrawn from a well of cross-sectional area πr_w^2 $(\equiv \Gamma_2)$. The sum of bulk fluxes $\bar{q}_b$ through that part of Γ_1 of a fixed regional closed surface Γ that does not include Γ_2 equals -Q. Incompressible constituents (liquid and solid) are assumed.

$$\int_{\Gamma} (\bar{q}_b \cdot \hat{n})dS = \int_{\Gamma_1} (\bar{q}_b \cdot \hat{n})dS + \int_{\Gamma_2} (\bar{q}_b \cdot \hat{n})dS = 0 \tag{25a}$$

The area of Γ_2 is specified for convenience to be πr_w^2. The corresponding $\hat{n}$ becomes the vertical upward normal $\hat{k}$. Γ_2 is thereby within a discharging well of finite radius r_w on a horizontal plane above a producing aquifer. Γ_1 is the remaining part of an arbitrarily specified bounding surface Γ (Figure 2). The upward bulk flux $\bar{q}_b \cdot \hat{k}$ through Γ_2 equals the local upward flow of water v_{wz}. The discharge Q from the producing well equals $\pi r_w^2 v_{wz}$. In other words, we have specified Γ_2 such that

$$\int_{\Gamma_2} (\bar{q}_b \cdot \hat{n})dS = \pi r_w^2 v_{wz} = Q$$

Substituting this known value of flux through specified Γ_2 into (25a) leads to

$$\int_{\Gamma_1} (\bar{q}_b \cdot \hat{n})dS = -Q \tag{25b}$$

It can be seen from taking the local time derivative of both sides of (25b) why condition (23a) requires constant Q.

Uncoupling and Solving the Governing Equation

If the principal directions of hydraulic conductivity are specified, (24) can be uncoupled. Appropriate orientation allows us to simplify the equation greatly. For horizontal bedding, it is reasonable to assume $\bar{\bar{K}}$ to be transversely isotropic with the minimum principal direction to be vertical. This means that the vertical component of $\bar{\bar{K}}$, namely, K_{zz}, is smaller than any horizontal component, namely, K_{xx}, K_{yy}, $K_{\theta\theta}$, or K_{rr}. It also means that $\bar{\bar{K}}$ is isotropic on a horizontal plane, namely,

$$K_{zz} < K_{xx} = K_{yy} = K_{rr} = K_{\theta\theta} \tag{26}$$

and that the off-diagonal terms of the hydraulic conductivity tensor can be ignored.

For axially symmetric movement, we write $\bar{v}_s$ as

$$\bar{v}_s = (v_r, v_\theta, v_z) = (v_r, 0, v_z) \tag{27}$$

Based on (26) and (27), governing equation (24) can be uncoupled to form two equations, one in terms of the horizontal component v_r and the other in terms of the vertical component v_z, namely,

$$\partial v_r/\partial t - (K_{rr}/S_{sk}) [\partial(v_r/r)/\partial r + \partial^2 v_r/\partial r^2] = 0 \tag{28}$$

where we assume

$$\partial/\partial r\ (\partial v_r/\partial r + v_r/r) >> \partial/\partial r\ (\partial v_z/\partial z) \tag{28a}$$

and

$$\partial v_z/\partial t - (K_{zz}/S_{sk})\ \partial^2 v_z/\partial z^2 = 0 \tag{29}$$

where we assume

$$\partial/\partial z\ (\partial v_z/\partial z) >> \partial/\partial z\ (\partial v_r/\partial r + v_r/r) \qquad (29a)$$

Inequalities (28a) and (29a) state that we essentially are ignoring the change in shear strains with respect to time. The remainder of the paper is largely an analytic solution and discussion of the solution of (28) and (29) followed by a comparison of analytic solutions to the more standard equations for transient flow within a leaky confined aquifer system. First, however, we must discuss initial and boundary conditions that will be used in the solution.

Initial and Boundary Conditions

It has been observed in the field [Wolff, 1970; Allen, 1971] that a zone of radial compression develops near a discharging well (Figure 3). Beyond this zone is a zone of radial extension. The zone of compression represents an area of decrease in porosity on a horizontal plane due to a combination of two types of relative motion of solids. One type of motion involves movement along a given radius. If a grain moves radially inward further from its initial position than another grain that is closer to the discharging well, there is net radial compression of space between these two grains. If over the same time interval it also moves inward a further distance than another grain does which is on the same radius at a greater distance from the discharging well, there is net radial extension between these latter two grains. There is a boundary that separates the zone of radial compression from the zone of radial extension. This boundary represents zero radial skeletal strain, namely, $\varepsilon_{rr} = 0$. Interestingly, it also represents maximum radial displacement of solids over a specified time span. With time this boundary of zero radial strain can be expected to migrate outward from a continuously discharging well.

The second type of motion involves two grains close to each other that essentially lie equidistant from the discharge well but on

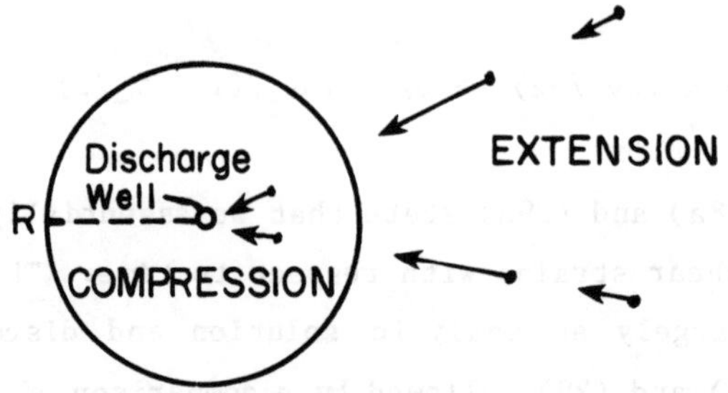

Fig. 3. A zone of skeletal compression develops near a discharging well in response to fluid withdrawal. An outer zone of skeletal extension develops beyond a radius R of no horizontal strain. Arrows schematically represent cumulative displacement of solids.

different radii. They will tend to move closer to each other as they move closer to the discharging well. This relative movement toward one another is true whether they lie in the zone of skeletal radial compression or in the zone of skeletal radial extension. This type of strain $\varepsilon_{\theta\theta}$ (hoop strain) is therefore always compressional in response to fluid withdrawal. For simplicity of specifying boundary conditions, we label the radial distance where the sum of horizontal normal strains, $\varepsilon_{rr} + \varepsilon_{\theta\theta}$, is zero valued by the letter R.

A distance r_0 is associated through a permeability ellipse (Figure 4) with an elevation b, namely,

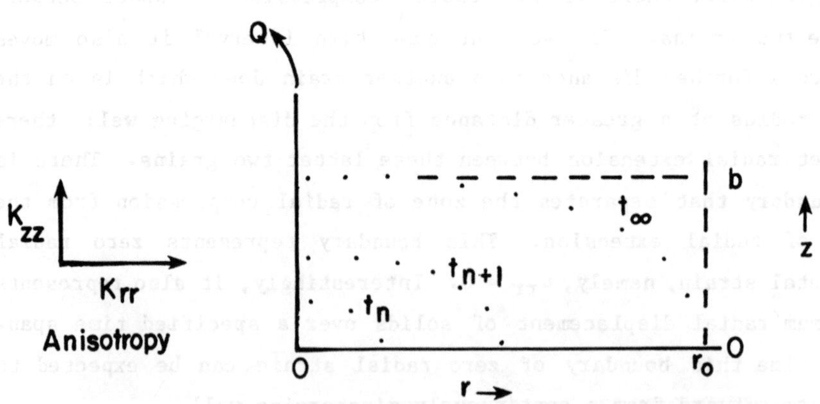

Fig. 4. Schematic showing the effect of a permeability ellipse, $K_{zz}/K_{rr} = b^2/r_0^2$, on a migrating pressure boundary within a confined aquifer with drawdown eventually arrested at z = b.

$$(b/r_o)^2 = K_{zz}/K_{rr} \tag{30}$$

If a physical process, such as a decrease in pressure, migrates upward and outward through an anisotropic confined aquifer and is arrested due to a vertical boundary condition at $z = b$, the radius r_o represents the distance that the physical process had reached radially before its further upward progress was arrested.

According to leaky aquifer theory [Hantush, 1960] there exists a no-drawdown boundary at a specified elevation. This can be interpreted to mean that a large source of water is available at this elevation from a very permeable stratum. Somewhat analogous to this, we require no vertical skeletal strain to occur at b. However, the lowest elevation Z above which no vertical strain occurs may be a function of r (namely, $Z = Z(r) \leq b$). The local value of Z is governed by regional b (Figure 5) and local steady state pressure gradients.

The initial and boundary conditions for vertical skeletal movement are

$$\partial\varepsilon_{zz}/\partial t = 0 \qquad t = 0,\ z > 0,\ r > 0 \tag{31a}$$

$$\partial\varepsilon_{zz}/\partial t = 0 \qquad t > 0,\ z = b,\ r > 0 \tag{31b}$$

$$v_z = 0 \qquad t > 0,\ z = 0,\ r > 0 \tag{31c}$$

Equation (31c) merely states that a depth exists (and is labeled $z = 0$) at which no vertical movement occurs. Comparable to equation set (31) are initial and boundary conditions for horizontal skeletal movement.

$$\partial\varepsilon_{rr}/\partial t + \partial\varepsilon_{\theta\theta}/\partial t = 0 \qquad t = 0,\ r > 0,\ z > 0 \tag{32a}$$

$$\partial\varepsilon_{rr}/\partial t + \partial\varepsilon_{\theta\theta}/\partial t = 0 \qquad t > 0,\ r = R,\ z > 0 \tag{32b}$$

$$v_r = 0 \qquad t > 0,\ r = 0,\ z > 0 \tag{32c}$$

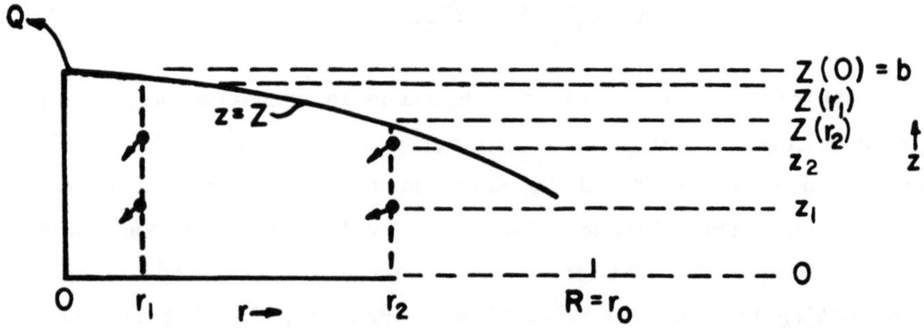

Fig. 5. Schematic showing the lowest elevation Z(r) where vertical skeletal strain does not change with time. Arrows schematically represent cumulative displacement of solids.

Boundary condition (32c) merely states that the radial movement of solids within the aquifer is constrained by the well casing of the producing well whose radius is mathematically collapsed to zero for computational convenience.

Keeping in mind that $\partial\varepsilon_{zz}/\partial t$ and $\partial\varepsilon_{rr}/\partial t$ equal $\partial v_z/\partial z$ and $\partial v_r/\partial r$, respectively, and that for axially symmetric movement $\partial\varepsilon_{\theta\theta}/\partial t$ equals v_r/r, we rewrite the initial and boundary conditions as

$$\partial v_z/\partial z = 0 \qquad t = 0,\ z > 0,\ r > 0 \tag{31a'}$$

$$\partial v_z/\partial z = 0 \qquad t > 0,\ z = b,\ r > 0 \tag{31b'}$$

$$v_z = 0 \qquad t > 0,\ z = 0,\ r > 0 \tag{31c'}$$

and

$$\partial v_r/\partial r + v_r/r = 0 \qquad t = 0,\ r > 0,\ z > 0 \tag{32a'}$$

$$\partial v_r/\partial r + v_r/r = 0 \qquad t > 0,\ r = R,\ z > 0 \tag{32b'}$$

$$v_r = 0 \qquad t > 0,\ r = 0,\ z > 0 \tag{32c'}$$

We are now in a position to solve (28) and (29).

Solution for the Radial Component

We shall first solve (28) for radial skeletal movement. According to the method of separation of variables, we write

$$v_r\ (r,z,t) = \psi \cdot \tau$$

If ψ_m is a typical solution for ψ, then linearity implies

$$\psi = \psi\ (r,z) = \sum_{m=1}^{\infty} \psi_m$$

Similarly,

$$\tau = \tau(t) = \sum_{m=1}^{\infty} \tau_m$$

Hence

$$v_r = \sum_{m=1}^{\infty} \psi_m \cdot \tau_m \tag{33}$$

A typical expression of (28) is

$$\psi_m \partial\tau_m/\partial t - c_{rr}\ \tau_m\ [\partial(\psi_m/r)/\partial r + \partial^2\psi_m/\partial r^2] = 0$$

or

$$(1/c_{rr}\tau_m)\partial\tau_m/\partial t = (1/\psi_m)\partial/\partial r\ (\partial\psi_m/\partial r + \psi_m/r) = -\lambda_m^2 \tag{34}$$

where λ_m is a constant [Bowman, 1958, p. 38] and

$$c_{rr} \equiv K_{rr}/S_{sk} \tag{34a}$$

Equation (34) is satisfied if both

$$\partial\tau_m/\partial t + c_{rr}\ \lambda_m^2\tau_m = 0 \tag{35}$$

with initial condition (32a') becoming

$$\tau(t=0) = \sum_{m=1}^{\infty} \tau_m\ (t=0) = f(r,z) \tag{35a}$$

$$\partial f/\partial r + f/r = 0 \tag{35b}$$

and also if

$$\partial^2\psi_m/\partial r^2 + (1/r)\ \partial\psi_m/\partial r + [\lambda_m^2\ (1/r)^2]\psi_m = 0 \qquad (36)$$

with boundary conditions (32c') and (32b') becoming

$$\psi_m\ (0,z) = 0 \qquad (36a)$$

$$\partial\psi_m\ (R,z)/\partial r + \psi_m\ (R,z)/R = 0 \qquad (36b)$$

The general solution to (35) is

$$\tau_m = C_1\ \exp(-\lambda_m^2\ c_{rr} t) \qquad (37)$$

The general solution of (36) is

$$\psi_m = C_2\ J_1(\lambda_m r) + C_3\ Y_1(\lambda_m r) \qquad (38)$$

where C_i ($i = 1, 2, 3$) are constants, J_1 is a first-order Bessel function of the first kind, and Y_1 is a first-order Bessel function of the second kind. Evaluating (36) is greatly simplified by requiring $r \gg r_w$. In effect the radius of a discharge (injection) well is reduced mathematically to a line sink (source) at $r = 0$. Because $Y_1\ (\lambda_m r) \to -\infty$ when $r \to 0$, condition (35a) requires $C_3 = 0$. Solution (38) reduces to

$$\psi_m = C_2\ J_1(\lambda_m r) \qquad (39)$$

A recurrence formula for Bessel functions is

$$\partial J_n(x)/\partial x = J_{n-1}(x) - (n/x)\ J_n(x) \qquad (40)$$

which implies for $n = 1$ and $x = \lambda_m r$ that

$$\lambda_m\ J_o(\lambda_m r) = \partial J_1(\lambda_m r)/\partial r + (1/r)\ J_1(\lambda_m r) \qquad (41)$$

Substituting (39) and (41) into (36") leaves

$$\lambda_m J_o(\lambda_m R) = 0 \tag{42}$$

According to (42), boundary condition (36") is satisfied if

$$J_o(\beta_m) = 0 \tag{43}$$

where β_m ($\equiv \lambda_m R$) is a typical positive root of (43). Combining (33), (37), and (39) and merging C_2 and C_3 into one constant C gives

$$v_r = C \sum_{m=1}^{\infty} J_1(\beta_m r/R) \exp(-\beta_m^2 c_{rr} t/R^2) \tag{44}$$

Let us now find the merged constant C: From orthogonal and

$$C = \sum_{m=1}^{\infty} \frac{\int_o^R r\, f(r,z)\, J_1(\beta_m r/R) dr}{\int_o^R r\, J_1^2(\beta_m r/R) dr} \tag{45}$$

where f (r,z) represents an initial value of v_r for r > 0. Let us discuss initial flow conditions.

It is reasonable to assume that initially ($t=0^+$) there is no immediate change in flow past solids (specific discharge) for r > 0, namely,

$$\bar{q} = -\bar{\bar{K}}\ \bar{\nabla} h = 0 \qquad t = 0^+ \qquad r > 0 \tag{46}$$

where for the remainder of this paper we use the terms $\bar{q}$ and h to indicate changes in flow and head from a background unstressed state ($t=0^-$). The latter state is expressed by (18b). Combining (4), (25b), and (46) gives

$$\int_{\Gamma_1} \bar{v}_{si} \cdot \hat{n}\, dS = -Q \qquad r > 0 \tag{47}$$

where $\bar{v}_{si}$ is the initial value ($t=0^+$) of $\bar{v}_s$. The surface Γ_1 is fixed in space and when combined with Γ_2 ($=\pi r_w^2$) completely surrounds a sink discharging at volume rate Q. Let this combined surface be a cylinder of height b at a perimeter distance r from a discharging well with top and bottom horizontal surfaces at z = b and z = 0 and a vertical axis at r = 0. In cylindrical coordinates, (47) becomes

$$-Q = 2\pi r \int_0^b v_{ri}\, dz + 2\pi \int_{r_w}^{r} r v_{bi}\, dr \tag{47a}$$

where v_{ri} is the initial value at r of axially symmetric v_r and v_{bi} is the initial value of v_z along the surface z = b. The value of v_z everywhere along the horizontal surface at z = 0 is zero for $t \geq 0$ (see equation (31c)). The integral sum taken over r of the initial value of v_z is thereby zero on the specified surface z = 0 and hence does not appear in (47a).

For the purposes of this discussion, constant Q can be broken conveniently into two parts, namely, Q_r and Q_b such that $-Q = Q_r + Q_b$. Q_r is the initial total horizontal bulk volume rate of incompressible material moving through the vertical surface of the specified cylinder at r. Q_b is the initial total vertical bulk volume rate moving through a horizontal surface at z = b between a radius r ($>r_w$) and r_w. Q_r equals the first integral in (47') and Q_b equals the second. For the case of vertically uniform v_{ri} within a column of aquifer material, we write

$$v_{ri} = Q_r/2\pi rb \tag{47b}$$

It is worth interpreting (46) and (47). Because there is initially no change in porosity for r > 0 (see (17a), (17b), (31a), and (32a)), the confined aquifer system responds initially as an undifferentiated incompressible bulk material (see (25)). Initially, solids and water move together toward a producing well. For constant Q, a zone within which porosity decreases spreads outward from the producing well. For such material, water does not begin to flow past the solids until there occurs a change in porosity. Perforations in the casing of the producing well are designed to impede the inward movement of solids. This is where differentiation between water and solid is introduced and relative flow of water past solids (specific discharge) is initiated. Initial conditions (46) and (47) are direct consequences of this physical dynamic. For constant Q the initial value of $\bar{v}_s$ is its maximum everywhere. The value of $\bar{v}_s$ at a point of interest gradually decreases to zero as a new stable porosity distribution or strain equilibrium is approached. For $\bar{v}_s$ to reach zero, this new equilibrium must be reached not only locally but everywhere between the producing well and the specified local point of interest. Relative flow (specific discharge) of water past solid can continue even after $\bar{v}_s$ locally decreases to zero so long as either porosity continues to decrease at some more distant point or when a source of leakage at a distant point is intercepted by an outward migrating decrease in fluid pressure.

Let us return to finding a solution for (28). Substituting v_{ri} of (47b) for f(r, z), (45) becomes

$$C = \frac{1}{2\pi} \frac{\int_0^R (Q_r/b) J_1(\beta_m r/R)\,dr}{\int_0^R r\, J_1^2\,(\beta_m r/R)\,dr} \tag{48}$$

For roughly uniform Q_r on a horizontal plane, (48) simplifies to

$$C = \frac{Q_r}{2\pi b} \frac{\int_0^R J_1(\beta_m r/R)\,dr}{\int_0^R r\, J_1^2\,(\beta_m r/R)\,dr} \tag{48a}$$

For the denominator in (48a) we write [Bowman, 1958, p. 101]

$$\int_0^R rJ_1{}^2(\beta_m r/R)\,dr = (R^2/2)\; J_1^2(\beta_m) \tag{49}$$

Completing the integration of the numerator in (48a) yields

$$\int_0^R J_1\;(\beta_m r/R)\,dr = R/\beta_m \tag{50}$$

Combining (44), (48a), (49), and (50) gives the dimensionless relation

$$v_{rD} = \sum_{m=1}^{\infty} [J_1(\beta_m r_D)/\beta_m J_1^2(\beta_m)]\; \exp(-\beta_m^2 T_R) \tag{51}$$

where

$$v_{rD} \equiv \pi R b v_r/Q_r \tag{51a}$$

$$r_D \equiv r/R \tag{51b}$$

$$T_R \equiv c_{rr} t/R^2 \tag{51c}$$

We now have a dimensionless solution for the radial component of $\bar{v}_s$ subject to boundary and initial conditions (32a), (32b), and (32c). Before finding the cumulative transient displacement and strain of the solid matrix, we must solve for the vertical component of $\bar{v}_s$ subject to boundary conditions (31a), (31b), and (31c). This is done in the following section.

Solution for the Vertical Component

The transformed expression of (29) is

$$\partial^2 v^*/\partial z^2 - s\, v^*/c_{zz} = -\, v_{zi}/c_{zz} \tag{52}$$

where v^* is the Laplace transform of v_z, v_{zi} is the initial value of v_z, and

$$c_{zz} = K_{zz}/S_{sk} \tag{52a}$$

In transform space, conditions (31a'), (31b'), and (31c') become

$$\partial v_{zi}/\partial z = 0 \tag{52a'}$$
$$\partial v^*/\partial z = 0 \qquad z=b \tag{52b}$$
$$v^* = 0 \qquad z=0 \tag{52c}$$

The general solution of (52) is

$$v^* = C_1 \cosh(\sqrt{s/c_{zz}}\, z) + C_2 \sinh(\sqrt{s/c_{zz}}\, z) + v_{zi}/s \tag{53}$$

Conditions (52c) and (52b) require

$$C_1 = - v_{zi}/s \tag{54a}$$

$$C_2 = (v_{zi}/s) \sinh(\sqrt{s/c_{zz}}\, b)/\cosh(\sqrt{s/c_{zz}}\, b) \tag{54b}$$

Substituting (54a) and (54b) into (53) and using hyperbolic identities yields

$$v^* = \frac{v_{zi}}{s}\left[1 - \frac{\cosh(\sqrt{s/c_{zz}}\,(b-z))}{\cosh(\sqrt{s/c_{zz}}\,b)}\right] \tag{55}$$

The inverse transform [Spiegel, 1965, p. 252] of (55) is

$$v_z = 2v_{zi} \sum_{n=1}^{\infty} \frac{1}{N} \sin(Nz_D)\exp(-N^2T_z) \tag{56}$$

where

$$N \equiv (2n - 1)\pi/2 \tag{56a}$$
$$z_D \equiv z/b \tag{56b}$$
$$T_z \equiv c_{zz}\, t/b^2 \tag{56c}$$
$$v_{zi} = v_{zi}(r) \tag{56d}$$

In dimensionless terms we write

$$v_{zD} = \sum_{n=1}^{\infty} \frac{2}{N} \sin(Nz_D) \exp(-N^2 T_z) \tag{57}$$

where

$$v_{zD} = v_z/v_{zi} \tag{57a}$$

Equations (51) and (57) are analytic solutions for transient skeletal flux within a confined aquifer system. We shall now find solutions for cumulative displacement and skeletal strain.

Skeletal Displacement and Strain

Keeping (8) in mind, we integrate (51) with respect to time to get

$$u_{cum\ rD} = \sum_{m=1}^{\infty} \frac{J_1(\beta_m r_D)}{\beta_m^3 J_1^2(\beta_m)} [1 - \exp(-\beta_m^2 T_R)] \tag{58}$$

where

$$u_{cum\ rD} = \frac{\pi b K_{rr}}{R S_{sk} Q_r} u_{cum\ r} \tag{58a}$$

Integrating (57) with respect to time gives

$$u_{cum\ zD} = \sum_{n=1}^{\infty} \frac{2}{N^3} \sin(Nz_D) [1 - \exp(-N^2 T_z)] \tag{59}$$

where

$$u_{cum\ zD} = (K_{zz}/S_{sk} b^2 v_{zi}) u_{cum\ z} \tag{59a}$$

In summary, equations (58) and (59) are nondimensional solutions for the transient cumulative displacement of sedimentary skeletal material in response to discharging a confined aquifer at a constant rate Q. We shall plot and discuss these equations later. For the remainder of this section we shall discuss volume strain.

Recall that volume strain is the divergence of the displacement field (equation (12)), namely,

$$\varepsilon_v = \bar{\nabla} \cdot \bar{u}_{cum}$$

In cylindrical coordinates, volume strain can be expressed by

$$\varepsilon_v = \varepsilon_{rr} + \varepsilon_{\theta\theta} + \varepsilon_{zz} \tag{60}$$

where for axially symmetric displacement

$$\varepsilon_{rr} = \partial u_{cum\ r} / \partial r \tag{60a}$$

$$\varepsilon_{\theta\theta} = u_{cum\ r}/r \tag{60b}$$

$$\varepsilon_{zz} = \partial u_{cum\ z}/\partial z \tag{60c}$$

$$\bar{u}_{cum} = (u_{cum\ r}, 0, u_{cum\ z}) \tag{60d}$$

Substituting (58) and (59) into equation set (60) gives

$$\varepsilon_v = A' \sum_{m=1}^{\infty} \frac{\partial J_1(\beta_m r_D)/\partial r + J_1(\beta_m r_D)/r}{\beta_m{}^3\ J_1{}^2(\beta_m)} [1 - \exp(-\beta_m{}^2 T_R)] \tag{61}$$

$$+ B' \sum_{n=1}^{\infty} \frac{1}{N^2} \cos(Nz_D)[1-\exp(-N^2 T_z)]$$

where

$$A' = Q_r S_{sk} R/\pi b K_{rr} \tag{61a}$$

$$B' = S_{sk} b v_{zi}/K_{zz} \tag{61b}$$

A recurrence formula for Bessel functions is

$$\partial J_n(x)/\partial x = J_{n-1}(x) - (n/x)\, J_n(x)$$

which for n=1 and $x=\beta_m r_D$ helps to simplify (61) to the following nondimensional form

$$E_v = 2\sum_{m=1}^{\infty} \frac{J_o(\beta_m r_D)}{\beta_m^2 J_1^2(\beta_m)} \left[1-\exp(-\beta_m^2 T_R)\right]$$

$$+ C_z \sum_{n=1}^{\infty} \frac{2}{N^2} \cos(Nz_D) \left[1-\exp(-N^2 T_z)\right] \tag{62}$$

where

$$E_v = 2\pi b\varepsilon_v K_{rr}/S_{sk}Q_r \tag{62a}$$

$$C_z = 2\pi b^2 v_{zi} K_{rr}/K_{zz}Q_r \tag{62b}$$

Alternatively, (62) can be expressed [Bowman, 1958, p. 19] by

$$E_v = \ln r_D \quad -2\sum_{m=1}^{\infty} \frac{J_o(\beta_m r_D)}{\beta_m^2 J_1^2(\beta_m)} \exp(-\beta_m^2 T_R)$$

$$+ C_z \sum_{n=1}^{\infty} \frac{2}{N^2} \cos(Nz_D) \left[1-\exp(-N^2 T_z)\right] \tag{63}$$

The value of C_z can be considered a type of weighting function that indicates the degree to which a dimensionless change in volume E_v is influenced by its vertical component. It should be noted by substituting (14) into (62a) that E_v can be transformed into a dimensionless change in mean normal effective stress σ_D', namely,

$$\sigma_D' = -2\pi bK_{rr}\sigma'/3\rho_w gQ_r \tag{64}$$

where dimensionless σ_D', as defined above, equals the right-hand side of (63).

<u>Comparison With Previous Solutions</u>

It is worth comparing (63) to published solutions to transient flow problems.

Muskat [1937], for example, solved transient flow equations mathematically for a number of boundary conditions. One solution [Muskat, 1937, p. 655] is traditionally associated with a leaky circular boundary. This solution can be expressed in nondimensional form [Witherspoon et al., 1967, p. 105] essentially as

$$P_D = \ln r_D - 2 \sum_{m=1}^{\infty} J_0(\beta_m r_D) \exp(-\beta_m^2 r_D^2 T_r)/\beta_m^2 J_1^2(\beta_m) \quad (65)$$

Except for nomenclature, the right-hand side of (65) is identical to the first two terms in the right-hand side of (63). These terms in (63) represent the horizontal component of dimensionless strain. Similarly, they represent the horizontal component of σ_D' of (64). Witherspoon's nomenclature specifies a time factor or dimensionless time $T_r(\equiv t\ c_{rr}/r^2)$ that is distinct from T_R of (51c). Witherspoon et al. [1967, Appendix L] tabulated the relation of P_D to T_r in (65) and plotted the resulting family of curves for designated values of $0 < r_D < 1$. Figure 6 shows their results. Because P_D, T_r, and r_D in (65) and Figure 6 are dimensionless terms, Witherspoon et al.'s values are used in Figure 7 to express how the first two terms in the right-hand side of (63) vary as a function of T_R $(=r_D^2 T_r)$ for specified values of r_D. The only difference between Figures 6 and 7 is the definition of dimensionless time along the horizontal axis. The two figures merely represent two alternative ways to plot (65).

Witherspoon et al. [1967] have shown that nondimensional curves (65) fall closer and closer to a nondimensional Theis curve as r_D gets smaller. For fixed r, the Theis curve can be considered the limiting case of (65) for $R \to \infty$. Alternatively, for fixed R, the Theis curve can be considered the limiting case of (65) for $r \to 0$. In the present paper, we require a fixed R associated with unstrained leakage at elevation b. Hence at the discharging well the horizontal component of strain in (63) reduces to Theis' [1935] nondimensional solution.

The primary difference between (63) and (65) is the physical

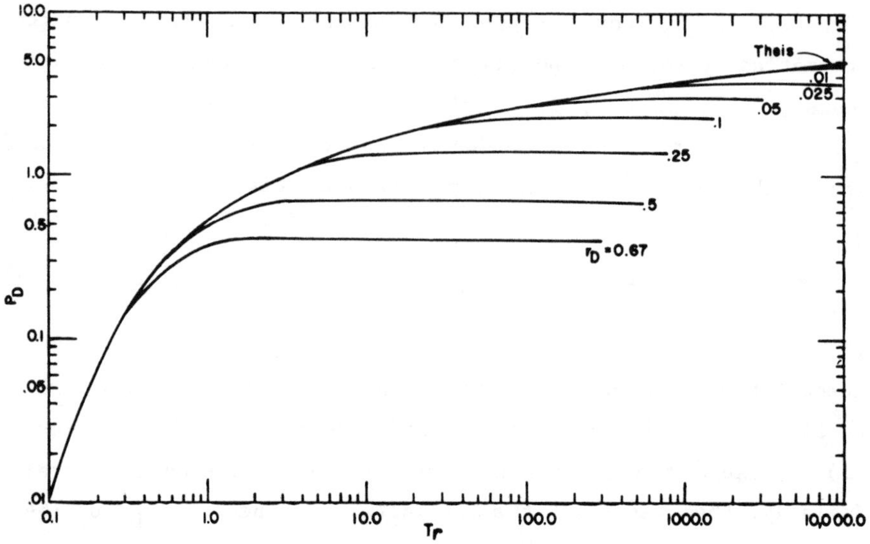

Fig. 6. P_D versus T_r for an aquifer with a constant-head circular boundary in accordance with (65). Modified from Witherspoon et al. [1967].

interpretation of the saturated flow system. The initial and boundary conditions which led Muskat [1937] to (65) as a solution for P_D are (1) uniform density at t=0, (2) constant density at r=R, and (3) the flow rate Q at the discharging well both remains constant and conforms to Darcy's law at r=0. In the present paper, density conditions 1 and 2 are interpreted to mean no change in strain at t=0 and r=R (see (32a) and (32b)). Witherspoon et al. [1967] interpreted Muskat's use of density condition 1 as uniform hydraulic head at t=0 and his use of condition 2 as no drawdown at r=R. Witherspoon et al.'s interpretation of Muskat's development is consistent with traditional physical interpretations that are common to both geohydrologists and petroleum reservoir engineers, namely, that the transient flow of fluid to a line sink is entirely horizontal within a confined aquifer and that the granular medium remains rigid. Witherspoon et al.'s [1967] interpretation of P_D as $2\pi bK_{rr}\Delta h/Q$ is entirely in keeping with Muskat's own interpretation. This interpretation is consistent with (64) though it contains more restrictive assumptions than (64). Because the radial

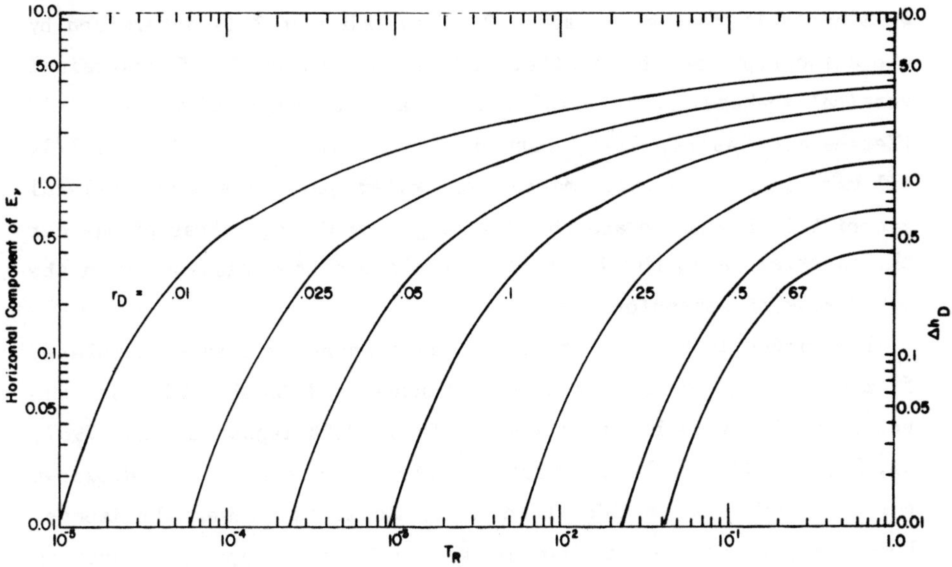

Fig. 7. Horizontal component of E_V (first two terms in the right-hand side of (63)) versus T_R. Dimensionless drawdown Δh_D versus T_R at the top of a confined aquifer (z_D = 1.0) in accordance with (68).

displacement field of skeletal material toward a discharging well is mistakenly ignored, the physical interpretation of boundary condition 2 is that an infinite or continuously replenished source of water surrounds the well field at a radius R from the pumping well. This is one possible interpretation of condition 2 but not the only one. Such an infinite source of water completely surrounding a well field at R seldom, if ever, occurs in nature. Hence because of this overly restrictive interpretation of condition 2, Muskat [1937, p. 655] himself considered (65) to be of little practical importance.

Contrastingly, for (63) to be valid there is no need to require an infinite or continuously replenished source of water at R. We need merely to distinguish physically a zone of lateral skeletal compression that lies near the pumping well from a zone of lateral extension that lies farther from the pumping well (Figure 3). The boundary between these two zones migrates outward from a discharging well until it reaches a radius R. The value of R is determined by the anisotropy of the aquifer and the elevation b of unstrained vertical leakage. Horizontal strain has been observed in the field [Yerkes and Castle, 1970; Hatton, 1970; Wolff, 1970; Lee, 1977]. All macroscopic lateral motion of water particles and skeletal material is inward toward the discharging well regardless of whether the material is in the inner zone of skeletal compression or in the outer zone of extension.

The curves in Figure 6 have a similar shape to curves calculated from the theory of leaky aquifers [Hantush and Jacob, 1955] and the theory of leaky faults [Stallman, 1952]. Witherspoon et al. [1967, pp. 55, 59, 107, 119] contrast the shapes of theoretical drawdown curves expected from idealized leaky aquifers (linear horizontal boundary), leaky faults (linear vertical boundary), and equation (65). For points near the discharge well (namely, for $r_D < 0.1$), the curves for a leaky fault and for a corresponding leaky aquifer are identical. For points further out (namely, $r_D > 0.1$) the curves deviate but are similar in shape. Curves based on (65) and shown in Figure 6 begin to deviate from the Theis curve at larger values of T_r for specified value of r_D than do those calculated from the theory of leaky aquifers or leaky faults. They also flatten sooner once they begin to deviate from the Theis curve. The ultimate value of flattened P_D for a specified r_D is identical for (65) as it is for a linear vertical boundary (theory of leaky faults).

The second summation in the right-hand side of (63) represents the vertical component of skeletal strain. At the base of the compression zone (z=0), this summation reduces to the average consolidation ratio U used by soil engineers [Taylor, 1948, p. 234], which is

$$U = 1 - \sum_{n=1}^{\infty} (2/N^2)\exp(-N^2 T_z) \qquad (66)$$

Briefly, the term U is used in the laboratory to predict the vertical movement of the top of a saturated clay specimen in response to water being squeezed out of it due to an incremental change in stress. It is curious that quantitatively U of (66) equals the vertical component of strain in (61) at the base ($z_D = 0$) of an idealized aquifer system when B' of (61b) equals unity.

The standard boundary conditions which are used in one-dimensional consolidation theory and lead to (66) can be interpreted to be $v_z=0$ at the base of the saturated soil specimen that is being compressed and $\partial\varepsilon_{zz}/\partial z = 0$ at the midplane. Symmetric strain is assumed by soil engineers above and below the midplane. If we compare these boundary conditions to conditions (31b) and (31c), we see that one is identical (namely, $v_z=0$), whereas the other condition is significantly different. The laboratory requires a zero-valued space derivative of strain at a boundary, whereas (31b) is a time derivative. Hence we cannot and shall not draw a direct comparison between results from laboratory consolidation theory, as expressed in (66), and the vertical component of theoretical displacement in the field, as expressed in (59). If the reader is interested in the direct application of the present theory to boundary value problems of one-dimensional consolidation, he is referred to the mathematical solutions plotted by Helm [1979a].

Calculated Transient Drawdown

At a point fixed in space, (14) and (21) can be combined to give the relation of change in hydraulic head to a corresponding amount of induced volume strain, namely,

$$\Delta h = \varepsilon_v/S_{sk} \tag{67}$$

where we assume no incremental change in total load. This last assumption of no change in total load is made in order to compare the computational results in this paper with similar results in the theory of transient groundwater flow. This assumption is stronger than assumption (23c). Substituting (63) into (67) gives

$$\Delta h_D = \ln r_D - 2 \sum_{m=1}^{\infty} \frac{J_o(\beta_m r_D)}{\beta_m^2 J_1^2(\beta_m)} \exp(-\beta_m^2 T)$$

$$+ C_z \sum_{n=1}^{\infty} \frac{2}{N^2} \cos(N z_D) [1 - \exp(-N^2 T)] \quad (68)$$

where we have used a generalized time factor T, as discussed below. Substituting (62a) into (67) gives

$$\Delta h_D = 2\pi b K_{rr} \Delta h / Q_r \quad (68a)$$

where we define Δh_D as the dimensionless drawdown equivalent of dimensionless volume strain E_v.

In order to introduce a generalized time factor T, boundary condition (32b) is now assumed to be associated with condition (31b) through the permeability ellipse (30). In other words, for computational convenience we let r_o of (30) roughly equal R of (32b), namely,

$$r_o = R \quad (68b)$$

R can be considered a type of radius of influence of horizontal and vertical strain that depends on degree of anisotropy and on the regional existence of zero vertical skeletal strain at z = b. Substituting (30), (34a), and (52a) into (51c) and (56c) leads to

$$T_z = T_R \quad (68c)$$

This allows use in (68) of a generalized time factor T for a transversely isotropic aquifer, where T equals

$$T = (K_{rr}/R^2)\ t/S_{sk} = (K_{zz}/b^2)\ t/S_{sk} \quad (68d)$$

We restrict our analysis to values of T between zero and one. For larger values of T, the influence of physical boundaries may cause the upward and outward progress of a physical process (Figure 4) not to follow a simple elliptic shape. Note that T of (68d) differs from dimensionless time T_r as generally used by hydrogeologists [Witherspoon et al., 1967, p. 6] by a factor r_D^2, namely,

$$T = r_D^2 T_r \tag{69}$$

As illustrated by comparing Figures 6 and 7, (69) merely shifts solution curves along the dimensionless time axis. The advantage of using (68d) as a definition of dimensionless time T is that the influence on drawdown Δh_D of both vertical and horizontal components of strain can be plotted together as one family of curves.

It is interesting to note that (68a) reduces to the standard definition of dimensionless drawdown [Witherspoon et al., 1967, p. 6] when Q_r equals Q and b is interpreted as thickness of the confined aquifer. This observation is somewhat surprising when contrasted to earlier interpretations of skeletal deformation. Previous mathematical solutions, such as (65), of transient groundwater flow equations are based on a more traditional approach [Jacob, 1940, 1950; Hantush, 1964; DeWiest, 1965] of deriving and solving equation (19d). These solutions and this equation have been interpreted [Verruijt, 1969, p. 343] to require matrix movement to be only vertical. This interpretation is due to an overly restrictive application of elastic theory. Our conclusion is that a condition of only horizontal matrix movement is fully consistent with solution (65). Note that S_{sk} of (14) can be considered an empirical term for relating volume strain to a change in mean normal effective stress.

Figures 7, 8, and 9 show families of type curves that indicate how dimensionless drawdown Δh_D varies with T at different dimensionless locations (r_D, z_D) in accordance with (68). Different values for the weighting factor C_z of (62b) have been selected. Figure 8 uses a value of 10 for C_z; Figure 9 uses 1; Figure 7 uses 0.0. For all values of C_z, type curves for drawdown at the top of a confined aquifer system (z_D=1) for different values of $0 < r_D < 1$ are plotted in Figure 7. Figures 8a and 9a are similar curves midway into the

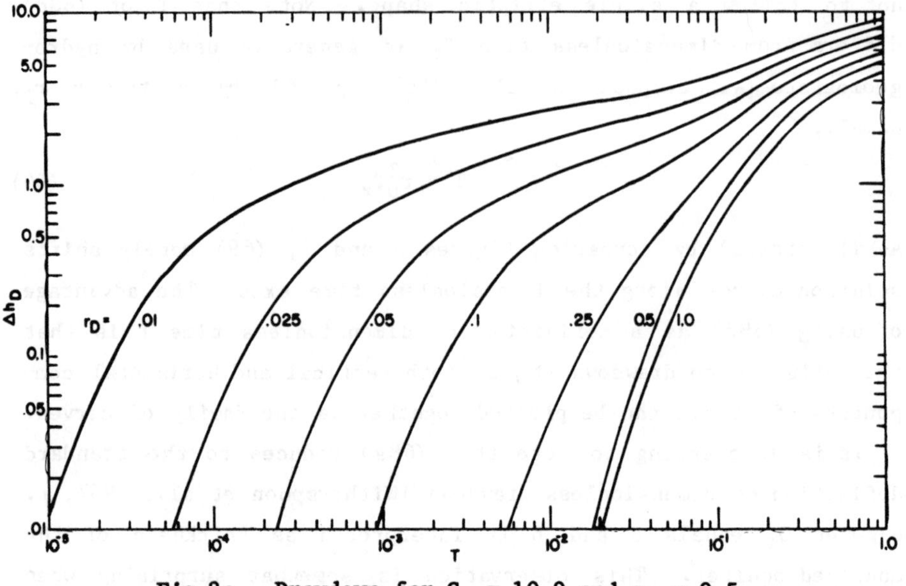

Fig 8a. Drawdown for C_z = 10.0 and z_D = 0.5.

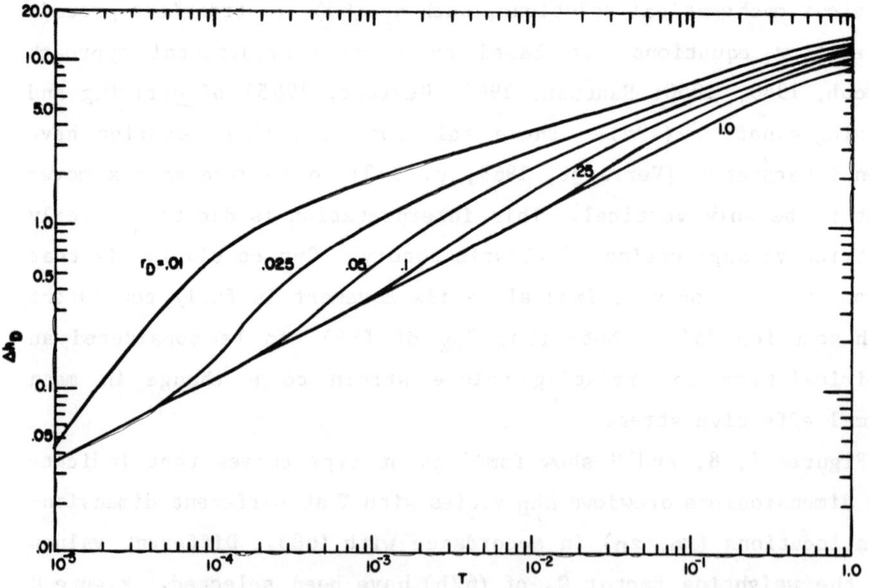

Fig 8b. Drawdown for C_z = 10.0 and z_D = 0.0.

Fig. 8. Dimensionless drawdown Δh_D versus dimensionless time T for a C_z value of 10.0 at (a) the midplane (z^d = 0.5) and (b) the base (z_D = 0.0) of a confined aquifer and (c) for a fully penetrating observation well.

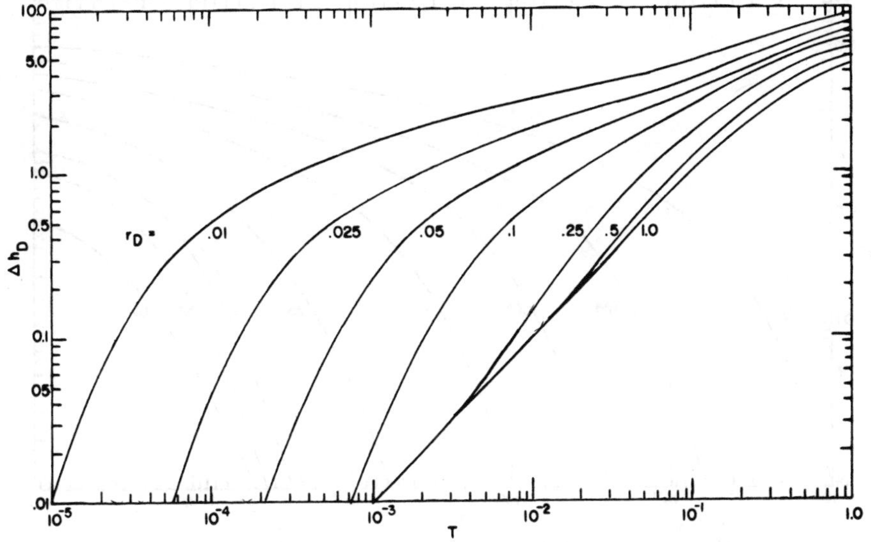

Fig. 8c. Drawdown for C_z = 10.0 and a fully penetrating observation well.

aquifer system (z_D = 0.5). Figures 8b and 9b are similar curves at the base of the aquifer system (z_D=0). Figures 8c and 9c are similar curves assuming that a fully penetrating observation well is open to the aquifer system between $0 < z_D < 1$. To plot Figures 8c and 9c, the last summation in the right-hand side of (68) is integrated over z_D. This integral mathematically equals the right-hand side of (59).

Vertical heterogeneity of transient drawdown within an aquifer column has been neglected by most previous investigators through invoking the Dupuit assumption (namely, dh/dz = 0). The main exception has been the calculated drawdown within a semiconfining bed in the theory of leaky aquifers. The method we have used in the present paper is to analyze skeletal deformation. The vertical component of decrease in porosity has been calculated to be greater in the lower part of an aquifer than in the upper part. The horizontal component is at a maximum near a discharging well. The time-dependent pattern of change in porosity is therefore significantly different in the upper part of an aquifer near a producing well from what it is in the lower part of the same confined aquifer at a radially

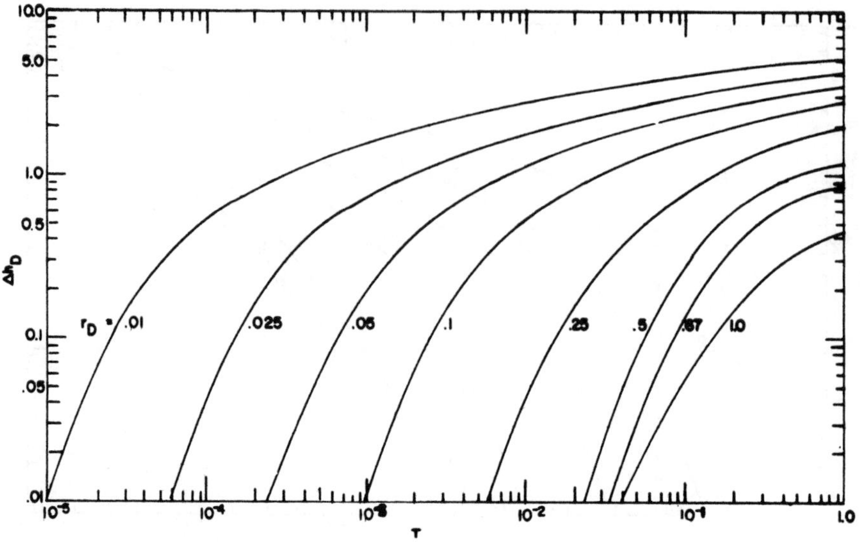

Fig. 9a. Drawdown for C_z = 1.0 and z_D = 0.5.

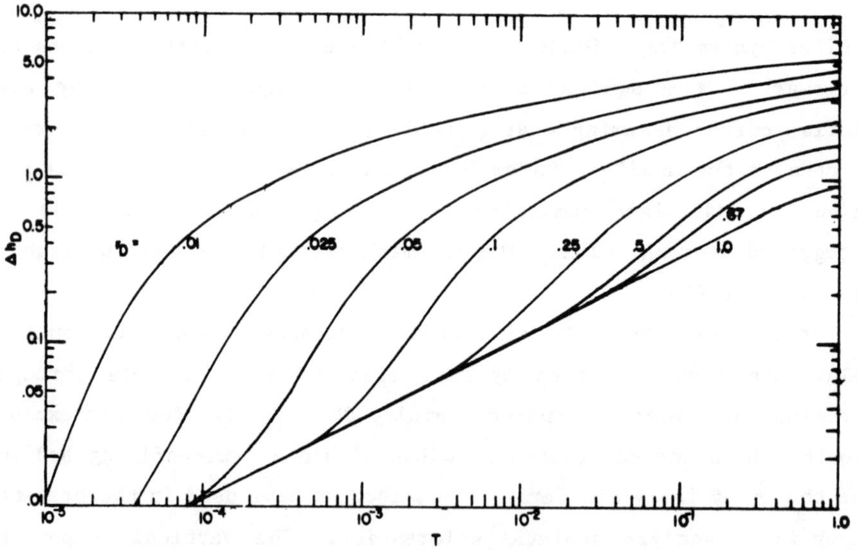

Fig. 9b. Drawdown for C_z = 1.0 and z_D = 0.0.

Fig 9. Dimensionless drawdown Δh_D versus dimensionless time for T for a C_z value of 1.0 at (a) the midplane (z_D = 0.5) and (b) the base (z_D = 0.0) of a confined aquifer and (c) for a fully penetrating observation well.

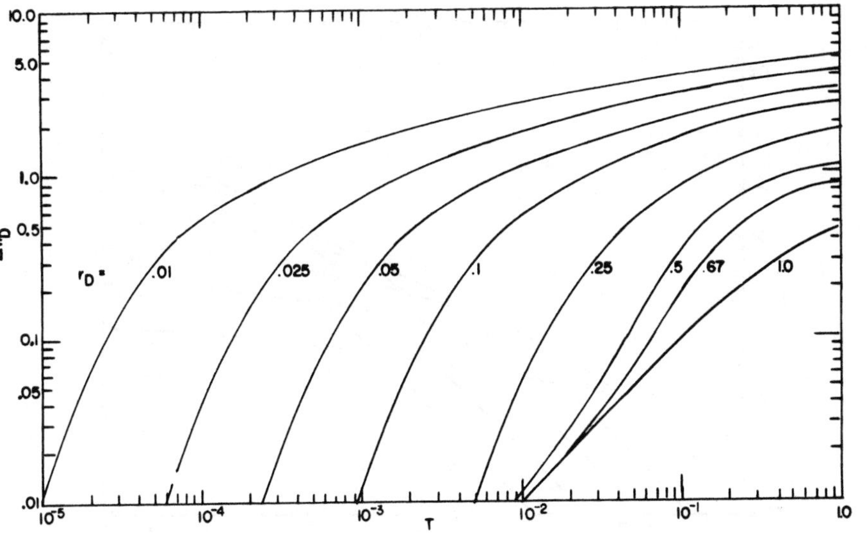

Fig. 9c. Drawdown for $C_z = 1.0$ and a fully penetrating observation well.

distant point. During transient flow, drawdown patterns in observation wells (Figures 7, 8, and 9) are predicted to reflect these distinctive local patterns of skeletal strain. Hence vertical heterogeneity of transient drawdown within an homogeneous confined aquifer need no longer be neglected.

It is surprising that S-shaped dimensionless drawdown curves (Figures 8 and 9) were calculated to occur within a homogeneous aquifer system. This shape resulted entirely from anisotropic displacement of solids. Through the reciprocal of anisotropic intrinsic permeability, the resistance to vertical movement of solids is distinct from the resistance to horizontal movement. Hence the horizontal component of time-dependent decrease in porosity can and probably does occur at a different rate from the vertical component. Theoretically, drawdown cannot help but reflect this lack of directional strain rates being synchronized with respect to time. A system of S-shaped drawdown curves are a result. Standard theory (even soil mechanics consolidation theory) essentially calculates an average or implicitly assumed isotropic pore-volume change. It thereby neglects the possibility of anisotropic

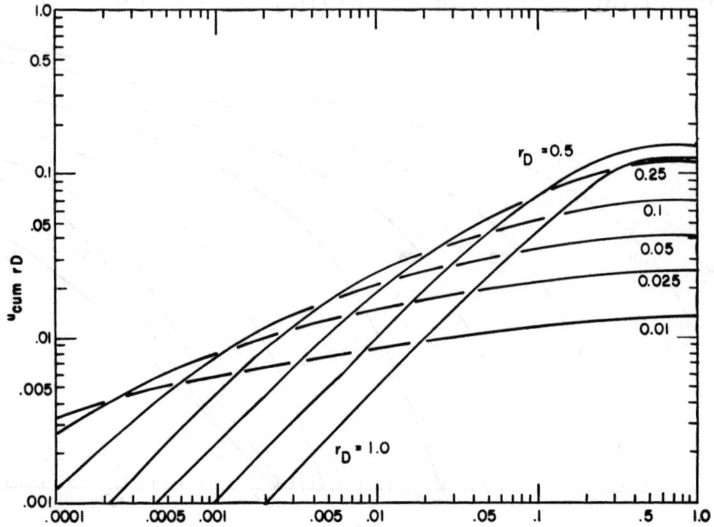

Fig. 10. Dimensionless radial displacement u_{cum} rD versus dimensionless time T_R for selected dimensionless distances r_D from a discharging well.

strain rates which have been laid bare for the first time by the present theoretical development.

Calculated Horizontal and Vertical Displacement

Equations (58) and (59) lie at the heart of the present paper. They are solutions to a boundary value problem (namely, equation sets (31) and (32)) that represents skeletal movement of an idealized confined aquifer system.

Figure 10 shows dimensionless horizontal displacement u_{cum} r D versus dimensionless time T_R for selected dimensionless distances r_D from the producing well. Figure 11 shows u_{cum} r D versus r_D for selected values of T_R. Both figures are plotted in accordance with (58).

As can be seen in Figure 10, during earliest plotted time the selected points near the discharging well have moved radially inward a further distance than other points that lie farther from the discharging well. This relative movement expresses radial extension.

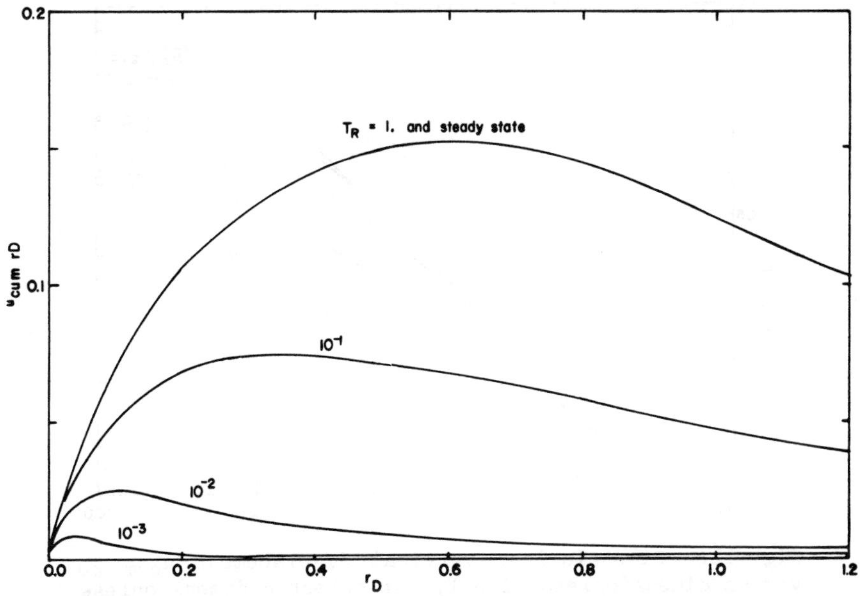

Fig. 11. Dimensionless radial displacement u_{cum} rD versus dimensionless distance r_D from a discharging well for selected values of dimensionless time T_R.

For example, at time T_R equal to 0.0001, all points $r_D > 0.01$ are experiencing radial extension. At all times $(t>0^+)$ there exists mathematically a zone of radial compression nearest the discharging well. This inner zone of radial compression expands outward with time. At dimensionless time T_R equal to 0.00023 the relative movement of solid particles at r_D equal to 0.01 and 0.025 changes from net radial expansion to net radial compression. At T_R equal to 0.0011 the relative movement of solid particles at r_D equal to 0.025 and 0.05 makes a similar change from net radial expansion to net radial compression. The outward migration of the boundary that separates the inner zone of radial compression from the outer zone of radial extension could be shown in Figure 10 by the envelope of maximum displacement. The envelope of maximum cumulative displacement, if plotted, would satisfy a zero radial strain condition, namely, $du_{cum\ rD}/dr = 0$.

Figure 11 also shows the same process. At any specified time

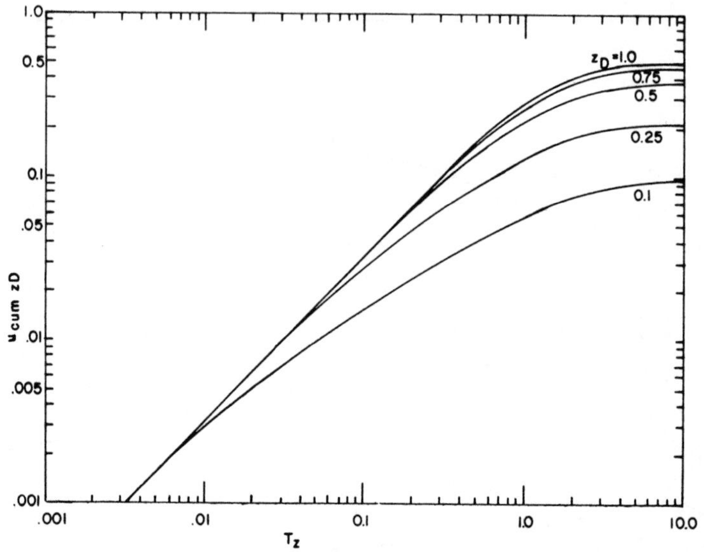

Fig. 12. Dimensionless vertical displacement $u_{cum\ zD}$ versus dimensionless time T_z for selected dimensionless elevations z_D above a depth of no vertical movement.

T_R, there is a point or particle that has moved further inward from its initial location than any other solid particle that lies on the same radial line. The locus of points of maximum displacement progressively moves outward with time and satisfies the condition $du_{cum\ rD}/dr = 0$. Because of our assumption that ultimately an infinite source of water is tapped and correspondingly that steady state flow conditions eventually prevail, the calculated values of $u_{cum\ rD}$ are arrested near T_R equal to 1 in Figure 11.

There are characteristic shapes of displacement versus time curves in Figure 10 as a function of r_D. At any specified radius, at early time the curve is a 45° line on a logarithmic scale. For late time, the line becomes horizontal. Between the initial and ultimate slopes is a transitional slope which is both flatter and longer on a logarithmic plot for points closer to a discharging well.

Figure 12 is plotted in accordance with equation (59). It shows dimensionless vertical displacement $u_{cum\ zD}$ versus dimensionless time T_z for selected dimensionless elevations z_D above a depth of no vertical movement. For early time there is no vertical compression except in a zone nearest the base as all other skeletal particles are initially moving downward at the same rate. As time progresses, a zone of vertical compression migrates upward from the base. Because we have assumed that ultimately an infinite source of water is tapped which leads eventually to steady state flow conditions, the calculated values of $u_{cum\ zD}$ are arrested near T_z equal to 10 in Figure 12. No zone of vertical extension is calculated to occur.

To compare Figures 10 and 12, we note that for a transversely isotropic aquifer, relation (68c) can be assumed; T_R equals T_z. A comparison of the vertical coordinates in Figures 10 and 12 is not so straightforward.

Rearranging terms in (58a) gives

$$u_{cum\ r} = (RS_{sk}Q_r/\pi bK_{rr})\ u_{cum\ rD} \tag{70}$$

Because 0.15 is the ultimate steady state value of $u_{cum\ rD}$ at r_D equal to 0.5 (see Figures 10 and 11), we express a near maximum value of $u_{cum\ r}$ as

$$u_{cum\ r\ max} \leq (0.15)RS_{sk}Q/\pi bK_{rr} \tag{70a}$$

where we have used (30) and (32d) for a transversely isotropic aquifer system and realize that Q_r cannot be larger than the discharge rate Q.

Rearranging terms in (59a) gives

$$u_{cum\ z} = (S_{sk}b^2 v_{zi}/K_{zz})u_{cum\ z\ D} \tag{71}$$

Because 0.5 is the ultimate steady state value of $u_{cum\ z\ D}$ at the top of the aquifer system (z_D = 1), we express a maximum value of $u_{cum\ z}$ as

$$u_{cum\ z\ max} = (0.5)S_{sk}b^2 v_{zi}/K_{zz} \tag{71a}$$

The degree to which $u_{cum\ z\ max}$ is greater than $u_{cum\ r\ max}$ can be found by taking the ratio

$$u_{cum\ z\ max}/u_{cum\ r\ max} = (3.3)\pi b^3 K_{rr}v_{zi}/Q_r RK_{zz} \quad (72)$$

Invoking the assumption of a permeability ellipse simplifies (72) to

$$u_{cum\ z\ max}/u_{cum\ r\ max} = (3.3)\pi bRv_{zi}/Q_r \quad (72a)$$

where we have used (30) and (68b).

Using (70a), we shall approximate a representative value for $u_{cum\ r\ max}$. Let us choose reasonable values for the parameters which appear in the right-hand side of (70a). For semicompressible aquifer material, S_{sk} can be expected to be approximately $1 \times 10^{-5}\ m^{-1}$. For a reasonably good aquifer, K_{rr} can be expected to be about 10 darcys and b to be about 30 m. For Q equal to $10^5\ cm^3/s$ and R equal to 10^4 m, we find

$$u_{cum\ r\ max} < 15\ cm \quad (73)$$

It would theoretically take about 4 months of pumping at this rate to reach this ultimate value. It is important to realize that although turning off the pump will allow the confined fluid pressures to recover, the horizontal ground movement will not rebound, not even in theory [Helm, 1979a]. Reversal of pressure gradients are required for solid particles to return to their initial configuration rather than merely a recovery of hydraulic head. Within a stress range greater than preconsolidation, S_{sk} of clayey material may be an order of magnitude larger than that estimated above or equal to roughly $1 \times 10^{-4}\ m^{-1}$ [Helm, 1978]. For this case, $u_{cum\ r\ max}$ is predicted to be nearly 1.5 m. However, for such material, the long-term maintaining of a large Q is unlikely. Permeable aquifer material is generally associated with smaller S_{sk} values. Less permeable interbeds of clayey material are associated with larger S_{sk} values [Helm, 1975, 1976]. It may be that horizontal skeletal movement is controlled by hydraulically connected aquifer material, whereas vertical movement is controlled by more compressible and slowly draining clay interbeds. The hydraulic properties of clay interbeds may partly control the initial vertical velocity of solids v_{zi} that appears in the right-hand side of (72) and (72a). Analysis of factors that control v_{zi} is beyond the scope of the present paper.

If we combine (72a) with the representative values that led to (73), we find that

$$u_{cum\ z\ max} < 4.5 \times 10^6 s \times v_{zi}$$

where at present we are unable to predict v_{zi}.

Conclusions

A governing equation in which $\bar{v}_s$ is the unknown has been derived (equation (24)), namely,

$$\partial \bar{v}_s / \partial t - (\bar{\bar{K}}/S_{sk}) \bar{\nabla}(\bar{\nabla} \cdot \bar{v}_s) = 0$$

Equation (24) was solved using initial and boundary conditions analogous to those used in the theory of leaky aquifers. Rather than introducing a semiconfining bed, a transversely isotropic confined aquifer was assumed. The cumulative displacement field of solids $\bar{u}_{cum}$ was found to equal

$$u_{cum\ r} = \frac{RS_{sk}Q_r}{\pi b K_{rr}} \sum_{m=1}^{\infty} \frac{J_1(\beta_m r_D)}{\beta_m^3 J_1^2(\beta_m)} [1 - \exp(-\beta_m^2 T)]$$

$$u_{cum\ z} = \frac{b^2 S_{sk} v_{zi}}{K_{zz}} \sum_{n=1}^{\infty} \frac{2}{N^3} \sin(Nz_D)[1 - \exp(-N^2 T)]$$

where

$$\bar{u}_{cum} = (u_{cum\ r},\ 0,\ u_{cum\ z})$$

The specific storage term S_{sk} in the present paper is related only to volume strain of the skeletal frame. Interstitial water is assumed incompressible as well as individual solids. If S_{sk} is smaller than roughly $1 \times 10^{-5}\ m^{-1}$, then the expansion of water should be included in the analysis. Anisotropy of skeletal deformation is contained entirely in the hydraulic conductivity tensor $\bar{\bar{K}}$.

Using the above solutions, we found the corresponding dimensionless drawdown Δh_D to equal (equation (68))

$$\Delta h_D = \ln r_D - 2 \sum_{m=1}^{\infty} \frac{J_o(\beta_m r_D)}{\beta_m^2 J_1^2(\beta_m)} e^{-\beta_m^2 T} + C_z \sum_{n=1}^{\infty} \frac{2}{N^2} \cos (N z_D) (1 - e^{-N^2 T})$$

where (equation 62b')

$$C_z = 2\pi R^2 v_{zi}/Q_r$$

Based on the plots of (68) (see Figures 7, 8, and 9), it became evident that the shape of the Theis curve is associated more with the horizontal component of skeletal compression and skeletal movement than with the vertical component. This conclusion runs counter to previous interpretations [Jacob, 1940; Verruijt, 1969] as to which component of skeletal deformation dominates the shape of the drawdown curve.

Equation (24) can be uncoupled and solved for boundary conditions distinct from those assumed in the present paper. For example, the effect of a semiconfining bed or other heterogeneities can be investigated in conjunction with the present analysis. The effect of an impermeable cap rock rather than an overlying infinite source of water needs to be studied. Such theoretical investigations are beyond the scope of the present paper.

It is clear from the present analysis that a useful equation for calculating transient movement of solids has been derived. Its link to drawdown has been established. By comparing results of this method to established theories of drawdown (Figures 6-9), the validity of (24) has to a great extent been corroborated.

Final corroboration must await comparing predicted movement to direct measurements in the field of transient movement of the skeletal frame in response to fluid withdrawal. Because in accordance with (73) the predicted maximum horizontal movement is so small and may take a long time ultimately to occur, the state of the art of field measurements of transient skeletal movement needs to advance significantly before direct field corroboration of (24) can take place.

The cumulative displacement of solids may quantitatively be small or large. The significance of the present paper is to demonstrate the intimate interconnection between the velocity of matrix material and the transient changes in hydraulic head. The dynamics of skeletal movement is not a computational addendum, but lies at the heart of the mechanics of aquifer systems.

Notation

b elevation near a discharge well of no change in vertical skeletal strain, equal to $Z(0)$, L.

c_{rr} radial coefficient of consolidation, equal to K_{rr}/S_{sk}, L^2/T.

c_{zz} vertical coefficient of consolidation, equal to K_{zz}/S_{sk}, L^2/T.

C_z weighting factor that relates horizontal and vertical components of strain.

E_v function of ε_v, dimensionless.

g gravitational acceleration, L/T^2.

h hydraulic head, L.

h_D hydraulic head, dimensionless.

h_i initial unstrained value of h, L.

$\bar{\bar{I}}$	identity matrix.
J_n	nth order Bessel function of the first kind.
$\hat{k}$	unit vertical vector.
$\bar{\bar{K}}$	hydraulic conductivity tensor, L/T.
$K_{xx}, K_{yy}, K_{rr}, K_{\theta\theta}$	horizontal components of $\bar{\bar{K}}$, L/T.
K_{zz}	vertical component of $\bar{\bar{K}}$, L/T.
p	fluid pressure, M/LT^2.
p_D	drawdown as generally used by hydrogeologists, dimensionless.
$\bar{q}$	specific discharge, L/T.
$\bar{q}_b$	bulk flux, L/T.
$\bar{q}_i$	initial unstrained value of $\bar{q}$, L/T.
Q	volume rate of fluid withdrawal, L^3/T.
Q_r	initial horizontal bulk volume rate moving through a vertical cylindrical surface at r whose an axis is at r=0, L^3/T.
r	radial distance from a discharge well, L.
r_o	radial distance associated with elevation b through the anisotropy of a permeability ellipse, L.
r_D	radial distance from a discharge well, equal to r/R, dimensionless.
R	radial distance from a discharge well denoting where no change in horizontal skeletal strain occurs, L.
S	surface, L^2.
S_s	specific storage, 1/L.
S_{sk}, S_{sk}^*	skeletal component of S_s, 1/L.
t	time, T.
T	generalized time factor.
T_r	time as generally used by hydrogeologists, equal to $c_{rr}t/r^2$, dimensionless.
T_R	time, equal to $c_{rr}t/R^2$, dimensionless.
T_z	time, equal to $c_{zz}\ t/b^2$, dimensionless.
$\bar{u}_{cum}$	displacement field of solids, L.
$u_{cum\ r}$, $u_{cum\ z}$	radial and vertical components of $\bar{u}_{cum}$, L.
$u_{cum\ r\ D}$	$u_{cum\ r}$, dimensionless.

$u_{cum\,z\,D}$ $u_{cum\,z}$, dimensionless.

U average consolidation ratio.

v_{bi} initial value of v_z on a horizontal plane at $z = b$, L/T.

v_r, v_θ, v_z components of v_s in the radial, horizontal tangential, and vertical directions, L/T.

v_{rD} radial velocity of solids, equal to $\pi Rbv_r/Q_r$, dimensionless.

$\bar{v}_s$ velocity of solids, L/T

$\bar{v}_{si}$ initial value $\bar{v}_s$, L/T.

$\bar{v}_w$ velocity of interstitial pore water, L/T.

v_{wz} vertical component of $\bar{v}_w$, L/T.

v_{zi} initial value of v_z, L/T.

v_{zD} vertical velocity of solids, equal to v_z/v_{zi}, dimensionless.

V volume, L^3.

z elevation above a specified datum, L.

z_D elevation, equal to z/b, dimensionless.

$Z, Z(r)$ elevation of no change in vertical skeletal strain, L.

β_m typical positive root that satisfied $J_0(\beta_m)$ equal to zero.

β_w compressibility of water, LT^2/M.

$\bar{\nabla}$ gradient, 1/L.

$\bar{\nabla}\cdot$ divergence, 1/L.

∇^2 Laplacian, $1/L^2$.

ε_{rr}, $\varepsilon_{\theta\theta}$ radial and horizontal tangential components of the skeletal strain tensor.

ε_v cumulative volume strain of the skeletal frame; trace of the skeletal strain tensor.

ε_{zz} vertical component of the skeletal strain tensor.

φ porosity.

ρ_s density of individual solids, M/L^3.

ρ_w density of water, M/L^3.

σ incremental change of σ_{tr}, M/LT^2.

σ_i	initial unstrained value of σ_{tr}, M/LT^2.
σ_{tr}	trace of $\bar{\bar{\sigma}}$, M/LT^2.
$\bar{\bar{\sigma}}'$	incremental change of σ'_{tr}, M/LT^2.
σ'_D	change in mean normal effective stress, dimensionless.
σ'_i	initial unstrained value of σ'_{tr}, M/LT^2.
σ'_{tr}	trace of $\bar{\bar{\sigma}}'$, M/LT^2.
$\bar{\bar{\sigma}}$	total stress tensor, M/LT^2
$\bar{\bar{\sigma}}'$	effective stress tensor, M/LT^2.

Acknowledgment. Work performed under auspices of U.S. Department of Energy.

References

Allen, D. R., Horizontal movement and surface strain due to rebound, report, Dep. of Oil Properties, Long Beach, Calif., 1971.

Bear, J., and G. F. Pinder, Porous medium deformation in multiphase flow, J. Eng. Mech. Div. Am. Soc. Civ. Eng., 104(EM4), 881-894, 1978.

Biot, M. A., General theory of three-dimensional consolidation, J. Appl. Phys., 12(2), 155-164, 1941.

Bowman, F., Introduction to Bessel Functions, 135 pp., Dover, New York, 1958.

DeWiest, R. J. M., Geohydrology, 326 pp., John Wiley, New York, 1965.

Ferris, J. G., D. B. Knowles, R. H. Brown, and R. W. Stallman, Theory of aquifer tests, U.S. Geol. Surv. Water Supply Pap. 1536-E, 69-174, 1962.

Gersevanov, N. M., The Foundation of Dynamics of Soils(in Russian), 3rd ed., Stroiizdat, Leningrad, 1937.

Hantush, M. S., Modification of the theory of leaky aquifers, J. Geophys. Res., 65(11), 3713-3725, 1960.

Hantush, M. S., Hydraulics of wells, Adv. Hydrosci., 1, 281-432, 1964.

Hantush, M. S., and C. E. Jacob, Nonsteady radial flow in an infinite leaky aquifer, EOS Trans. AGU, 36(1), 95-100, 1955.

Hatton, J. W., Ground subsidence of a geothermal field during exploitation, Geothermics (Spec. Issue 2), 2(2), 1294-1296, 1970.

Helm, D. C., One-dimensional simulation of aquifer system compaction near Pixley, California, 1, Constant parameters, Water Resour. Res. 11(3), 465-478, 1975.

Helm, D. C., One-dimensional simulation of aquifer system compaction near Pixley, California, 2, Stress-dependent parameters, Water Resour. Res., 12(3), 375-391, 1976.

Helm, D. C., Field verification of a one-dimensional mathematical model for transient compaction and expansion of a confined aquifer system, in Verification of Mathematical and Physical Models in Hydraulic Engineering, pp. 189-196, American Society of Civil Engineers, New York, 1978.

Helm, D. C., A postulated relation between granular movement and Darcy's law for transient flow, in Evaluation and Prediction of Subsidence, edited by S. K. Saxena, pp. 417-440, American Society of Civil Engineers, New York, 1979a.

Helm, D. C., Comment on Governing equations for geothermal reservoirs, by D. H. Brownell, S. K. Garg, and J. W. Pritchett, Water Resour. Res., 15(3), 723-726, 1979b.

Helm, D. C., Conceptual aspects of subsidence due to fluid withdrawal, Recent Trends in Hydrogeology, edited by T. N. Narasimhan and R. A. Freeze, Spec. Pap. Geol. Soc. Am., 189, 103-139, 1982.

Hubbert, M. K., The theory of ground-water motion, J. Geol., 48(8), 785-944, 1940.

Jacob, C. E., The flow of water in an elastic artesian aquifer, EOS Trans. AGU, 21, 574-586, 1940.

Jacob, C. E., Flow of groundwater, in Engineering Hydraulics, edited by H. Rouse, pp. 321-386, John Wiley, New York, 1950.

Lambe, T. W., and R. V. Whitman, Soil Mechanics, 553 pp., John Wiley, New York, 1969.

Lee, K. L., Calculated horizontal movements at Baldwin Hills, California, Land Subsidence, edited by A. I. Johnson, IAHS-AISH Publ., 121, 299-308, 1977.

Mikasa, M., The consolidation of soft clay--A new consolidation theory and its applications, in Civil Engineering in Japan, pp. 21-26, Japan Society of Civil Engineers, location, 1965.

Muskat, M., The Flow of Homogeneous Fluids Through Porous Media, 763 pp., McGraw-Hill, New York, 1937.

Spiegel, M. R., Theory and Problems of Laplace Transforms, 261 pp., Schaum, New York, 1965.

Stallman, R. W., Nonequilibrium type curves modified for two-well systems, U.S. Geol. Surv. Ground Water Note, 3, 1952.

Taylor, D. W., Fundamentals of Soil Mechanics, 700 pp., John Wiley, New York, 1948.

Theis, C. V., The relation between the lowering of the piezometric surface and the rate and duration of discharge of a well using ground-water storage, EOS Trans. AGU, 16, 519-524, 1935.

Verruijt, A., Elastic storage of aquifers, in Flow Through Porous Media, edited by R. J. M. DeWiest, pp. 331-376, Academic, New York, 1969.

Witherspoon, P. A., I. Javandel, S. P. Neuman, and R. A. Freeze, Interpretation of Aquifer Gas Storage Conditions for Water Pumping Tests, 273 pp., American Gas Association, New York, 1967.

Wolff, R. G., Relationship between horizontal strain near a well and reverse water level fluctuation, Water Resour. Res., 6(6), 1721-1728, 1970.

Yerkes, R. F., and R. O. Castle, Surface deformation associated with oil and gas field operations in the United States, Land Subsidence, edited by L. J. Tison, IAHS-AISH Publ., 88, 55-66, 1970.

Averaged Regional Land Subsidence Equations for Artesian Aquifers

J. Bear
Technion-Israel Institute of Technology, Haifa, Israel

M. Y. Corapcioglu
Department Civil Engineering, University of Delaware, Newark, Delaware 19711

Introduction

In recent years, in many areas of the world (for example, San Joaquin Valley, California; Houston, Texas; Mexico City; Venice), extensive withdrawal of groundwater has been accompanied by significant settling of the land surface due to the severe reduction in pore pressure. This phenomenon is known as land subsidence. Simultaneously with compaction in the vertical direction, horizontal displacement also takes place in the aquifer. They may reach significant values and cause damage to engineering structures. Horizontal ground displacement, sometimes of significant and damage-causing magnitudes, has actually been observed. In the course of developing a groundwater management plan which calls for extensive lowering of groundwater levels, below their original elevations, the planner should take into account also the nature and extent of land subsidence and horizontal displacements which might occur. The long-term prediction of the latter as a result of planned groundwater withdrawal, in turn, requires the use of mathematical models. In this paper, we review the mathematical models developed previously by the authors [Bear and Corapcioglu, 1981a, 1981b] for regional land subsidence due to pumping from aquifers. Here we shall consider only a confined aquifer. The reader is referred to these papers for some of the computational steps not repeated here.

For the sake of simplicity, in this paper we shall assume that the entire compaction of the aquifer manifests itself as land subsidence, although in reality the presence and response of the soil overlying the aquifer should be taken into account when compaction is translated into land subsidence. In reality, the compaction that takes place is mainly that of relatively soft layers and lenses imbedded within the aquifer, and (in the form of continuous semipermeable layers) between leaky aquifers. Here we shall assume that the material comprising the aquifer is deformable everywhere, i.e., has some average behavior that permits deformation at each point. We shall assume that for the range of pressure changes considered here, the aquifer material is perfectly elastic, although the general model presented here may be modified to accommodate other types of materials.

In order to determine land subsidence, we have first to determine the (effective) stress distribution in the solid skeleton and then employ some assumed stress-strain relationship to determine the distribution of strain. In determining the stress distribution, we rely on the relationship between the effective stress and the pore pressure.

Based on the effective stress concept introduced by Terzaghi [1925], two basic approaches may be found in the literature on land subsidence in addition to empirical relations between water level decline and compaction [e.g., Lofgren and Klausing, 1969]. In the first approach, originally presented by Biot [1941], a simultaneous solution is sought for the pressure in the water and for the strain in the solid matrix. Actually, Biot's theory describes the strain in the solid matrix in a three-dimensional space. Verruijt [1969], who employs Biot's approach, shows that when only vertical displacement is being considered, Biot's formulation reduces to that presented by Jacob [1940], who assumed vertical consolidation only. In the second approach, following Terzaghi [1925], water pressure is first obtained by solving a mass conservation equation in terms of pore pressure in the aquifer as the only dependent variable. In deriving this equation it is assumed that pressure changes produce changes in the effective stress, which, in turn, produce changes in the porosity of the solid skeleton. Once the water pressure distribution has been derived, the effective stress and the resulting strain distribution are determined. Finally, the latter is used to determine land subsidence. Thus Terzaghi's theory is implemented as a two-step procedure [e.g., Helm, 1975]. Pore pressure distribution is either obtained from field measurements or calculated independently by solving the fluid flow equation. In this approach the land subsidence is assumed to be one-dimensional (vertical) only. Gambolati and Freeze [1973] and Narasimhan and Witherspoon [1976] made use of this approach.

The first approach stated above, based on Biot's [1941] fully coupled three-dimensional formulation in terms of pore pressure and displacements, has been further developed by several researchers. In a coupled three-dimensional model, one fluid flow equation in terms of pressure, or piezometric head, and three equilibrium equations in terms of vertical and horizontal displacements are the governing equations. Consequently, numerical methods for the simultaneous solution of the coupled partial differential equations have been utilized [e.g., Ghaboussi and Wilson, 1973; Schrefler et al., 1977; Lewis and Schrefler, 1978; Safai and Pinder, 1979]. An alternative approach to developing a three-dimensional stress-strain approach has been proposed by Helm [1982]. It related the velocity field for solids in three-dimensions to appropriate force fields. After calculating the transient velocity field of saturated granular medium, the cumulative displacement field is found by integrating velocity over time.

An early attempt by Carrillo [1949] to arrive at a mathematical analysis of land subsidence in the Long Beach-San Pedro area constitutes another kind of approach. Carrillo represented the earth as a homogeneous, isotropic, elastic half space with a free flat upper surface. Carrillo and, later, McCann and Wilts [1951] presented 'tension center' and 'vertical pincer' models, respectively, both based on elastic continuum mechanics.

Based on the major two approaches reviewed above, two models are developed and presented in this study: one in which it is assumed, as an approximation, that the entire deformation takes only the form of vertical compaction and the other where three-dimensional soil movements are considered and both vertical and horizontal displacements of the solid skeleton take place as a result of changes in the effective stress.

Our main interest in the present paper are deformations, expressible as vertical and horizontal displacements as produced by pumping from a confined aquifer. Hence we have introduced the term 'regional' in the title to emphasize that we are not interested in a 'local' problem, where horizontal distances of interest are comparable in magnitude to the thickness of the considered aquifer. Instead, the considered problem of land subsidence and horizontal displacements, when the latter are also considered, is such that horizontal distances of interest, say between wells and points at which subsidence is determined, are much larger than the thickness of the aquifer, and corrections in values of state variables of interest, here pressure or head, along the horizontal direction are much larger than the (vertical) aquifer's thickness. This approach is used in all aquifer models for determining drawdown. To obtain these models, the three-dimensional description of flow is integrated over the thickness of the aquifer. We then have the case where flow in the aquifer may be considered as essentially horizontal, and the hydraulic approach is applicable [Bear, 1979]. In a similar way we may integrate Jacob's [1940] model of vertical displacement only, or Biot's [1941] three-dimensional consolidation model, over the thickness of an aquifer and obtain a model in which the dependent variables (averaged piezometric head in the aquifer, averaged horizontal displacement, and land subsidence) are functions of plane coordinates and of time only. The two-dimensional model thus derived is simpler than the three-dimensional one. One should note that the considerations of an aquifer system as a thin, two-dimensional slab have been reported by a number of authors [e.g., Corapcioglu and Brutsaert, 1977], but their models are different from that obtained by averaging over the aquifer's thickness. Obviously, the integrated approach should not be used for problems of a local nature.

The work presented here is divided into two parts. In the first one, a regional subsidence model is obtained by assuming vertical soil compressibility only. First, the three-dimensional mass conservation equation is developed for compressible fluid and deformable solid matrix. This equation is then integrated over the aquifer's thickness, taking into account conditions on the top and bottom surfaces bounding the aquifer. The result is a flow equation in terms of averaged piezometric head. By relating changes in head to land subsidence a single equation is then obtained for land subsidence as a single dependent variable. In this way, land subsidence is obtained as a single-step procedure. In the second part, the starting point is Biot's [1941] theory of consolidation, but the objective is the same, namely, to derive integrated equations for averaged pressure, land subsidence, and this time also average horizontal displacements, all as functions of plane coordinates and time.

Mass Conservation Equations

We start from the three-dimensional equation of mass conservation for a saturated porous medium [e.g., Bear, 1979]

$$\nabla \cdot \rho q + \frac{\partial(\rho n)}{\partial t} = 0 \qquad q = nV \tag{1}$$

where V and q are the velocity and the specific discharge of water, respectively, ρ is the density of the water, and n is the porosity of the porous medium. In deriving (1), we have neglected both the dispersive flux and molecular diffusion due to spatial variations in water density. In a deforming porous medium it is the specific discharge relative to the moving solid, q_r, that is expressed by Darcy's law:

$$q_r = q - nV_s = -K \cdot \nabla\phi^* \qquad \phi^* = z + \int_{p_o}^{p} \frac{dp}{g\rho(p)} \tag{2}$$

where V_s is the velocity of the solid, ϕ^* is Hubbert's [1940] piezometric head for a compressible fluid and K is the hydraulic conductivity tensor.

Definition of material derivatives with respect to the flowing water (at velocity V) and the deforming solids (at velocity V_s) are denoted by $d_w(\ )/dt$ and $d_s(\ )/dt$, respectively. The conservation of mass equation for solid with incompressible grains can be written as

$$\nabla.(1-n)V_s + \frac{\partial(1-n)}{\partial t} = 0 \tag{3}$$

At this point we introduce Terzaghi's [1925] concept of intergranular stress:

$$\sigma = \sigma_s - pi \tag{4}$$

where σ is the total stress, σ_s is the intergranular stress, p is the pressure in the water (positive for compression). I is the unit second rank tensor, and it is assumed that no shear stresses exist in the water.

If we assume that $d\sigma = 0$, i.e., no change in total stress, then

$$d\sigma_s = d(pi)$$

Assuming vertical compressibility only

$$d\sigma_s = dp \tag{5}$$

We may now return to (3), and in view of (5), based on the fact that $d\sigma = 0$, rewrite it as

$$\nabla.V_s = \frac{1}{1-n} \frac{d_s n}{dt} = \alpha^* \frac{d_s \sigma S}{dt} = \alpha^* \frac{d_s p}{dt} \tag{6}$$

where α^* is the coefficient of matrix compressibility of a moving solid.

The compressiblity of water (in motion) is defined by

$$\beta^* = \frac{1}{\rho}\frac{d_w p}{dt} \tag{7}$$

where we focus our attention on a fixed mass of water.

By inserting (2) into (1), employing the definition of material derivates, and making use of (6) and (7), we obtain

$$\nabla \cdot q_r + n\beta^* \frac{d_w p}{dt} + \alpha^* \frac{d_s p}{dt} = 0 \tag{8}$$

By assuming that $|n\partial p/\partial t| \gg |q.\nabla p|$ and $|\partial p/\partial t| \gg |V_s.\nabla p|$, we may approximate (8) by

$$\nabla \cdot q_r + (\alpha^* + n\beta^*)\partial p/\partial t = 0 \tag{9}$$

It can easily be verified that in writing (9), we have neglected the term $[n\beta^* V + \alpha^* V_s].\nabla p$ on the left-hand side. We note that the term $q.\nabla p$ is proportional to $(\nabla p)^2$ and hence is negligible small. Similar considerations justify the other approximations which are given by Bear and Corapcioglu [1981a].

The relative specific discharge q_r in the continuity equation can be written in the form

$$q_r = -K \cdot \nabla\phi^* = -\frac{k}{\mu} \cdot (\nabla p + \rho g \nabla z) \tag{10}$$

where k is the medium's permeability and μ is the dynamic viscosity of the water.

By assuming that $|n\partial p/\partial t| \gg |q.\nabla p|$ and (as an approximation) defining the solid and fluid compressibilities in (6) and (7) by partial derivatives rather than by material ones (using the symbols α and β, respectively, to indicate this approximation), (1) becomes

$$\nabla \cdot q + S_{op}\frac{\partial p}{\partial t} = 0 \qquad S_{op} = (1-n)\alpha + \beta n \tag{9'}$$

where S_{op} may be interpreted as the specific storativity with respect to pressure changes (equal to volume of water added to storage per unit volume of porous medium per unit rise in pressure).

Boundary Conditions on Top and Bottom of Aquifer Boundaries

To obtain equations of flow and subsidence in an aquifer, we integrate the three-dimensional equation (9) or (9') over the aquifer's thickness. By this procedure we obtain an integrated equation where the dependent variables (e.g., average pressure or effective stress) depend only on the planar coordinates x and y and on time. The bottom and top boundary surfaces of the aquifer cease to serve as boundaries of the regional model. The conditions on these boundaries become source/sink terms in the corresponding two-dimensional equations.

In order to perform the integration along the vertical, we have to know the boundary conditions on the top and bottom bounding surfaces of the aquifer. We shall consider an impervious boundary for top and bottom aquifer bounding surfaces. A more detailed discussion of the integrated aquifer equations, taking into account the conditions on the top and bottom bounding surfaces, is given by Bear [1977, 1980].

Denoting the elevation of a point on a surface by $b = b\,(x,y,t)$, the shape of this surface can be described by the function

$$F = F(x,y,z,t) \quad z - b(x,y,t) = 0 \tag{11}$$

The thickness of the aquifer is given by $B = b_2 - b_1 = F_1 - F_2$ (Figure 1). For any moving boundary, we also have

$$\frac{dF}{dt} = \frac{\partial F}{\partial t} + u \cdot \nabla F = 0 \tag{12}$$

where u is the speed of displacement of the boundary.

When we consider the flow of water, the condition to be satisfied at the upper boundary surface is

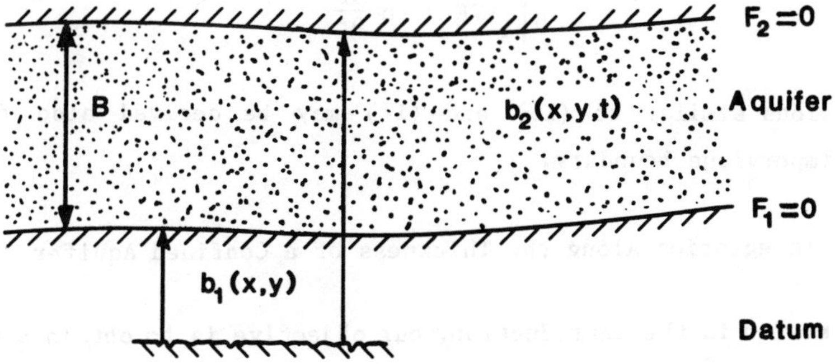

Fig. 1. Nomenclature for an artesian aquifer.

$$\begin{aligned} &[q - nu]_{u,1} \cdot \nabla F = 0 \\ &[q_r + n\,(V_s - u)]_{u,1} \cdot \nabla F = 0 \\ &[q]_{u,1} \cdot \nabla F = -[n]_{u,1} \frac{\partial F}{\partial t} \end{aligned} \tag{13}$$

where $[A]_{u,1} = A|_u - A|_1$ denotes a jump in A from the upper side (subscript u) to the lower (aquifer) side (subscript 1) of the boundary.

At an impervious boundary, $n|_u = 0$, $q|_u = 0$, $n|_1 = n$ = porosity of aquifer, hence

$$(q - nu)\Big|_1 \cdot \nabla F \equiv [q_r + n(V_s - u)]\Big|_1 \cdot \nabla F = 0 \tag{14}$$

If we also assume that F is a material surface with respect to the solid, then

$$(V_s - u)\Big|_1 \cdot \nabla F = 0 \tag{15}$$

Hence (14) written for an upper boundary F = 0 becomes

$$q_r\Big|_1 \cdot \nabla F = 0 \tag{16}$$

An alternative form of the boundary condition for an impervious (possibly moving) boundary can be obtained in terms of q from the third equation of (13)

$$q\big|_1 \cdot \nabla F = -n \frac{\partial F}{\partial t} \tag{17}$$

Expressions similar to (16) and (17) may be derived also for a lower impervious boundary.

Integration Along the Thickness of a Confined Aquifer

As stated in the introduction, our objective is to obtain a field equation to be used for predicting subsidence as a function of time and of the plane coordinates x and y only. To achieve this goal, we start by integrating (9') and describing saturated flow in a three-dimensional space, along the vertical thickness of a considered aquifer, taking into account the various boundary conditions on its top and bottom surfaces. This will yield equations written in terms of dependent variables which are averaged values. The latter are functions of time and the plane coordinates x and y only. If the procedure outlined by Bear [1977, 1979, p. 522] is followed, the integration of (9') by making use of (17) and assuming vertical equipotentials, i.e., $\bar{\phi}^* \cong \phi^*\big|_{F_2} \cong \phi^*\big|_{F_1}$ and $\bar{n} = n\big|_{F_1} \simeq n\big|_{F_2}$ (or accurately so, if the aquifer is homogeneous) would yield

$$-\nabla' \cdot B\bar{q}' = \bar{n}\frac{\partial B}{\partial t} + \overline{S_{op}}\bar{\rho} Bg \frac{\partial \bar{\phi}^*}{\partial t} \qquad \overline{(\ )} = \frac{1}{B}\int_{(B)} (\ )dz \tag{18}$$

where we have assumed a stationary aquifer bottom, $b_1 = b_1(x,y)$, and the overbar $(\bar{\ })$ symbol indicates an average over the vertical thickness $B = B(x,y,t)$ and the prime symbol, $(\ )'$, on a vector or a vector operation indicates that the vector or the operation is in the xy plane only.

Bear and Corapcioglu [1981a] have shown that $\partial B/\partial t$ could be approximated by $\bar{\alpha}^* B\, \bar{\rho}\, g\, (\overline{\partial\phi^*}/\partial t)$. Then (18) becomes

$$-\nabla' B\bar{q}' = \overline{S_o}^{**} B \frac{\partial \bar{\phi}^*}{\partial t} \tag{19}$$

where $\overline{S_o}^{**} = \bar{\rho}\, (\overline{n\alpha^*} + \overline{S_{op}})$. With approximations $\overline{q'} \cong \overline{q'_r}$, and both (18) and (19) may be approximated by

$$\nabla' . (\bar{K}' B \cdot \nabla' \bar{\phi}) = \bar{S}_o \; B \; \partial\bar{\phi}/\partial t \tag{20}$$

where $\overline{S_o} \equiv \overline{S_o}^{**}$ is the average specific storativity and we have approximated ϕ^* by ϕ.

If net withdrawal takes place at a rate of $Q = Q(x,y,t)$ (in terms of volume of water per unit area per unit time), we should add $-Q(x,y,t)$ on the left-hand side of (20).

We note that in (20) the transmissivity of the aquifer $T = \overline{K'}\,B$ and its storativity $S = \overline{S_o}\,B$ vary with B. In fact, $\bar{K}$ and $\overline{S_o}$ also depend on the porosity n, which varies continuously during the consolidation process. Usually this effect is neglected in aquifers (but not in clays).

Subsidence in a Confined Aquifer

At this point the usual procedure for determining land subsidence $\delta(x,y,t)$ in an aquifer due to pumping is to start by using (20) for determining the piezometric head distribution $\bar{\phi} = \bar{\phi}(x,y,t)$ or the drawdown $s = \bar{\phi}(x,y,0) - \bar{\phi}(x,y,t)$ in the aquifer, as produced by the pumping distribution described by $-Q(x,y,t)$. Then the total subsidence, $\delta(x,y,t) = B^o(x,y) - B(x,y,t)$ is determined by using

$$\delta(x,y,t) = B\bar{\alpha}^*\bar{\rho}\; g(s - \Delta\bar{z}) \simeq B\bar{\alpha}^*\bar{\rho}gs \begin{cases} z = b_1 + B/2 \\ \Delta\bar{z} = \bar{z}(x,y,0) - \bar{z}(x,y,t) \end{cases} \tag{21}$$

where, usually, B is taken as its initial value, B^0. One may also adjust B and n as subsidence and soil compaction progresses. Actually, (21) is valid only for our present assumption of $b_1 = b_1\,(x,y)$. Otherwise the land subsidence should be defined only as $\delta(x,y,t) = b_2^0\,(x,y) - b_2(x,y,t)$.

Instead, let us try to state the problem directly in terms of $\delta = \delta(x,y,t)$ as the dependent variable. We shall continue to assume that $b_1 = b_1(x,y)$, i.e., independent of time. We shall assume, as is common in consolidation studies, that some initial steady state exists and that pumping produces incremental effective stresses and pressures which cause subsidence. Accordingly, with ϕ replacing ϕ^*, we write for the deviation from steady state produced by pumping

$$\nabla'\cdot(\overline{K'B}\cdot\nabla'\overline{\phi}^e) - Q(x,y,t) = \overline{S}_o B\,\frac{\partial\overline{\phi}^e}{\partial t} \tag{22}$$

where superscript e denotes the stress producing incremental head.

Now

$$\frac{\partial\overline{\phi}^e}{\partial t} \equiv \frac{1}{\overline{\rho} g}\frac{\partial\overline{p}^e}{\partial t} + \frac{\partial\overline{z}^e}{\partial t} = \frac{1}{\overline{\rho} g}\frac{\partial}{\partial t}\left(\frac{\delta}{\overline{\alpha}^* B}\right) + \frac{\partial}{\partial t}\left(\frac{\delta}{2}\right)$$
$$= \left(\frac{1}{\overline{\rho} g\overline{\alpha}^* B} + \frac{1}{2}\right)\frac{\partial\delta}{\partial t} + \frac{1}{\overline{\rho} g\overline{\alpha}^*}\,\delta\,\frac{\partial}{\partial t}\left(\frac{1}{B}\right) \simeq \frac{1}{\overline{\rho} g\overline{\alpha}^* B}\frac{\partial\delta}{\partial t} \tag{23}$$

i.e., the effect of changes in the elevation of the aquifer's axis is negligible, and

$$\nabla\overline{\phi}^e \cong \frac{1}{\overline{\rho} g}\nabla\overline{p}^e + \nabla\overline{z}^e = \frac{1}{\overline{\rho} g}\nabla\left(\frac{\delta}{\overline{\alpha}^* B}\right) + \nabla\left(\frac{\delta}{2}\right) =$$
$$= \left(\frac{1}{\overline{\rho} g\overline{\alpha}^* B} + \frac{1}{2}\right)\nabla\delta + \frac{1}{\overline{\rho} g}\,\delta\nabla\left(\frac{1}{\overline{\alpha}^* B}\right) \approx \frac{1}{\overline{\rho} g\overline{\alpha}^* B}\nabla\delta \tag{24}$$

where we have assumed that $\overline{\alpha}^*B$ may vary in the xy plane; yet, we have assumed that

$$\left(\frac{1}{B}\frac{\partial B}{\partial t} + \frac{1}{\overline{\alpha}^*}\frac{\partial\overline{\alpha}^*}{\alpha t}\right) << \frac{1}{\delta}\frac{\partial\delta}{\partial t} \qquad \frac{1}{\delta}\nabla\delta >> \frac{1}{B}\nabla B + \frac{1}{\overline{\alpha}^*}\nabla\overline{\alpha}^* \qquad \frac{1}{\overline{\rho} g\overline{\alpha}^* B} >> \frac{1}{2} \tag{25}$$

With these approximations, we obtain from (20)

$$\nabla\cdot\left(\frac{\overline{k'}}{\mu\overline{\alpha}^*}\cdot\nabla\delta\right) - Q(x,y,t) = \overline{S}_v\,\frac{\partial\delta}{\partial t} \tag{26}$$

where

$$\overline{S}_v = \frac{\overline{S}_o}{\overline{\rho} g\overline{\alpha}^*} = \left(1 + \overline{n}\,\frac{\beta}{\overline{\alpha}^*}\right) \qquad \overline{K'} = \overline{k'}\,\overline{\rho} g/\mu$$

In soil mechanics the coefficient of consolidation for an isotropic medium is defined by

$$C_v = k/\mu\overline{\alpha}^* = k(1+e)/\mu a_v \simeq k(1 + e^o)/\mu a_v$$

where e and e^0 denote the void ratio and its initial value, respectively.

For clay, $\bar{n}\beta \ll \bar{\alpha}*$. Then $\overline{S_v} = [1 + \bar{n}(\beta/\bar{\alpha}*)] \simeq 1$ and (26), with $\overline{C_v} = \bar{k}'/\bar{\alpha}*\mu$, may be written in the form

$$\nabla \cdot (\bar{C}_v \cdot \nabla\delta) - Q = \frac{\partial\delta}{\partial t} \tag{27}$$

In order to solve (27) for $\delta = \delta(x,y,t)$, with $\delta(x,y,0) = 0$, in a given aquifer domain, we have still to provide boundary conditions in terms of δ on the boundaries of the flow domain in the xy plane. For example, at a sufficiently large distance from the zone of pumping, subsidence vanishes, i.e., $\delta = 0$.

Thus we have achieved our objective of constructing a mathematical model in terms of vertical land subsidence, δ. The advantage lies in determining $\delta(x,y,t)$ in one step rather than by first determining drawdown.

It is of interest to note that the thickness B is eliminated from both the partial differential equation (27) and the boundary conditions. Puzzling as this may seem at first, this is a reasonable consequence of the linearization and of the fact that for the same Q, drawdown is inversely proportional to thickness while subsidence is proportional to drawdown.

Bear and Corapcioglu give the example of a single well pumping from an infinite confined aquifer. The solution of (27) for an extensive aquifer with a single pumping well in radial coordinates is

$$\delta = \frac{Q_w}{4\pi\bar{C}_v} W\left(\frac{r^2}{4\bar{C}_v t}\right) \tag{28}$$

Equation (28) is analogous to the classical Theis solution for drawdown in a single pumping well: W( ) is the well function for a confined aquifer. Hence by averaging the three-dimensional equation of water mass conservation over the thickness of an aquifer and introducing a relationship between changes in averaged piezometric head and land subsidence, a single equation in terms of land subsidence has been obtained for pumping from a confined aquifer. The development has been based on (1) Terzaghi's concept of effective stress, (2) an assumption of essentially horizontal flow in the aquifer, and (3) an assumption of vertical aquifer compressibility only and no horizontal displacements.

As an alternative model for regional subsidence, Biot's [1941] three-dimensional equation could be averaged over the thickness of an aquifer, and such a model would provide also the estimates of averaged horizontal displacements. This approach is presented in the following section.

The Integrated Mass Conservation Equation With a Three-Dimensional Displacement Field

In the first part of this paper, equation (6) actually served as a definition for α^*. Instead, for a three-dimensional displacement field, u, one can write $\nabla . V_s = \nabla . \partial u/\partial t \cong \partial\varepsilon/\partial t$, where ε denotes the volumetric strain. Then, the mass conservation equation can be written in the form

$$\nabla \cdot q_r + \frac{\partial \varepsilon}{\partial t} + \rho g n \beta \frac{\partial \phi^*}{\partial t} = 0 \tag{29}$$

Because we are interested only in subsidence, we separate the flow into (1) initial steady flow, including possibly steady pumping, without subsidence and (2) excess flow producing subsidence. Denoting variables of part 2 by superscript e and integrating along the vertical thickness with impervious top and bottom boundaries and with the presence of distributed sinks of magnitude $\bar{Q}^e$ (x,y,t), we obtain

$$\nabla' \cdot B\bar{q}_r{}'^e + B\,\overline{\frac{\partial \varepsilon}{\partial t}} + \bar{\rho} g \bar{n} \beta B \frac{\partial \bar{\phi}^{*e}}{\partial t} + \overline{Q^e} = 0 \tag{30}$$

where $B\bar{q}_r{}'^e$ is expressible by

$$B\bar{q}_r{}'^e = -B\bar{K}' \cdot \nabla' \bar{\phi}^{*e} \tag{31}$$

In separating $\bar{q}_r{}'$ into two parts, we have neglected the effect of consolidation on k, ρ, etc. Equation (30) may be rewritten in terms of $\bar{p}^e$ by employing

$$\nabla'\bar{\phi}^* \simeq \nabla'\bar{\phi} = \frac{1}{\bar{\rho}g} \nabla'\bar{p} + \nabla'\bar{z} \qquad \phi = z\frac{p}{\bar{\rho}g} \quad \bar{z} = \frac{b_1 + b_2}{2} \tag{32}$$

$$\frac{\partial\bar{\phi}^*}{\partial t} \approx \frac{\partial\bar{\phi}}{\partial t} \approx \frac{1}{\bar{\rho}g}\frac{\partial\bar{p}}{\partial t} + \frac{\partial\bar{z}}{\partial t} \tag{33}$$

The averaged temporal rate of change of ε, $\overline{\partial\varepsilon/\partial t}$ can be obtained as follows

$$B\overline{\frac{\partial\varepsilon}{\partial t}} = B\,\overline{\nabla\cdot V_s} = \nabla'\cdot B\bar{V}'_s + V_s\Big|_{F_2}\cdot\nabla F_2 - V_s\Big|_{F_1}\cdot\nabla F_1 \tag{34}$$

Since the top and bottom surfaces are assumed to be material surfaces with respect to the solid, and also by assuming no variation in horizontal displacements along the vertical; $\bar{U}' \cong U'\Big|_{F_2} \cong \bar{U}'\Big|_{F_1}$, (34) reduces to

$$B\overline{\frac{\partial\varepsilon}{\partial t}} = \nabla'\cdot B\frac{\partial\overline{U'}}{\partial t} + \frac{\partial B}{\partial t} = \frac{\partial\overline{U'}}{\partial t}\cdot\nabla' B + B\frac{\partial}{\partial t}\nabla'\cdot\overline{U'} + \frac{\partial B}{\partial t} \tag{35}$$

The continuity equation (30) can therefore be rewritten for a confined aquifer in terms of variables $\bar{\phi}^{*e}$(or $\bar{p}^e$) and $\bar{U}'$ (or components $\bar{U}_x$ and $\bar{U}_y$) as

$$\nabla'\cdot B\bar{q}_r{}'^e + \nabla'\cdot B\frac{\partial\overline{U'}}{\partial t} + \frac{\partial B}{\partial t} + \bar{\rho}g\bar{n}\beta B\frac{\partial\bar{\phi}^{*e}}{\partial t} + \overline{Q^e} = 0 \tag{36}$$

where $B\bar{q}_r{}'^e$ can be expressed in terms of $\overline{\phi^{*e}}$ following (31). Note that

$$B(x,y,t) = \beta^o(x,y) + \Delta_z \qquad \Delta_z = U_z\Big|_{F_2} - U_z\Big|_{F_1} \tag{37}$$

which can be written by neglecting second-order terms.

The second and third terms on the left-hand side of (36) may also be written in a linearized form:

$$\nabla' \cdot B \frac{\partial \overline{U'}}{\partial t} + \frac{\partial B}{\partial t} \approx B^{o} \frac{\partial \bar{\varepsilon}}{\partial t} \qquad (36')$$

where $B = B_0 + \Delta_z$, $\Delta_z << B_0$, and we have assumed that $\overline{V}'_s \cdot \nabla' B << \partial B/\partial t$.

The Integrated Equilibrium Equations

The components of the consolidation producing incremental total stress tensor σ at a point within the flow domain satisfy the following equilibrium equations

$$\frac{\partial \sigma_{ij}'^{e}}{\partial x_j} - \frac{\partial p^e}{\partial x_i} = 0 \qquad i,j = x,y,z \qquad (38)$$

where σ'^{e} and p^e denote incremental effective stress and excess pore pressure, respectively. The summation convention is to be invoked in (38) as in subsequent equations written in indicial notation. In writing the last equation we have assumed that the initial body force remains unchanged, i.e., $f_i^e = 0$, $i = x,y,z$.

We now assume that the solid is isotropic and (for the relatively small deviations considered here) perfectly elastic. For such a solid, the stress-strain relationship is expressed by

$$\sigma_{ij}'^{e} = G\left(\frac{\partial U_i}{\partial x_j} + \frac{\partial U_j}{\partial x_i}\right) + \lambda\left(\frac{\partial U_k}{\partial x_k}\right) \delta_v \qquad (39)$$

where $\varepsilon = \partial U_k/\partial x_k$, $G = E/2(1 + v)$ is the shear modulus, E is the modulus of elasticity, ν is Poisson's ratio, and $\lambda = E/(1 + \nu)(1 - 2\nu)$; λ and G are the Lamé constants. Relationships other than (39) may also be used.

Following the methodology of deriving integrated aquifer equations, before continuing to integrate (38) and (39), we have to introduce the conditions on the top and bottom boundary surfaces. The conditions for total stress at any boundary $F = 0$ is

$$[\sigma]_{u,l} \cdot \nabla F = 0 \qquad (40)$$

or, in terms of effective stress and pressure,

$$[\sigma' - p\mathrm{i}]_{u,l} \cdot \nabla F = 0 \tag{41}$$

Separating into initial steady state and displacement-producing excess effective stress and pressure, assuming that the total stress remains unchanged, i.e., $\sigma^e = 0$, we obtain the condition

$$(\sigma'^e - p^e \mathrm{i})\Big|_{F_i} \cdot \nabla F_i = 0 \tag{42}$$

for the lower (i=1) and upper (i=2) surfaces.

Next, we integrate the first equation in (38), i.e., for i = x,

$$\begin{aligned}&\frac{\partial}{\partial x}(B\bar{\sigma}_{xx}'^e) + \sigma_{xx}'^e\Big|_{F_2}\frac{\partial F_2}{\partial x} - \sigma_{xx}'^e\Big|_{F_1}\frac{\partial F_1}{\partial x} + \frac{\partial}{\partial y}(B\bar{\sigma}_{xy}'^e)\\ &+ \sigma_{xy}'^e\Big|_{F_2}\frac{\partial F_2}{\partial y} - \sigma_{xy}'^e\Big|_{F_1}\frac{\partial F_1}{\partial y} + \sigma_{xz}'^e\Big|_{F_2} - \sigma_{xz}'^e\Big|_{F_1}\\ &- \frac{\partial}{\partial x}(B\bar{p}^e) - p^e\Big|_{F_2}\frac{\partial F_2}{\partial x} + p^e\Big|_{F_1}\frac{\partial F_1}{\partial x} = 0\end{aligned} \tag{43}$$

However, from the first equations of (42), we obtain

$$\sigma_{xx}'^e\Big|_{F_i}\frac{\partial F_i}{\partial x} + \sigma_{xy}'^e\Big|_{F_i}\frac{\partial F_i}{\partial y} + \sigma_{xz}'^e\Big|_{F_i}\frac{\partial F_i}{\partial x} - P\Big|_{F_i}\frac{\partial F_i}{\partial x} = 0 \quad i = 1,2 \tag{44}$$

Since $\partial F/\partial z = 1$ for i = 1,2 by inserting (44) into (43), we obtain

$$\frac{\partial}{\partial x}(B\bar{\sigma}_{xx}'^e) + \frac{\partial}{\partial y}B\bar{\sigma}_{xy}'^e) - \frac{\partial}{\partial x}(B\bar{p}^e) = 0 \tag{45}$$

In a similar way, from the two remaining equations of (38) we obtain

$$\frac{\partial}{\partial x}(B\bar{\sigma}_{yx}'^e) + \frac{\partial}{\partial y}(B\bar{\sigma}_{yy}'^e) - \frac{\partial}{\partial y}(B\bar{p}^e) = 0 \tag{46}$$

$$\frac{\partial}{\partial x}(B\bar{\sigma}_{xz}'^e) + \frac{\partial}{\partial y}(B\bar{\sigma}_{yz}'^e) = 0 \tag{47}$$

in which all averaged values are functions of x, y and t only.

We now average the excess effective stress components to obtain expressions in terms of averaged displacements by making use of (39) and the assumption of uniform horizontal displacements along the vertical.

By inserting these expressions in (45) through (47), we obtain three equations in the four variables $\bar{p}^e$, $\bar{U}_x$, $\bar{U}_y$, and $\bar{U}_z$, all functions of x, y and t.

$$\frac{\partial}{\partial x}\left\{B\left[(\lambda + 2G)\frac{\partial \bar{U}_x}{\partial x} + \lambda\left(\frac{\partial \bar{U}_y}{\partial y} + \frac{\Delta z}{B}\right)\right]\right\} + \frac{\partial}{\partial y}\left\{BG\left(\frac{\partial \bar{U}_x}{\partial y} + \frac{\partial \bar{U}_y}{\partial x}\right)\right\} - \frac{\partial}{\partial x}(B\bar{p}^e) = 0 \tag{48}$$

$$\frac{\partial}{\partial x}\left\{BG\left(\frac{\partial \bar{U}_x}{\partial x} + \frac{\partial \bar{U}_y}{\partial y}\right)\right\} + \frac{\partial}{\partial y}\left\{B\left[\frac{\partial \bar{U}_x}{\partial x} + (\lambda + 2G)\frac{\partial \bar{U}_y}{\partial y} + \lambda\frac{\Delta z}{B}\right]\right\} - \frac{\partial}{\partial y}(B\bar{p}^e) = 0 \tag{49}$$

$$\frac{\partial}{\partial x}\left\{BG\frac{\partial \bar{U}_z}{\partial x} + G\left(\bar{U}_r\frac{\partial B}{\partial x} + U_z\Big|_{F_2}\frac{\partial F_2}{\partial x} - U_z\Big|_{F_1}\frac{\partial F_1}{\partial x}\right)\right\} + \frac{\partial}{\partial y}\left\{BG\frac{\partial \bar{U}_z}{\partial y} + G\left(\bar{U}_z\frac{\partial B}{\partial y} + U_z\Big|_{F_2}\frac{\partial F_2}{\partial y} - U_z\Big|_{F_1}\frac{\partial F_1}{\partial y}\right)\right\} = 0 \tag{50}$$

For constant λ and G (actually $\bar{\lambda}$ and $\bar{G}$) and with $B(x,y,t) = B^0(x,y) + \Delta_z(x,y,t), \Delta_z \ll B^0$, we obtain, by linearizing (48) and (49)

$$G\nabla'^2\bar{U}_x + (\lambda + G)\frac{\partial \bar{\varepsilon}}{\partial x} - G\frac{\partial(\Delta z/B^0)}{\partial x} - \frac{\partial \bar{p}^e}{\partial x} = 0 \tag{51}$$

$$G\nabla'^2\bar{U}_y + (\lambda + G)\frac{\partial \bar{\varepsilon}}{\partial y} - G\frac{\partial(\Delta z/B^0)}{\partial y} - \frac{\partial \bar{p}^e}{\partial y} = 0 \tag{52}$$

In (50) we have used $U_x\Big|_{F_2} = U_x\Big|_{F_1}$ and $U_y\Big|_{F_2} = U_y\Big|_{F_1}$. We note that in (50) we have only $\bar{U}_z$. In principle, together with (36) we have four equations for the four dependent variables. However, in these four equations we have $U_z\Big|_{F_1}$ and $U_z\Big|_{F_2}$ (and $\Delta_z = U_z\Big|_{F_2} - U_z\Big|_{F_1}$, $B = B^0 + \Delta_z$). These are actually conditions on the two boundaries, $F_1 = 0$ and $F_2 = 0$, for which we have no information. In fact, the land subsidence $U_z\Big|_{F_2}$ is the very unknown for which a solution is sought in most subsidence problems.

At this point we may continue by introducing certain simplifying assumptions instead of the missing information. For example, we may assume that the bottom of the aquifer is stationary, i.e., $U_z|_{F_1} = 0$ and that U_z varies linearly with z, i.e., $\bar{U}_z = \frac{1}{2} U_z|_{F_2} = -\delta/2$, where δ is the land subsidence (positive downward). Then we end up with four equations for $\bar{U}_x$, $\bar{U}_y$, δ, and $\bar{p}^e$ (or $\bar{\phi}^{*e}$).

Another approach is to assume that consolidation occurs under the condition of plane incremental total stress, as suggested by Verruijt [1969, p. 347]. This means that

$$\sigma_{zz}{}^{e} = 0 \quad \sigma_{xz}{}^{e} = \sigma_{zx}{}^{e} = 0 \quad \sigma_{yz}{}^{e} = 0 \; \sigma_{zy}{}^{e} = 0 \tag{53}$$

As indicated by Verruijt, this assumption is justified when the aquifer is between two soft, confining layers (e.g., clay) which cannot resist shear stress. Furthermore, this assumption also justifies $\bar{U}' = U'|_{F_2} = U'|_{F_1}$, since in a relatively thin aquifer, as implied by the plane stress assumption, lateral deformation is more or less uniform throughout the relatively small thickness. From (53), (38) reduces to two equations (i = x,y) with the boundary conditions (40) also written in xy coordinates only. Following the integration procedure which led above to (45) through (47), we now obtain (45) and (46) only. Equation (47) drops out. We could have obtained the same result from (45) through (47) by assuming that deformation occurs under conditions of plane incremental averaged total stress described by (53).

Accordingly, (50) vanishes, but (48) and (49), or (51) and (52), remain unchanged. We now have to solve (36), (48), and (49) for $\bar{p}^e$, $\bar{U}_x$, $\bar{U}_y$, and Δ_z. The needed fourth equation is now obtained from the first condition in (53), which leads to

$$\sigma_{zz}{}^{'e} = p^{e} \tag{53'}$$

and therefore, from the constitutive equation for σ'^{e}_{zz} (i.e., averaged excess effective stress displacement)

$$\bar{p}^{e} = \lambda\left(\frac{\partial \bar{U}_x}{\partial x} + \frac{\partial \bar{U}_y}{\partial y}\right) + (\lambda + 2G)\,\frac{\Delta_z}{B} = \lambda\bar{\varepsilon} + 2G\,\frac{\Delta_z}{B} \tag{54}$$

This completes the formulation of the problem where $\bar{p}^e\,(x,y,t)$ $\bar{U}_x(x,y,t)$, $\bar{U}_y(x,y,t)$, and $\Delta_z(x,y,t)$ are the sought unknowns. Usually, we assume $U_z|_{F_1} = 0$; hence $\Delta_z = U_z|_{F_2} = -\delta(x,y,t)$, equal to the land subsidence. We note that this result is identical to that obtained by assuming that U_z varies linearly along the vertical.

Initially, the values of these variables are zero. As external boundary conditions we shall usually assume vanishing values of all these variables at a distance sufficiently remote from the zone of pumping (at a rate $\bar{Q}_w^{\,e}\,(x,y,t)$).

Example: Displacements due to Pumping From a Single Well

Bear and Corapcioglu [1981b] considered consolidation in the vicinity of a single well pumping from an infinite aquifer. They started by writing all flow and equilibrium equations in cylindrical coordinates, then assumed axial symmetry, and integrated over the aquifer's thickness. Then these equations are solved simultaneously by employing the Laplace transformation. Bear and Corapcioglu [1981b] have obtained for $\bar{\phi}^e$ and δ the expressions

$$\bar{\phi} \cong \frac{\bar{p}^{e}}{\bar{\rho}^{o} g} = -\frac{Q_w}{4\pi T}\,W(u) \qquad u = \frac{r^2}{4C_v t} = \frac{Sr^2}{4Tt} \tag{55}$$

$$\delta = \frac{Q_w}{8\pi C_v}\,W(u) \tag{56}$$

This is approximately half the value obtained as (28) in the first part, where it was assumed that only vertical consolidation takes place. The latter assumption thus leads to an overestimation of subsidence as the part of the volumetric strain due to horizontal displacements has been overlooked. The ratio is not exactly one half because of the difference between the definitions of C_v

(with α'') and in the first part (with α^*); $\alpha^*/\alpha'' = (\lambda + G)/(\lambda + 2G) \neq 1$.

For the radial horizontal displacement, the solution is

$$\bar{U}_r = -\frac{Q_w r}{16\pi C_v B^o}\left[W(u) + \frac{1 - e^{-u}}{u}\right] \qquad (57)$$

which satisfied $\bar{U}_r \to 0$ as $r \to r_w$ and $z \to \infty$, and shows a maximum of U_r at $r = 1.1367(C_v t)^{1/2} = 1.136\ (Tt/S)^{1/2}$.

Summary and Conclusions

By averaging the three-dimensional equation of water mass conservation over the thickness of an aquifer and introducing a relationship between changes in averaged piezometric head and land subsidence, assuming no horizontal displacements to occur, a mathematical model in the form of a single equation in terms of land subsidence has been obtained for pumping from an artesian aquifer. Then a second model for regional subsidence is developed, based on averaging Biot's [1941] coupled three-dimensional model. Such a model also provides estimates of averaged horizontal displacements. The development is based on (1) Terzaghi's concept of effective stress, (2) an assumption of essentially horizontal flow in the aquifer, (3) elastic stress-strain relations, and (4) an assumption of plane stress and shear free boundaries. The resulting model consists of a set of averaged mass conservation and equilibrium equations, with averaged pressure, volumetric strain, and average vertical and horizontal displacements as dependent variables for which a solution is sought.

Closed form analytical solutions are derived by solving both models, using the Laplace transformation, for an example of radial flow to a single pumping well in an infinite homogeneous aquifer. The solutions for changes in piezometric head (for pressure) obtained for this example in the first model, where the flow equilibrium equations are uncoupled by introducing a coefficient of aquifer compressibility, and the solution obtained in the second model, where flow and strains are coupled, are identical. In fact, this result is obtained due to the boundary conditions of the problem. The values of land subsidence obtained in the second model are smaller than those obtained by the uncoupled approach in the first model. It is interesting to note that in a pumped aquifer, horizontal displacements are of the same order of magnitude as vertical ones, a fact which was also noted by Verruijt [1970].

Obviously, because of the various approximate assumptions involved, the results should be viewed only as an estimate of the true values. Such estimates should be sufficient for most engineering purposes. The justification for averaging over the vertical stems from the fact that we are interested here only in regional land subsidence problems, i.e., problems in which horizontal lengths of interest are much smaller than the aquifer's thickness.

Notation

b_1, b_2	elevation of bottom and top aquifer bounding surfaces.
B	aquifer thickness, equal to $b_2 - b_1$.
C_v	consolidation coefficient tensor.
ds()/dt	total derivative of () with respect to the moving solid.
dw()/dt	total derivative of () with respect to the moving water.
e	void ratio.
F = 0	equation of a surface, where $F(x,y,z,t) = z - b(x,y,t)$.
G	shear modulus.
i	unit second rank tensor.
k	permeability tensor.
K	hydraulic conductivity tensor.
n	porosity.
0,e	as superscripts, denote steady initial values and incremental, or excess, unsteady ones, causing consolidation.
p	pressure in water.
q	specific discharge (vector) of water.
q_r	specific discharge (vector) relative to solids.
Q	distributed withdrawal from an aquifer, L/T.

Q_w rate of withdrawal from a well, L^3/T.

r radial coordinate.

s drawdown.

S storativity, equal to S_oB.

S_o $\rho g(\alpha' + n\beta)$.

$S_o{**}$ $\rho g(n\alpha' + S_{op}) = \rho g(\alpha + n\beta)$.

S_{op} specific storativity with respect to pressure changes, equal to $(1 - n)\alpha + n\beta$.

T transmissivity tensor.

u,l as subscripts, denote the upper and lower side of surface $F(x,y,z,t) = 0$.

u speed of displacement of the boundary.

U solid displacement.

U_r radial displacement.

U_{os} volume of solid.

U_{ow} volume of water.

V,V_s water and solid velocities, respectively.

$W()$ well function.

x,y horizontal Cartesian coordinate.

z vertical Cartesian coordinate.

$\bar{z}$ average elevation of the midthickness of the aquifer.

$\alpha*$ coefficient of matrix compressibility, equal to $(\lambda + 2G)^{-1}$.

α'' coefficient of matrix compressibility, equal to $(\lambda + G)^{-1}$.

Δ_z $B - B_o$.

β water compressibility.

δ land subsidence.

ε volume dilatation.

λ a Lamé constant.

μ dynamic viscosity of water.

ρ density of water.

σ,σ_s total and effective stresses, respectively.

$\phi*$ Hubbert's potential, equal to $z + \int [dp/g\rho(p)]$.

$\bar{\phi}^*$ $1/B\int_{(B)}\phi^*\,dz$, and similarly for other variables and parameters.

ϕ piezometric head, equal to $z + p/\rho g$.

' over a vector or an operator (∇'), denotes vector components or operators in the xy plane only (i.e., not in α'', β', and σ').

References

Bear, J., On the aquifer's integrated balance equation, Adv. Water Resour., 1, 15-23, 1977.

Bear, J., Hydraulics of Groundwater, McGraw-Hill, New York, 1979.

Bear, J., On the aquifer balance equations, technical comment, Adv. Water Resour., 3, 141-142, 1980.

Bear, J., and M. Y. Corapcioglu, Mathematical model for regional land subsidence due to pumping, 1, Integrated aquifer subsidence equations based on vertical displacement only, Water Resour. Res., 17, 937-946, 1981a.

Bear, J., and M. Y. Corapcioglu, Mathematical model for regional land subsidence due to pumping, 2, Integrated aquifer subsidence equations for vertical and horizontal displacements, Water Resour. Res., 17, 947-958, 1981b.

Biot, M. A., General theory of three-dimensional consolidation, J. Appl. Phys., 12, 155-164, 1941.

Carrillo, N., Subsidence in the Long Beach-San Pedro area, report, pp. 67-69, 227-242, Stanford Res. Inst., Stanford, Calif., 1949.

Corapcioglu, M. Y., and W. Brutsaert, Viscoelastic aquifer model applied to subsidence due to pumping, Water Resour. Res., 13, 597-604, 1977.

Gambolati, G., and R. A. Freeze, Mathematical simulation of the subsidence of Venice, 1, Theory, Water Resour. Res., 9, 721-733, 1973.

Ghaboussi, J., and E. L. Wilson, Flow of compressible fluid in porous elastic media, Int. J. Numer. Math. Eng., 5, 419-442, 1973.

Helm, D. C., One-dimensional simulation of aquifer system compaction near Pixley, California, 1, Constant parameters, Water Resour. Res., 11, 465-478, 1975.

Helm, D. C., Conceptual aspects of subsidence due to fluid withdrawal, Recent Trends in Hydrogeology, edited by T. N. Narasimhan and R. A. Freeze, Spec. Pap. Geol. Soc. Am., 189, 103-139, 1982.

Hubbert, M. K. The theory of ground-water motion, J. Geol., 48, 785-944, 1940.

Jacob, C. E., The flow of water in an elastic artesian aquifer, Eos Trans. AGU, 21, 574-586, 1940.

Lewis, R. W., and B. Schrefler, A fully coupled consolidation model of the subsidence of Venice, Water Resour. Res., 14, 223-230, 1978.

Lofgren, B. E., and R. L. Klausing, Land subsidence due to groundwater withdrawal, Tulare-Wasco area, California, U.S. Geol. Surv. Prof. Pap., 437-B, 103 pp., 1969.

McCann, G. D., and C. M. Wilts, A mathematical analysis of the subsidence in the Long Beach-San Pedro area, California, report, 119 pp., Calif. Inst. of Technol., Pasadena, 1951.

Narasimhan, T. N., and P. A. Witherspoon, Numerical model for land subsidence in shallow groundwater systems, IAHS-AISH Publ. 121, 133-143, 1976.

Safai, N. M., and G. Pinder, Vertical and horizontal land deformation in a desaturating porous medium, Adv. Water Resour., 2, 19-25, 1979.

Schrefler, R. A., R. W. Lewis, and V. A. Norris, A case study of surface subsidence of the Polesine Area, Int. Numer. Anal. Math. Geomech., 1, 377-386, 1977.

Terzaghi, K., Erdbaumechanik auf bodenphysikalischer Grundlage, Franz Deuticke, Vienna, 1925.

Verruijt, A., Elastic storage of aquifers, in Flow Through Porous Media, edited by R. J. M. DeWiest, pp. 331-376, Academic, New York, 1969.

Verruijt, A., Horizontal displacements in pumped aquifers, Eos Trans. AGU, 51, 284, 1970.

Multiple-Well Systems in Layered Soils

Shragga Irmay
Technion-Israel Institute of Technology, Haifa, Israel

Introduction

Multiple-well systems or batteries of wells occur in water supply, oil fields, and drainage by well points. If we drive a cylindrical well down to the impervious horizontal base of a phreatic aquifer of initial head H in a homogeneous water-bearing stratum of constant hydraulic conductivity K, for small drawdowns it is customary to apply the Dupuit [1863] approximation:

$$H^2 - h^2 = (Q/\pi K)\ \ln(L/r) \tag{1}$$

Q is the well discharge; L its radius of influence, i.e., the distance where the drawdown $s=H-h$ is negligible; and $h(r)$ water depth above base at distance r from the well.

N wells W_k $(k=1,2,\ldots,N)$ of discharge Q_k and influence radius L_k each, spaced at distances $c \ll L$, affect each other's drawdown for a given total discharge Q. Forchheimer [1924] shows that in extended aquifers $(h^2/2)$ is a harmonic (or Laplace) potential:

$$(h^2/2)_{,xx} + (h^2/2)_{,yy} = 0 \tag{2}$$

The water depth h(M) at point M at distance r_k from well W_k is

$$H^2 - h^2 = (\pi K)^{-1} \sum_{k=1}^{N} Q_k\ \ln(L_k/r_k) \qquad Q = \Sigma Q_k \tag{3}$$

For $Q_1=\ldots=Q_k=\ldots=Q_N=Q/N$, and $L_1=\ldots=L_k\ldots=L_N\equiv L$,

$$H^2-h^2 = (Q/\pi K)\ln(L/R) \qquad R = (r_1 r_2 \ldots r_N)^{1/N} \tag{4}$$

The drawdown is that of a single well of discharge Q located at the equivalent distance R (= geometrical mean of the N distances r_k from M).

Equivalent Trench

Irmay (Szeps) [1935] and Irmay [1960] show that a battery of N uniformly spaced (spacing c) well points W_k of equal discharge Q/N along a line W_1W_N, (Figure 1) can be approximated by an equivalent trench AB of length $\lambda = Nc$, so that each well point is at the midpoint of a segment c. Its uniform linear discharge is Q/λ. Equation (4) may be used with R(M) at any point M given by

$$\ln R = \left(\sum_{k=1}^{N} \ln r_k\right)/N = \sum_{k=1}^{N} \ln r_k \cdot c/Nc \qquad (4')$$

Assuming N to be very large ($N \to \infty$), (4) tends to the curvilinear integral

$$\ln R(M) = \int_A^B \ln r(M,P)ds(P)/\lambda \qquad (5)$$

Where $r>c$, the sum of (4') converges rapidly to (3), even for N = 1, 2,3. When M is far from the line AB (e.g., $r>\lambda$), $r(M) = r(s)$ varies little about a mean value r_m:

$$r = r_m[1+a(s)] \qquad \int_A^B a(s)ds = 0 \qquad (6)$$

Developing into series:

$$\lambda \ln R = \int_A^B \ln r \cdot ds = \lambda \ln r_m + \int_A^B a(s)ds - \ldots$$

As $a(s) \ll 1$, $R \approx r_m$, and the equivalent distance R is almost the mean distance r_m.

Plotting lnr as function of s, lnR of (5) is the area under the curve divided by λ.

When the point M is in the vicinity of a well W_k ($r<c$), then the equivalent trench AB may be replaced by two equivalent trenches:

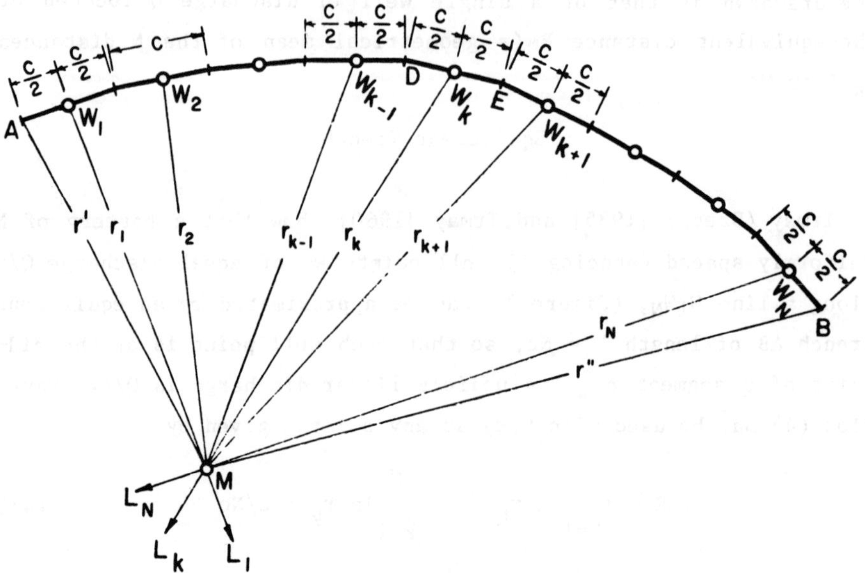

Fig. 1. Equivalent ditch.

one AD of length $\lambda' = (k-1)c$ and discharge $Q' = Q(k-1)/N$; another trench EB of length $\lambda'' = (N-k)c$ and discharge $Q'' = Q(N-k)/N$; and well point W_k of $Q_k = Q/N$. Equation (5) is then replaced by

$$\ln R(M) = \left(\int_A^D \ln r\cdot ds\right)/\lambda' + \left(\int_E^B \ln r\cdot ds\right)/\lambda'' + (\ln r_k)/N \tag{7}$$

For small drawdowns ($s<<H$),

$$H^2-h^2 = (H-h)(H+h) \approx s\cdot 2H \tag{8}$$

$$s \approx (Q/2\pi T)\ln(L/R) \tag{9}$$

$T = KH$ is the aquifer transmissivity.

In the case of an artesian aquifer of constant thickness D, the former results are valid, with the piezometric head h replacing $(h^2/2)$ as potential. Equation (4) is replaced by

$$H-h = (Q/2\pi KD)\ln(L/R) \qquad \nabla^2 h = 0 \tag{10}$$

The streamlines ψ = const are orthogonal to the equipotentials h = const of (4) or (5). Introducing the complex potential w,

$$w = \ln(\zeta+i\eta) = \frac{1}{2}\ln(\zeta^2+\eta^2) + i\tan^{-1}(\eta/\zeta) = \ln R(M,P) + i\Theta(M,P) \tag{11}$$

$$\zeta = x(P)-x(M) \qquad \eta = y(P)-y(M)$$

The streamlines obey the equation

$$\psi(M) = \int_A^B \Theta(M,P)ds(P)/\lambda = \text{const} \tag{12}$$

and $\Theta(x,y)$ can be obtained directly from $\ln R(x,y)$ by equation

$$\Theta(x,y) = \int^{(x,y)} -(\ln R)_{,x}\,dx + (\ln R)_{,y}\,dy \tag{13}$$

The main weakness of all the above formulas is the vagueness of the influence radius L. Lembke [1886, 1887] proposed

$$L = H\sqrt{K/2\varepsilon} \tag{14}$$

where ε is the rate of accretion (= rate of rain infiltration - rate of evapotranspiration). Others [Dachler, 1936; Koussakin, 1935] suggested that

$$L = \alpha\sqrt{HtK/m} \tag{14'}$$

where t is the pumping time, m drainage coefficient, and α = 3 to 2. This indeterminacy of L can be removed in the neighborhood of a shoreline, where the potential ($h^2/2$ or h) is constant. In the case of a straight-line shoreline, one has to add to each well point W_k of discharge Q_k, its image W_k' of discharge $(-Q_k)$. In the case of a battery of well points (Figure 2), (4) becomes

$$H^2-h^2 = (Q/\pi K)\ln(R/R') \qquad R' = (r_1'r_2'\ldots r_N')^{1/N} \tag{15}$$

When the battery is far from the shoreline we have $R' \approx 2g$, where g is the distance from shoreline of battery's center of gravity.

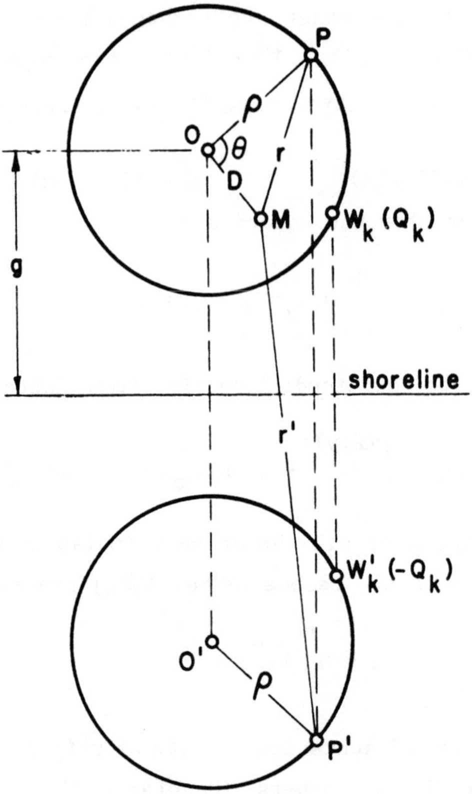

Fig. 2. Circular ditch near waterline.

Circular Battery

When the well points are on a circle of radius r at a point M at distance D from the center 0 (Figure 2), we have

$$\lambda = 2\pi\rho \qquad ds = \rho \; d\Theta$$

$$R = \phi\lambda r(M,P)ds/ = 2\pi\rho = \lambda \; (D^2 + \rho^2 - 2D\rho \; \cos\Theta)^{\frac{1}{2}} = \begin{cases} \ln D & (D>\rho) \\ \ln \rho & (D<\rho) \end{cases} \tag{16}$$

The drawdown is constant inside the battery and equal to that of a single well point of radius ρ. Outside the battery, the drawdown is the same as in the case of a single well point at center discharge Q.

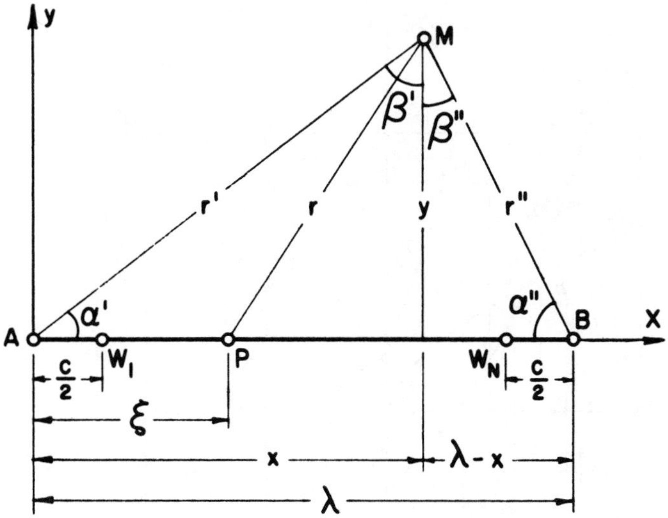

Fig. 3. Linear ditch AB.

Rectilinear Battery

For a point M(x,y) (Figure 3)

$$\lambda \ln R(M) = \int_A^B \ln r \cdot ds = \int_{\zeta=0}^{\lambda} \ln[(x-\zeta)^2+y^2]^{\frac{1}{2}} d\zeta \tag{17}$$

$$= x\ln r' + (\lambda-x)\ln r'' - \lambda + \beta y$$

$$\beta = \beta' + \beta'' = \tan^{-1}(x/y) + \tan^{-1}[(\lambda-x)/y]$$

The streamlines are defined by

$$\lambda\psi(M) = \int_{\zeta=0}^{\lambda} \tan^{-1}[y/(x-\zeta)]d\zeta = \alpha' x - \alpha''(\lambda-x) + y\ \ln(r'/r'') \tag{18}$$

$$\alpha' = \tan^{-1}(y/x) = \pi/2-\beta' \qquad \alpha'' = \pi/2-\beta''$$

when M is to the left of A, the signs of x and β' are changed. Figure 4 gives some cross sections through the aquifer. The lowest points are along AB. Drawdown is felt more on both sides than on the line outside AB. Introducing

$$\phi \equiv \lambda\ln(R/L) = (h^2-H^2)\pi K\lambda/Q \tag{19}$$

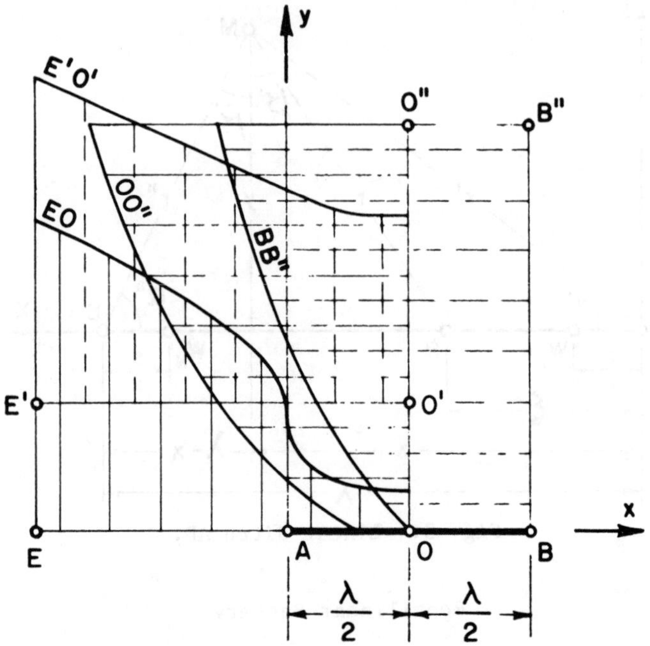

Fig. 4. Profile of water table for linear ditch AB.

the slopes of the water table in direction x and y are

$$-h,_x = -(Q/2\pi Kh\lambda)\phi,_x \qquad \phi,_x = \ln(r'/r'')$$
$$-h,_y = -(Q/2\pi Kh\lambda)\phi,_y \qquad \phi,y = \beta \tag{20}$$

See Table 1.

Polygonal Battery

In the case of an open or closed polygon of m sides, (17) can be applied to each side k:

$$\lambda \ln R(M) = \Sigma x_k \ln r_k' + \Sigma(\lambda_k - x_k)\ln r_k'' + \Sigma\beta_k y_k - \lambda$$
$$\lambda = \Sigma\lambda_k \qquad \beta_k = \beta_k' + \beta_k'' \tag{21}$$

x_k and β_k' (or $\lambda_k - x_k$ and β_k'') are negative when the normal from M to side k touches its continuation on the side of r_k' (or r_k'').

TABLE 1. Rectilinear Battery

Point	x/λ	y/λ	ln(R/L)	$-\phi,_x$	$\phi,_y$
A	0	0	1.000-2	∞	1.571
0	0.5	0	0.307-2	0	3.142
0'	0.5	0.5	1.438-2	0	1.572
E	-1.0	0	2.386-2	0.693	0

At the center of an open or closed n-polygon of m sides ($m \lesssim n$),

$$x_k = \lambda_k \quad -x_k = \lambda/2m \qquad r_k' = r_k'' \qquad \beta_k = 2\pi/n$$
$$\ln R = \ln r_k' + (\pi/m)\cot(\pi/m) - 1 \tag{22}$$

For $m \to \infty$, the polygon becomes a circle, $r_k' \to \rho$, so at the center of a circular arc of radius ρ, $R = \rho$.

In the case of a rectangle a x b (Figure 5),

$$\phi = \lambda \ln R \qquad \lambda = 2(a+b) \tag{23}$$
$$\phi,_x = \ln(r_1 r_4/r_2 r_3) + \beta_4 - \beta_2 \qquad \phi,_y = \ln(r_1 r_2/r_3 r_4) + \beta_1 - \beta_3$$

$\phi,_x$ and $\phi,_y$ are proportional to the surface slopes in directions x and y.

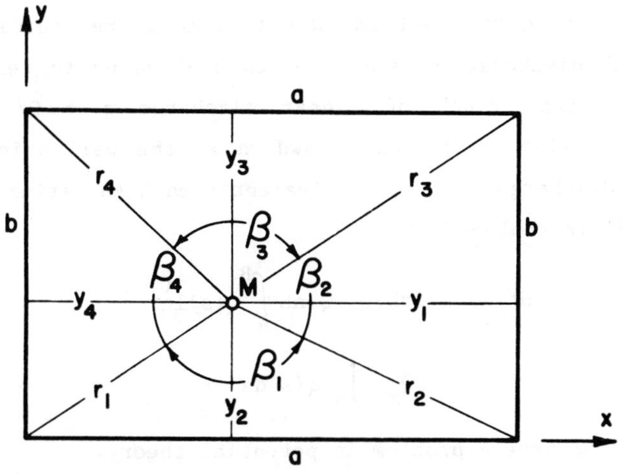

Fig. 5. Rectangular ditch.

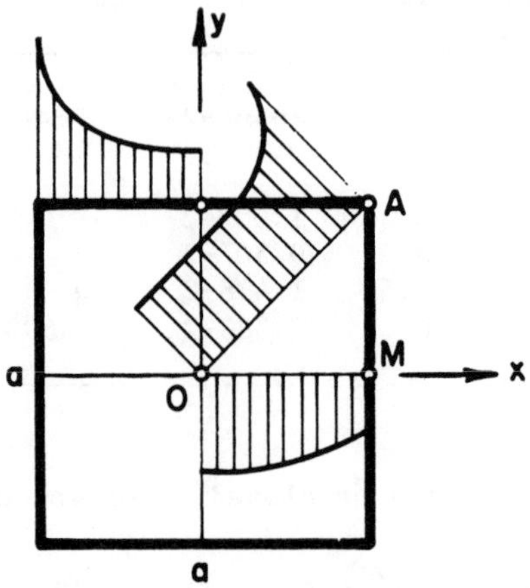

Fig. 6. Profile of water table for square ditch.

In the case of a square a x a (Figure 6) see Table 2. The water table has the form of a double saddle. The lowest points are at M (side midpoints), the highest at A (corners). The drawdown at 0 is approximately equal to that of a single well point of radius R at 0 of cross section a^2: $\pi R^2 = a^2$; $R/a = 0.566$, very close to 0.568 at 0 (see Table 2).

Constant Drawdown

We have assumed that all the N well points are equidistant (c) and of equal discharge $Q_k = Q/N$, which enabled us to replace them by an equivalent trench of linear discharge $q = Q/\lambda = Q/Nc$.

Often we prefer a constant drawdown at the well points. Then the linear discharge q of the equivalent trench AB varies along it. Equation (3) is replaced by

$$\varphi \equiv H^2-h^2 = (\pi K)^{-1}[Q\ln L-\int_A^B q(s)\ln r\cdot ds] \tag{24}$$

$$Q = \Sigma Q_k = \int_A^B q(s)ds$$

As $\nabla^2\varphi = 0$, we have a problem in potential theory.

For example, given a straight segment AB of length $\lambda = 2a$, con-

TABLE 2. Square Battery

Point	x/a	y/a	ln(R/a)	R/a	$\phi,_x$	$\phi,_y$
0	0	0	0.439-1	0.568	0	0
M	0.5	0	0.418-1	0.558	-0.604	0
A	0.5	0.5	0.566-1	0.650	∞	∞

stant potential $\phi = \phi_0$ in an infinite domain; find the equipotentials (φ = const), streamlines (ψ = const), and the linear discharge q(s)(Figure 7). The solution is well known [see, e.g., Irmay et al., 1968]: the equipotentials are confocal ellipses with foci at A and B:

$$x^2/a^2\cosh^2\varphi + y^2/a^2\sinh^2\varphi = 1 \tag{25}$$

The streamlines are confocal hyperbolas with the same foci:

$$x^2/a^2 \cos {}^2\psi \; - y^2/a^2 \sin {}^2\psi = 1 \tag{26}$$

The corresponding complex potential w(z) is

$$w(z) \equiv \varphi + i\psi = \cosh^{-1}(z/a) \qquad z = x + iy \tag{27}$$

Along AB, y = 0, $h = h_0$, $\varphi = \varphi_0 = H^2 - h_0^2$, and equation (25) gives

$$x/a = \pm \cos\psi \qquad \psi = \cos^{-1}(\pm x/a) \tag{28}$$

The specific discharge q(x) between two neighboring streamlines is proportional to $\delta\psi/\delta x \to \psi,_x$. In view of (25) we have

$$q = (Q/\pi)/(a^2-x^2)^{\frac{1}{2}} \tag{29}$$

At the vicinity of the midpoint 0 (x = 0), $q \approx Q/\pi a$, much less than in the case of uniform discharge. At the end points A and B, q is infinite. We may either use equidistant well points (distance c) of variable discharge ($Q_k = cq_k$) or well points of equal discharge

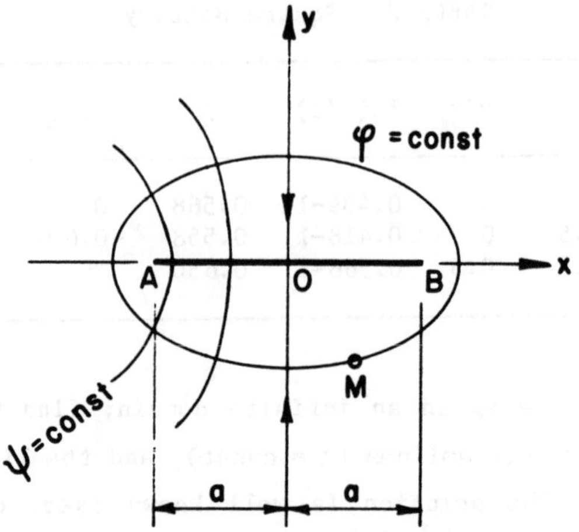

Fig. 7. Constant-head ditch AB.

($Q_k = Q/N$) but distributed nonuniformly. The graph q(x) is divided into N portions of equal area, and each well point is placed at the projection of the corresponding center of gravity (Figure 8).

At any point M on the equipotential φ, we have

$$\begin{aligned}(h^2-h_0^2)/(H^2-h_0^2) &= \ln[\mathrm{Ch}\varphi+\mathrm{Sh}\varphi)/\ln[(L+\sqrt{L^2-a^2})/a] \\ (H^2-h_0^2) &= (Q/\pi K)\cdot\ln[(L+\sqrt{L^2-a^2})/a]\end{aligned} \tag{30}$$

For L >> a = λ/2;

$$H-h_0^2 \approx (Q/\pi K)\cdot \ln(4L/\lambda) \tag{31}$$

when compared with (4), we have here R = λ/4 or ln(R/λ) = 0.614-2, which is less than in the case of uniform discharge at A (1.000-2) but more than at O (0.307-2).

Inside a closed trench the potential is constant. Outside a closed or open-ended trench we may use the usual methods of the theory of functions [e.g., Irmay et al., 1968]: conformal mapping, analytical, graphical, numerical, and other methods. An approximate method in the case of an open polygon is to divide the total discharge Q proportionally to each side's length λ_k:

$$Q_k = Q\lambda_k/\lambda \qquad \lambda = \Sigma\lambda_k \tag{32}$$

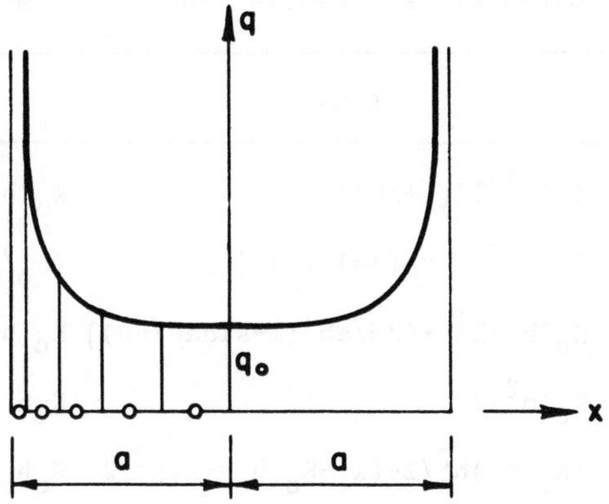

Fig. 8. Distribution of discharge along constant-head ditch.

At each side we apply (38)

$$q_k = (Q_k/\pi)/a_k{}^2 - x_k{}^2)^{\frac{1}{2}} \qquad a_k = \lambda_k/2 \tag{33}$$

Girinskii's Potential

When the subsoil consists of horizontal strata of K = K(z), variable vertically, Girinskii [1946] (and Irmay et al., [1968]) defined a new potential F(h) in steady state flow by

$$F[h(x,y)] \equiv \int_{z=0}^{h} (h-z)K(z)dz = h \int_{z=0}^{h} K(z)dz - \int_{z=0}^{h} zK(z)dz \tag{34}$$

$$= \eta\bar{K}\cdot h^2/2$$

$\bar{K}(h)$, mean value of K along a vertical, and $\eta(h)$, numerical coefficient (η = 1 for constant K), are given by

$$\int_{z=0}^{h} K(z)dz = h\bar{K}(h) \qquad \int_{z=0}^{h} zK(z)dz = \gamma\cdot h^2\bar{K}(h) \qquad \eta = 2(1-\gamma) \tag{35}$$

The derivative of this potential is the mean transmissivity $\bar{T}$:

$$F'(h) = h\bar{K}(h) = \bar{T}(h) \tag{36}$$

The total flux Q_1 in direction x per unit width $\Delta y = 1$ and Q_2 in

TABLE 3. Girinskii's Potential and Mean Transmissivity

K(z)	F(h)	$\bar{T}(h)$
$K_o(1+az)$	$K_o \cdot h^2/2 \cdot (1+ah/3)$	$K_o \cdot h \cdot (1+ah/2)$
$K_o \exp(az)$	$(K_o/a^2)[\exp(ah)-1-ah]$	$(K_o/a([\exp(ah)-1]$
$K_o(1+\delta \sin az)$	$K_o \cdot h^2/2 [1+(2\delta/ah)(1-\sin ah/ah)]$	$K_o[h+(\delta/a)(1-\cos ah)]$
$K_o(\text{for } 0<z<h_o)$	$K_o \cdot h^2/2$	$K_o h$
$K_1(\text{for } h_o<z<h_1)$	$(K_1-K_o)h_o^2/2-(K_1-K_o)h_o h+K_1 h^2/2$	$K_o h_o+K_1(h-h_o)$

direction y per unit length $\Delta x = 1$, may be derived from Darcy's formula relating the specific flux q (assumed, following Dupuit [1863], to be constant along a vertical) to the piezometric gradient (-grad h) at the surface; in Cartesian coordinates:

$$q = -K \,\text{grad}\, h \qquad q_1 = -Kh,_x \qquad q_2 = -Kh,_y \tag{37}$$

$$Q_1 = \int_{z=0}^{h} q_1 dz = -h,_x \int_{z=0}^{h} K(z)dz = -F,_x = -F'(h)h,_x = -\bar{T}(h)h,_x$$

$$Q_2 = \int_{z=0}^{h} q_2 dz = -h,_y \int_{z=0}^{h} K(z)dz = -F,_y = -F'(h)h,_y = -\bar{T}(h)h,_y \tag{38}$$

By continuity

$$Q_{1,x} + Q_{2,y} = 0 \qquad \nabla^2 F \equiv F,_{xx} + F,_{yy} = 0 \tag{39}$$

F[h(x,y)] is a harmonic function whose solutions can be obtained by analytical, function-theoretical, graphical, numerical, model, analogic, and approximate methods [Irmay et al., 1968].

Table 3 gives the values of F(h) and $\bar{T}(h)$ for several simple relationships K(z).

Battery of Well Points in Layered Soils

A battery of N uniformly spaced (spacing c) and uniformly discharging (total discharge Q) well points situated along some line, in a layered vertically inhomogeneous soil of K(z), can be approximated by an equivalent trench of total length $\lambda = Nc$ along that line, with a uniform specific discharge $q = Q/\lambda$. Applying Girinskii's potential (16), (15), and Forchheimer's [1924] procedure, one gets

$$F(H)-F(h) = (Q/2\pi)\ \ln(L/R)$$
$$\ln R(M) = \int_A^B \ln r(M,P)\, ds(P)/\lambda \tag{40}$$

In homogeneous soils, $F(h) = Kh^2/2$. For small drawdowns, $s = H-h$:

$$F(H)-F(h) \approx F'(H)\cdot(H-h) = sF'(H) \tag{41}$$

so that by (18) we get a result similar to (9):

$$s = (Q/2\pi\bar{T})\ln(L/R) \tag{42}$$

Unsteady Flow

Unsteady confined flow in a horizontal layer of constant thickness D, hydraulic conductivity K, transmissivity $T = KD$, and storativity S is [Jacob, 1950], if h is the piezometric head,

$$\nabla^2 h = (S/T)h,_t \tag{43}$$

In a phreatic aquifer of depth h(x,y,t),

$$\nabla^2(h^2/2) = (S/K)h,_t \tag{44}$$

For small drawdowns $s = H-h \ll H$, (44) becomes

$$\nabla^2 s = (S/T)s,_t \qquad T = KH \tag{45}$$

In the case of a single well, when $r^2S/4Tt<0.01$, i.e., after a long time t of pumping a constant discharge Q:

$$s \equiv H-h \approx (Q/4\pi T)\ \ln(2.25Tt/r^2S) \qquad (46)$$

Equations (43) and (45) are linear and additive, so for a battery of N wells,

$$s = H-h \approx \sum_{k=1}^{N} (Q_k/4\pi T)\ln(2.25Tt/r_k{}^2S)$$

For $Q_1 = \ldots = Q_k = \ldots = Q_N = Q/N$; $R = (r_1r_2\ldots r_N)^{1/N}$ and using an equivalent trench of length λ and uniform linear discharge $q = Q/\lambda$, we get

$$s \approx (Q/4\pi T)\ln(2.25Tt/R^2S)$$

$$\lambda\ln R = \int_A^B \ln r\cdot ds/\lambda \qquad (47)$$

In layered soils of $K = K(z)$ and constant porosity m, we may apply Girinskii's potential (16) [Polubarinova-Kochina, 1962]. Equation (21) is replaced by

$$Q_{1,x} + Q_{2,y} = -F,_{xx} - F,_{yy} = -(mh),_t = -F,_t \cdot m/F(h) = -F,_t \cdot m/\bar{T}(h)$$

When $m/\bar{T}(h)$ varies little, it may be assumed to be constant:

$$\nabla^2F = (m/\bar{T})F,_t$$

$$F(H)-F(h) \approx (Q/4\pi)\ln(2.25Tt/R^2S) \qquad (48)$$

Final Remarks

1. At a distant point $(r>\lambda)$, Q is practically independent of the number of well points and their distribution, so one should prefer simple assumptions as to the form of the polygonal trench.

2. Any closed trench of a regular shape may be replaced by a single well point of the same area.

3. For inhomogeneous nonlayered soils, the method of Irmay [1980] may be used.

Notation

a coefficient, L.

c well spacing.

D layer thickness, distance.

F Girinskii's potential.

g distance from shoreline.

H depth at L.

h depth at M(x,y).

K hydraulic conductivity.

L influence radius.

m Porosity, drainage coefficient.

N number of wells.

n number of polygon sides.

Q discharge.

q linear discharge.

$\underline{q}$ specific flux vector.

R equivalent distance.

r distance.

S storativity.

s drawdown, path.

ds element of length.

T transmissivity.

t time.

x,y,z Cartesian coordinates.

z=x+iy imaginary variable.

α coefficient, angle.

γ,δ coefficients.

ε accretion rate.

η coefficient.

θ angle.

ζ,η Cartesian coordinates.

λ trench length.

ρ circle radius.

ϕ function.

φ potential.

ψ stream function.

References

Dacher, R., Grundwasserströmung, 141 pp., Wien, Springer, 1936.

Dupuit, J., Études théoriques et pratiques sur le mouvement des eaux, 364 pp., Dunod, Paris, 1863.

Forchheimer, Ph., Hydraulik, 566 pp., Teubner, Leipzig-Berlin, 1924.

Girinskii, N. K., Complex potential of flow with free surface in a stratum of relatively small thickness and K = f(z), DAN, 51(5), 337-338, 1946.

Irmay, S. (Szeps, F.), Note sur le calcul du rabattement des nappes aquifères pare une batterie de puits filtrants, Bull. Sci. Assoc. Eleves Ecoles Spec. Univ. Liège, 32(7), 241-249, 1935.

Irmay, S., Calcul du rabattement des nappes aquifères, J. Hydraul. Nancy, 6th, 1(7), 61-70, 1960.

Irmay, S., Piezometric determination of inhomogeneous hydraulic conductivity, Water Resour. Res., 16(4), 691-694, 1980.

Irmay, S., J. Bear, and D. Zaslavsky, Physical principles of water percolation and seepage, Arid Zone Res., 29, 465 pp., 1968.

Jacob, C. E., Flow of groundwater in Engineering Hydraulics, edited by H. Rouse, pp. 321-386, John Wiley, New York, 1950.

Kussakin, I. P., Artificial lowering of groundwater level, ONTI, Moscow, 1935.

Lembke, K. E., Groundwater flow and theory of water collectors (in Russian), Zhurnal Ministerstva Putey soobshcheniya, (2), 1886.

Lembke, K. E., Groundwater flow and theory of water collectors (in Russian), Zhurnal Ministerstva Putey soobshcheniya, (17,18, 19), 1887.

Polubarinova-Kochina, P. Ya., Theory of Ground-Water Movement, translated from Russian, 613 pp., Princeton University Press, N.J., 1962.

Unsteady Drawdown in the Presence of a Linear Discontinuity

Paul R. Fenske
Water Resources Center, Desert Research Institute
Reno, Nevada 89506

Introduction

Drawdown in the presence of a linear discontinuity for the end cases, where the region beyond the discontinuity is either infinitely permeable or impermeable, has been derived using image theory, and the analytical method has been known for some time [Theis, 1941]. The steady state drawdown for cases where the region beyond the discontinuity is neither infinitely permeable nor impermeable has been presented by Verruijt [1970] and Bear [1972]. Complex unsteady state equations have been developed by Bixel et al. [1963] for pressure buildup and drawdown in the presence of linear discontinuities in oil-producing reservoirs. Nind [1965] presented both steady state and unsteady state equations for drawdown in the presence of partial hydrologic barriers (linear discontinuities). Nind provides a rigorous proof of the application of image theory to linear discontinuities under conditions of nonequilibrium flow and uniform diffusivities.

In this paper, the equations for unsteady drawdown in the presence of linear discontinuity are developed by superposition (image theory). The resultant equations are relatively simple and include analysis of diffusivity contrasts. The approach to the problem is unconventional, and the proposed equations are not shown to be exact solutions to the appropriate partial differential equation. However, if not exact, the proposed equations are considered to be good approximations and to provide insight into a well hydraulics problem that was heretofor not available.

Analysis

The components of the problem are illustrated in Figure 1, where two regions of differing transmissivities and diffusivities are separated by a boundary which, for simplicity of analysis, coincides with the y axis. The discharging well is located in region 1, $x > 0$. The usual assumptions of the Theis equation apply. The problem is solved when the unsteady drawdown at any location within region 1, $x > 0$, and region 2, $x < 0$, is described.

In the general case, one can consider the drawdown to consist of three components: the direct drawdown, or drawdown directly caused by the discharging well; the reflected drawdown, or drawdown caused by reflection into region 1 from the boundary; and the transmitted drawdown, or the drawdown which occurs in region 2 as a result of the discharging well in region 1. In region 1, the total drawdown will consist of the sum of the drawdown caused by the discharging well and the drawdown reflected by the boundary. The reflected drawdown, caused by the presence of the boundary, will be represented in region 1 by an image well located in region 2, the same distance from the boundary as the real well in region 1. The transmitted drawdown in region 2 will be represented by a component of the drawdown caused directly by the discharging well in region 1. Three conditions must be satisfied along the boundary.

1. The drawdown at the boundary must be equal on both sides of the boundary,

2. The specific discharge normal to the boundary at any point must be equal along both sides of the boundary, that is, the specific discharge into the boundary from one region must equal the specific recharge out of the boundary into the other region,

3. The tangent law of refraction of equipotential lines must be obeyed.

If the first two conditions are met, the third condition is satisfied. The law of refraction of equipotential lines derived from this analysis is a generalized form applicable to both steady state and transient flow conditions.

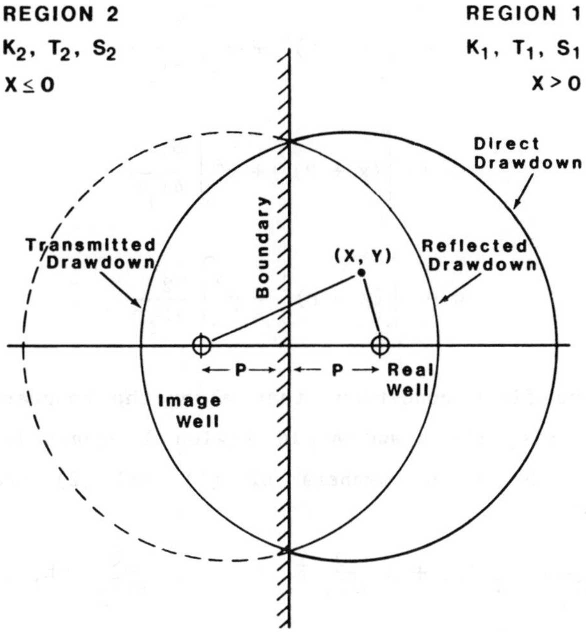

Fig. 1. Aquifer with linear discontinuity.

Drawdown in Region 1, x > 0

Drawdown in region 1 is the sum of the direct drawdown created by the discharging well and the portion of that drawdown reflected from the discontinuity:

$$s_1 = \frac{Q}{4\Pi T_1} E_1 \ (u) + \alpha \frac{Q}{4\Pi T_1} E_1 \ (v) \tag{1}$$

Drawdown in Region 2, x < 0

The drawdown in region 2 is the transmitted drawdown and is a proportion of the drawdown caused by the discharging well in region 1:

$$s_2 = \beta \frac{Q}{4\Pi T_2} E_1 \ (w) \tag{2}$$

The arguments of the exponential integrals used in (1) and (2) are defined in (3), (4), and (5). Along the boundary, x = 0, (3) and (4) are equal:

$$u = \left[(x - P)^2 + y^2\right] \frac{S_1}{4T_1 t} \tag{3}$$

$$v = \left[(x + P)^2 + y^2\right] \frac{S_1}{4T_1 t} \tag{4}$$

$$w = \left[(x - P)^2 + y^2\right] \frac{S_2}{4T_2 t} \tag{5}$$

To satisfy the first condition, that along the boundary, that is along the y axis, the drawdown in region 1 equals the drawdown in region 2, the right members of (1) and (2) are equated:

$$\frac{Q}{4\Pi T_1} E_1 (u) + \alpha \frac{Q}{4\Pi T_1} E_1 (v) = \beta \frac{Q}{4\Pi T_2} E_1 (w) \tag{6}$$

Then for the condition $x = 0$, the relationship between α and β is

$$1 + \alpha = \beta \frac{T_1}{\gamma T_2} \tag{7}$$

where

$$\gamma = \frac{E_1 (u')}{E_1 (w')} \tag{8}$$

To meet the requirements of condition 2, that the specific discharge normal to the boundary on either side of the boundary be equal, the partial differential of the drawdown parallel to the x axis has to be determined for each region and multiplied by the hydraulic conductivity for each region to obtain the specific discharges:

$$K_1 \frac{\partial s_1}{\partial x} = - \frac{Q}{4\Pi b} \frac{e^{-u}}{u} \frac{(2x-2P)}{4T_1 t} S_1 - \alpha \frac{Q}{4\Pi b} \frac{e^{-v}}{v} \frac{(2x+2P)}{4T_1 t} S_1 \tag{9}$$

$$K_2 \frac{\partial s_2}{\partial x} = - \beta \frac{Q}{4\Pi b} \frac{e^{-w}}{w} \frac{(2x-2P)}{4T_2 t} S_2 \tag{10}$$

The right member of (9) is set equal to the right member of (10). For condition 2 the relationship between α and β is

$$1 - \alpha = \frac{\beta}{\tau} \tag{11}$$

where

$$\tau = \frac{e^{-u'}}{e^{-w'}} \tag{12}$$

Equations (7) and (11) can then be solved simultaneously to obtain α and β in terms of the hydraulic conductivity of region 1 and region 2, or the transmissivity, assuming no discontinuity in aquifer thickness between region 1 and region 2:

$$\alpha = \frac{\tau T_1 - \gamma T_2}{\tau T_1 + \gamma T_2} \tag{13}$$

$$\beta = \frac{2\tau\gamma T_2}{\tau T_1 + \gamma T_2} \tag{14}$$

Since condition 3 requires that the tangent law of refraction of flow lines [Hubbert, 1940] be obeyed in the immediate vicinity of the linear discontinuity, the unsteady generalization of this law can be derived using these values for α and β. For convenience the refraction of equipotential lines orthogonal to the flow lines will be considered. The slope of the tangent of an equipotential line is the differential of y with respect to x. This tangent slope for (1), region 1, for any point along the y axis, x = 0, is

$$\left(\frac{dy}{dx}\right)_1 = \frac{(1-\alpha)p}{(1+\alpha)y} \tag{15}$$

Similarily, for region 2, x = 0, the tangent slope at any point along the y axis is

$$\left(\frac{dy}{dx}\right)_2 = \frac{p}{y} \tag{16}$$

The ratio between the two slopes, substituting for α, is the tangent law of refraction of equipotential lines:

$$\frac{\left(\frac{dy}{dx}\right)_2}{\left(\frac{dy}{dx}\right)_1} = \frac{1+\alpha}{1-\alpha} = \frac{\tau T_1}{\gamma T_2} \tag{17}$$

This is a form of the law of refraction of equipotential lines generalized to cover unsteady as well as steady state conditions. If diffusivities are the same in both regions, $\gamma = \tau$, and the law of refraction of equipotential lines reduces to the familiar equation.

The forgoing analyses are based upon unconventional application of image theory in that α and β are functions of time and the diffusivity ratio. For the case of constant α and β the equations are exact solutions to partial differential equations for groundwater movement, and Nind [1965] has shown that the requirements of the physical problem are met. It has not been shown that the equations as used here are exact solutions to the partial differential equations. If not exact solutions, however, the writer believes the solutions presented are good and useful approximations on the basis of the following rationale:

1. The conditions are met that drawdown at infinity is zero, drawdown and specific discharge are equal to both sides of the linear discontinuity, a singularity exists at which the discharge occurs, only one value of drawdown is found at any point at any time, and the limiting cases for recharging or discharging boundaries are satisfied exactly.

2. After initial transients both τ and γ and therefore α and β tend to change slowly with time. The straight-line segments of the curves therefore should be good approximations.

3. An infinite number of curves each with constant α or β and each of these curves representing an exact solution for the condition of uniform diffusivity can be constructed from (1) or (2). A family of curves for an infinite number of constant values of α,

for example, will also contain at one point on each curve the solution to (13) for differing diffusivities. This infinite number of points each satisfying (13) will represent the curve satisfying (1) with α dependent upon time and diffusivity ratio.

4. Reflection indicates that the dimensionless time drawdown curves derived from this analysis are qualitatively correct. If diffusivity in the region of the discharging well is higher, the drawdown will propagate more rapidly on the discharging well side of the linear discontinuity, temporarily causing steep hydraulic gradients on the other side of the discontinuity and high specific discharge through the discontinuity. This will cause the discontinuity to appear initially as a recharging boundary to an observation well in region 1, $x > 0$. The reverse of this argument applies when diffusivity is higher beyond the linear discontinuity, $x < 0$. In this case drawdown propagates more rapidly in region 2, $x < 0$, creating a relatively low gradient toward the discontinuity and low specific discharge across the discontinuity, which appears to an observation well in region 1, $x > 0$, as a discharging boundary. Figures 2 and 4 respectively, exhibit these expected phenomena.

Observation Well in Region 1

To illustrate the effect of linear discontinuities, semilogarithmic type curves are presented for region 1 in Figures 2, 3, and 4 for a K value [Stallman, 1963] of 20, selected values of transmissivity ratios between 0 and ∞, and diffusivity ratios from 10^{-6} to 10^{+6}. One would probably not encounter these extreme diffusivity contrasts in nature, but they are used here to illustrate the effect of diffusivity contrasts. For simplicity the observation well is assumed to be on the x axis, $y = 0$, and appropriate values of the ratio u'/u are used to correspond with $K = 20$.

As might be expected, if the transmissivity in region 2 is as much as 10 times greater than the transmissivity in region 1, the effect is sufficiently close to an infinite ratio to be, in practice, indistinguishable from a recharging boundary in the usual pumping

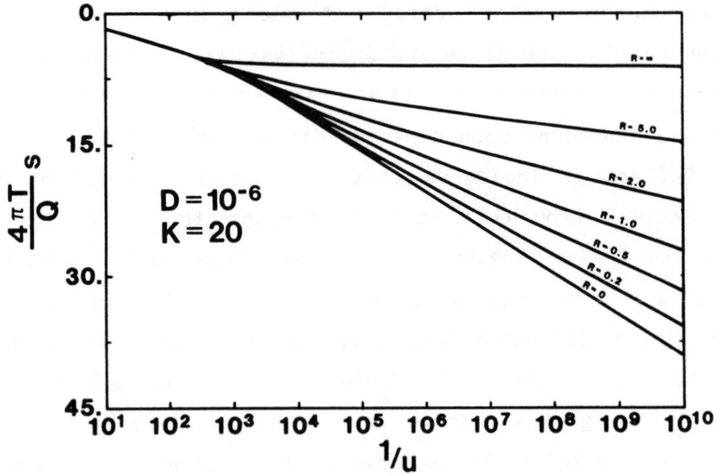

Fig. 2. Well response in region.

test data. Similarly, if the transmissivity in region 2 is any less than one tenth of the transmissivity of region 1, region 2 would essentially appear as impermeable in pumping test data. The distribution of the type curves in Figures 2, 3, and 4 shows that, as the transmissivity of region 2 approaches zero or infinity, the system becomes insensitive to diffusivity contrasts.

In Figure 2, the diffusivity in region 2 is much greater than

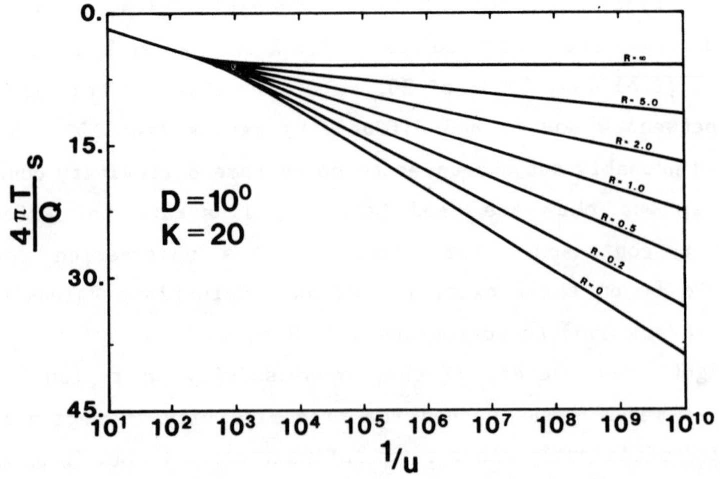

Fig. 3. Well response in region 1, uniform diffusivity.

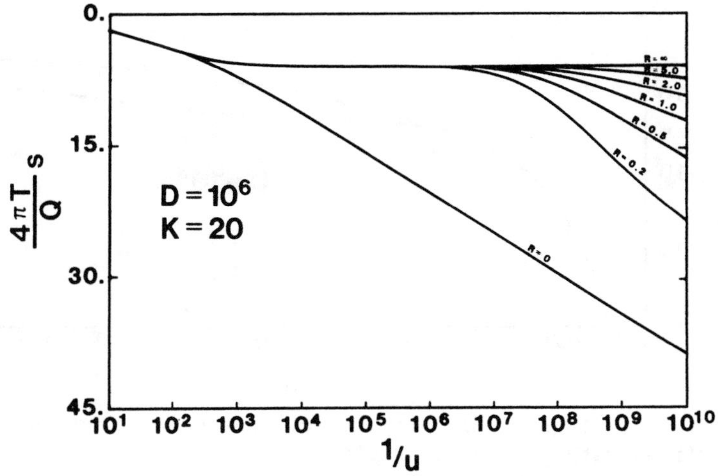

Fig. 4. Well response in region 1, diffusivity higher in region.

the diffusivity in region 1. This results in an early discharging boundary segment in the well hydraulics data. The extreme case of zero transmissivity in region 2 is indicated to show the slope of the discharging boundary curve.

Figure 3 illustrates the condition of uniform diffusivity. The cases of zero and infinite transmissivities in region 2 are presented for comparison. These two extremes are the end-members for all cases.

In Figure 4, the diffusivity in region 2 is much less than the diffusivity in region 1. Here again, as the transmissivity of region 2 approaches zero, the sensitivity to diffusivity contrasts disappears. When the transmissivity in region 2 is greater than zero to larger than the transmissivity of region 1, an early apparent recharging boundary segment appears in the well hydraulics data. This recharging segment appears even though the transmissivity in region 2 is significantly smaller than the transmissivity in region 1.

The similarity between delayed yield through a linear discontinuity and Boulton's [1963] delayed yield for an unconfined aquifer is evident for the case of uniform transmissivity. In both cases

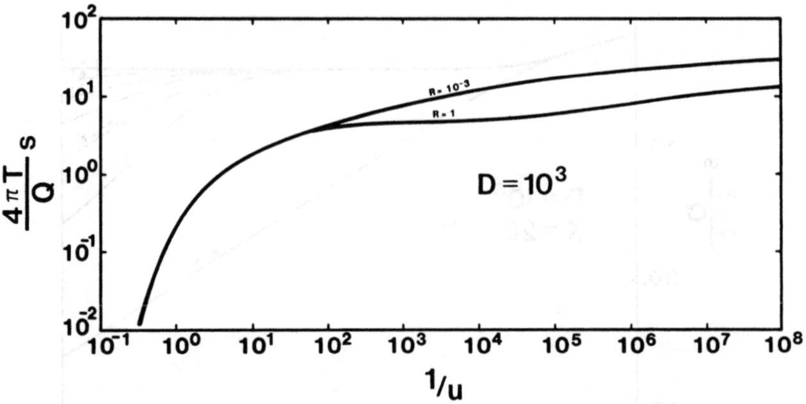

Fig. 5. Logarithmic plot of well response in region 1, diffusivity greater in region 1.

there is a shift during discharge from a smaller storage coefficient (higher diffusivity in region 1) to a larger storage coefficient (lower diffusivity in region 2) without a change in transmissivity. Since uniform transmissivities are not required for this response, the general case is that the effective diffusivity be higher early and lower later in the discharge period. Care should be exercised in interpreting well hydraulics data to differentiate between vertical delayed yield and lateral leakage through a vertical boundary.

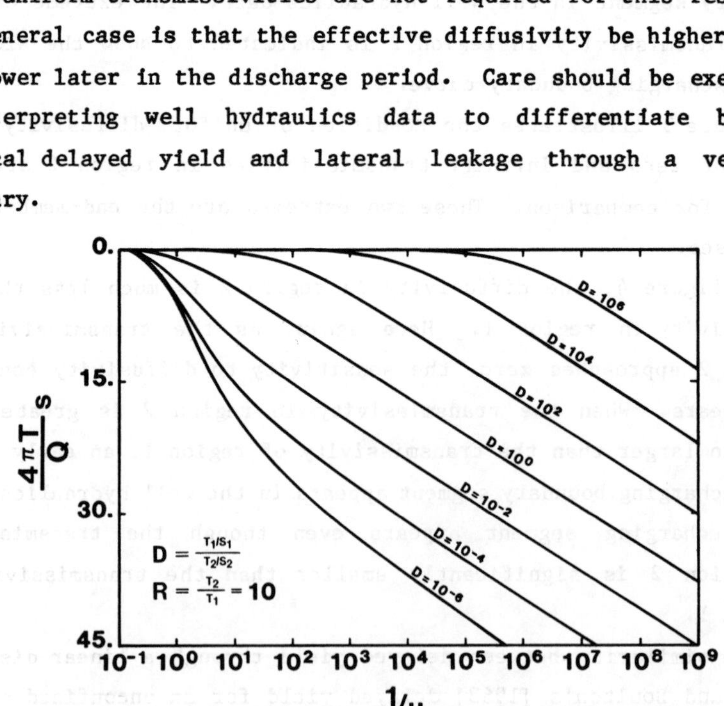

Fig. 6. Well response in region 2.

Figure 5 shows several logarithmic type curves plotted for a diffusivity ratio of 10^3, that is, the diffusivity in region 1 is 10^3 times the diffusivity in region 2. The different curves are obtained for K = 20, the ratio of the distance to the image well to the distance to the pumping well, $T_1 = T_2$, for the lower curve, and $T_1 = 10^3 T_2$ for the upper curve. The upper curve represents drawdown in the vicinity of a nearly impermeable boundary. The lower curve illustrates the similarity to Boulton's delayed yield curves and could represent a vertical linear discontinuity caused by a change from confined to unconfined conditions in the vicinity of the discharging well.

Observation Well in Region 2

Beyond the linear discontinuity, the drawdown of the well appears to obey the Theis equation when the diffusivity on the far side of the linear discontinuity is equal to or less than the diffusivity in the region of the discharging well. For the condition of higher diffusivity beyond the linear discontinuity, the drawdown versus time data at the observation well has an initial period of more rapid drawdown reflecting the lower diffusivity in the region of the discharging well. Figure 6 illustrates the condition at the observation well for a wide range of diffusivity contrasts when the transmissivity on the far side of the linear discontinuity in the region of the observation well is 10 times greater than the transmissivity in the region of the discharging well.

For uniform diffusivity, the apparent transmissivity obtained from a standard straight-line solution of observation well data will be the arithmetic average of the transmissivity in the region of the discharging well and the transmissivity in the region of the observation well. For nonuniform diffusivity the apparent transmissivity will be approximately the arithmetical average. Depending upon the differences in diffusivities, the apparent storage coefficient calculated at the observation well may be higher than, equal to, or lower than the true storage coefficient by a factor proportional to the diffusivity ratio. For example, if the diffusivity beyond the linear discontinuity is lower than the diffusivity in the vicinity of the discharging well, the apparent storage coefficient could be orders of magnitude higher than the real storage coefficient and could, in fact, exceed one. This situation could occur, for example, if an aquifer changes from a confined aquifer in the vicinity of the discharging well to an unconfined aquifer in the vicinity of the observation well. It is evident that a complete analysis of well hydraulics data requires both discharging and observation well data.

Straight-Line Solution

If equation (1) is plotted on semilog paper (Figures 2, 3, and 4) with the logarithmic axis representing dimensionless time, two straight-line segments connected by a curved line appear. For high diffusivity contrasts a recharging or discharging boundary straight-line segment will also appear. The early straight-line segment represents drawdown versus time in the aquifer before the boundary is felt. The late straight-line segment represents the effect of region 2 on the drawdown. The connecting segment is a function of diffusivity contrasts. For the case of uniform diffusivity, Figure 3, the transmissivities of both regions 1 and 2 can be calculated using a variation of Jacob's straight-line solution [Cooper and Jacob, 1946]. By substituting the long-time approximation of the exponential integrals in (1) the drawdown per log cycle of the first straight-line segment is

$$\Delta s' = \frac{Q}{4\Pi/T_1} \tag{18}$$

The drawdown per log cycle for the second straight-line segment is:

$$\Delta s'' = \frac{Q}{4\Pi T_1}(1 + \alpha) \tag{19}$$

For the analyses of well data, the transmissivity of the region in

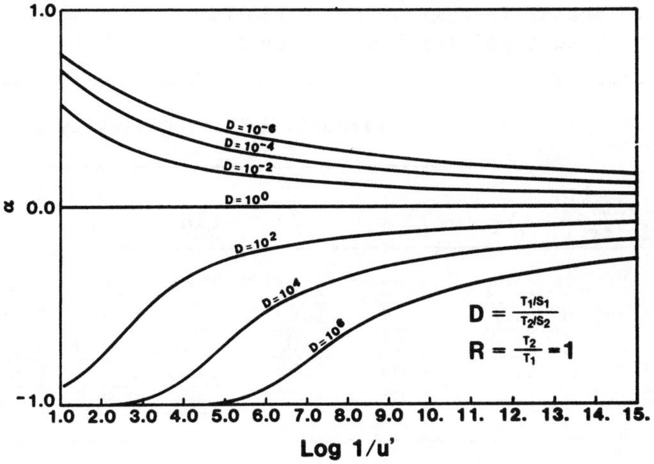

Fig. 7. Variation of o with dimensionless time.

which the well is discharging can be calculated from the first straight-line segment of the drawdown versus time curve on semilogarithmic paper, and the transmissivity of the second region can be determined by using the ratio between (18) and (19):

$$T_2 = T_1 \left(2 \frac{\Delta s'}{\Delta s''} - 1\right) \tag{20}$$

Examination of the figures presented shows that for a wide range of diffusivity ratios, the well hydraulics data will eventually plot as apparent straight lines regardless of the diffusivity ratios. If the ratio is larger than 1, as has been pointed out earlier, a recharging boundary segment appears on the semilogarithmic time-drawdown graph. If the ratio is less than 1, a discharging boundary segment appears on the semilogarithmic time-drawdown graph.

To investigate the error involved in calculating the transmissivity ratio from (20) for cases other than uniform diffusivity, the slopes of the second straight-line segments in the time drawdown graphs were calculated for a wide range of diffusivity ratios and a wide range of transmissivity ratios. Using the equation for the straight-line solution, it was found that in most cases, the transmissivity ratio for other than for uniform

TABLE 1. Actual Transmissivity Ratio Divided by Calculated Transmissivity Ratio Using Equation (20)

	Diffusivity 1/Diffusivity 2						
Transmissivity 2 / Transmissivity 1	10^{-6}	10^{-4}	10^{-2}	10^{0}	10^{2}	10^{4}	10^{6}
5	1.05	1.03	1.01	1.00	0.988	0.978	0.971
2	1.06	1.04	1.02	1.00	0.990	0.985	0.990
1	1.07	1.04	1.02	1.00	0.990	0.990	1.01
0.5	1.08	1.05	1.02	1.00	0.992	1.00	1.04
0.2	1.08	1.05	1.02	1.00	0.995	1.01	1.08

Underlined numbers delineate conditions in which diffusivity is directly proportional to transmissivity.

diffusivity was within 5% of the actual transmissivity ratio. This is because, at long times, the value of α does not change significantly over one log cycle, as indicated by Figure 7. The second straight-line segment is therefore displaced without an excessive change in slope. Table 1 presents the ratio between the calculated values and the actual values of transmissivity ratio for a large range of diffusivity ratios. In any but the less probable cases, it would appear that the straight-line solution of (20) for transmissivity rates would be useful. This is consistent with Nind's [1965] conclusion. The second straight-line segment, however, must be identified with care.

Application

During the summer of 1979 a series of constant discharge pumping tests were conducted in six aquifers overlying Tatum Salt Dome about 10 miles (16 km) west of Purvis, Mississippi [Fenske and Humphrey, 1980]. These aquifers were all in Miocene sand separated by clays. Six wells, one to each aquifer, were spaced around a circle of approximately 20-ft (6m) radius. The wells were drilled and tested sequentially so that during initial testing the aquifers

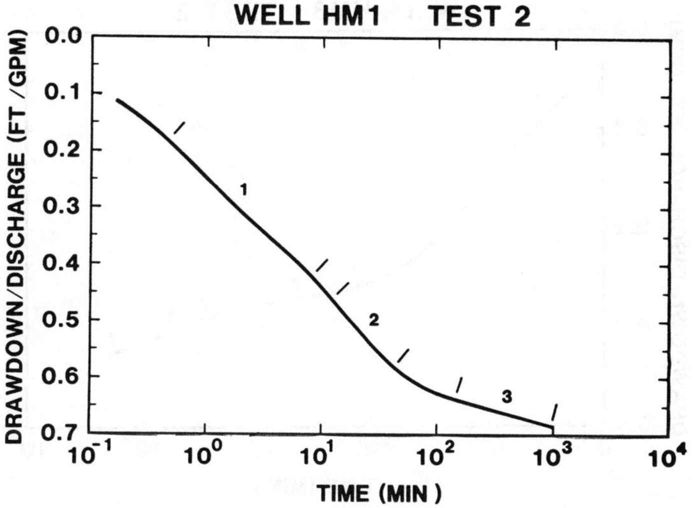

Fig. 8. Well response in aquifer 1, Tatum Dome, Mississippi.

overlying the aquifer being tested could be monitored. Subsequently, the aquifers were retested after recovery, and both overlying and underlying aquifers were monitored in all but the lowest aquifer, where only the overlying aquifer could be monitored. No effects from testing were noted in adjacent aquifers, so boundary effects are probably due to discontinuities in the aquifers. Two pumping tests are discussed here to illustrate the method of analysis in the presence of linear discontinuities. The assumption is made that the boundary is well defined and linear.

Figure 8 is a plot of drawdown versus time for a well in aquifer 1, 340 to 400 ft (104-122 m) below land surface. Aquifer 1 is apparently confined and isolated from the overlying and underlying aquifers. The aquifer was tested at a constant discharge of 60.1 gpm (227.5 l/min). The drawdown per log cycle for the first straight-line portion of the curve (segment 1, Figure 8) is 11.79 ft (3.6 m). The drawdown per log cycle for the second straight-line portion of the curve (segment 3, Figure 8) is 3.24 ft (0.99 m). Using (20), the transmissivity beyond the discontinuity in aquifer 1 is 6.3 times higher than the transmissivity in aquifer 1

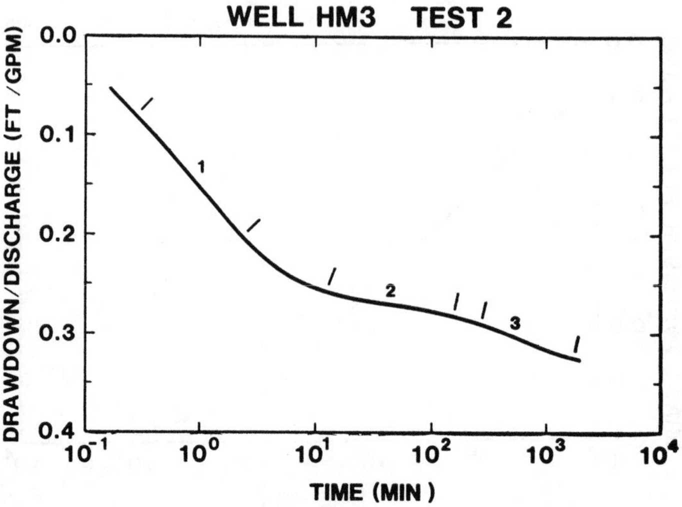

Fig. 9. Well response in aquifer 3, Tatum Dome, Mississippi.

near the discharging well. The transmissivity near the discharging well is 1344 gpd/ft (16.7 m^2/d), therefore the transmissivity beyond the discontinuity should be about 8437 gpd/ft (105 m^2/d). This is in the range of transmissivities measured elsewhere in aquifer 1. In this test the first and second straight-line segments are separated by a short discharging boundary segment (downward deflection of time-drawdown curve, segment 2, Figure 8), suggesting that the diffusivity is also greater in region 2 than in region 1.

Figure 9 is a semilog plot of the drawdown versus time for a well test in aquifer 3, 775 to 875 ft (236 to 267 m) below land surface. The aquifer was tested at a constant discharge of 65.8 gpm (249 l/min). This plot shows two straight-line segments (segments 1 and 3, Figure 8) separated by a recharging boundary segment (upward deflection of time-drawdown curve, segment 2, Figure 8). Although one cannot rule out leakage of water from confining clay layers, it would appear that this is an example of a linear discontinuity. The drawdown per log cycle for the first straight-line segment of the drawdown versus time curve is 8.94 ft (2.72 m). The drawdown per log cycle for the second straight-line portion of the curve is

3.06 ft (0.93 m). The short data base for the second straight-line segment of the drawdown versus time curve makes measurement of the drawdown per log cycle difficult, and some error might be expected. The first segment of the curve indicates a transmissivity for aquifer 3 of 1940 gpd/ft (24 m^2/d). Using (20), the transmissivity of region 2 is 9412 gpd/ft (117 m^2/d), 4.8 times higher than the transmissivity in region 1. The two straight-line segments are separated by a recharging boundary segment (segment 2, Figure 8), suggesting lower diffusivity in region 2, although the transmissivity is higher.

Application of the criterion of Papadopulos et al. [1967] for evaluating well storage effects ($t = 250\ r_c^2/T$) suggests that the first straight-line segments for both of these tests may be influenced by well storage effects. Examination of full logarithmic plots for the two tests, however, indicates no significant well storage effects within the area of the plots analyzed, and transmissivities derived from standard curve matching techniques are comparable.

Conclusion

The effect of linear discontinuities within an aquifer has been investigated for a wide range of transmissivity and diffusivity contrasts. Observation wells may be located at any point on either side of the discontinuity. Although not presented, the formulation permits projecting well interference effects for any number of arbitrarily located discharging wells. Representative dimensionless drawdown versus time curves are presented for an observation well in region 1, the same region as the discharging well and in region 2, the region on the other side of the discontinuity from the discharging well. Conclusions are as follows:

1. In the case of uniform diffusivity a plot of well hydraulics data in region 1 will consist of two straight lines on the conventional semilogarithmic plot. The first segment represents the transmissivity of region 1, and the second segment is related to the transmissivity of region 2.

2. In the case where the diffusivity of region 1 is greater than the diffusivity of region 2 a plot of well hydraulics data in region 1 will consist of two straight-line segments, as above, but these segments are separated by a recharging boundary segment. If the diffusivity contrast is great enough, the middle segment will also be a straight-line segment with zero drawdown per log cycle. Type curves are similar, but not identical, to Boulton's delayed yield curves.

3. In the case where the diffusivity of region 1 is less than the diffusivity of region 2, a plot of well hydraulic data in region 1 will consist of the two straight-line segments, as above, separated by a discharging boundary segment. For a large diffusivity contrast the middle discharging segment will become a straight-line segment with twice the drawdown per log cycle, indicating an apparent discharging boundary.

4. Transmissivity for region 1 can be calculated from the first straight-line segment on the semilogarithmic drawdown versus time plot. In the case of uniform diffusivity the transmissivity of region 2 can also be calculated using the ratio of the drawdown per log cycle for the two straight-line segments. For other than uniform diffusivities the equation for calculating the transmissivity of region 2 is a close approximation in most cases.

5. For an observation well in region 2, well hydraulics data will have a straight-line segment on semilogarithmic plots. For uniform diffusivity, transmissivity determined by conventional straight-line methods will be the arithmetic average of transmissivities of regions 1 and 2. For contrasting diffusivities this is approximately true. When the diffusivity of region 2 is less than the diffusivity of region 1, the presence of linear discontinuity is not obvious. In the opposite case an initial rapid drawdown may be seen reflecting lower region 1 diffusivity. Apparent storage coefficients calculated from region 2 data could be in error by orders of magnitude.

Two pumping tests on the top of Tatum Salt Dome near Purvis, Mississippi, appear to have linear discontinuities and are analyzed using the concept presented in this paper. The data plot for the first test contains two straight-line segments separated by a short discharging boundary segment, indicating that both the transmissivity and diffusivity is higher in region 2 than region 1. The second test contains a recharging boundary segment between the two straight-line segments, indicating a higher diffusivity in the region around the discharging well. The transmissivity, however, is higher on the far side of the linear discontinuity, region 2, for both tests.

In both tests, the differences in diffusivities on either side of the discontinuity cause an error in determining the transmissivity of region 2, the region on the opposite side of the discontinuity from the discharging well. Investigation of this potential error indicates that it is less than 5% in most instances.

Although this paper implicitly considers discontinuities caused by facies changes in the aquifer skeleton, discontinuities may also be caused by changes in fluid density, viscosity, or compressibility. The first two parameters effect the fluid conductivity of the aquifer. The third parameter effects storativity. In some of the tests at Tatum Dome, not analyzed in this paper, gas was present in the aquifers and had a significant impact on the well hydraulics data.

Notation

- b thickness of the aquifer.
- D diffusivity ratio (diffusivity in region 1/diffusivity in region 2).
- E_1 (u) exponential integral, with argument u, describing direct drawdown due to discharging well.
- E_1 (v) exponential integral, with argument v, describing reflected drawdown due to image well.
- E_1 (w) exponential integral, with argument w, describing transmitted drawdown in region 2, $x < 0$, on the other side of the discontinuity from the discharging well.

K ratio of distance to image well to distance to discharging well.

K_1 hydraulic conductivity in region 1, $x > 0$.

K_2 hydraulic conductivity in region 2, $x < 0$.

P distance from discharging or image well to linear discontinuity.

Q constant discharge of well.

R transmissivity ratio, T_2/T_1.

s_1 drawdown in region 1, $x > 0$, where the discharging well is located.

s_2 drawdown in region 2, $x < 0$.

S_1 storage coefficient in region 1, $x > 0$.

S_2 storage coefficient in region 2, $x < 0$.

$\Delta s'$ drawdown per log cycle for first straight-line segment of drawdown versus time curve on semi-logarithmic plot.

$\Delta s''$ drawdown per log cycle for second straight-line segment of drawdown versus time curve on semi-logarithmic plot.

T_1 transmissivity in region 1, $x > 0$.

T_2 transmissivity in region 1, $x < 0$.

u' argument of exponential integral along the discontinuity.

w' argument of exponential integral along the discontinuity.

α parameter describing amount of drawdown in region 1 reflected by linear discontinuity.

β parameter describing amount of drawdown transmitted across linear discontinuity into region 2.

Acknowledgments. This paper is the outgrowth of a hydrological program at Tatum Dome, Mississippi. Both the field work and this paper were supported by the Department of Energy under contract DE-AC08-81NV10162. Ron Sheen and Dennis Ghiglieri of Water Resources Center of the Desert Research Institute did the computer programming, plotting of analytical results, and data management for the hydrological program and for this paper.

References

Bear, J., Dynamics of Fluids in Porous Media, Elsevier, New York, 1972.

Bixel, H.C., B.K. Larkin, and H.K. Von Poollen, Effect of linear discontinuities on pressure build-up and drawdown behavior, J. Pet. Technol., 15, 885-895, 1963.

Boulton, N.S., Analysis of data from non-equilibrium pumping tests allowing for delayed yield from storage, Proc. Inst. Civ. Eng., 26, 469-482, 1963.

Cooper, H.H., and C.E. Jacob, A generalized graphical method for evaluating formation constraints and summarizing well-field history, EOS Trans. AGU, 27(4), 526-534, 1946.

Fenske, P.R., and T.M. Humphrey, Jr., The Tatum Dome Project Lamar County, Mississippi, Rep. NVO-225, 135 pp, U.S. Dep. of Energy, Nev. Oper. Office, Las Vegas, Aug. 1980.

Hubbert, M.K., The theory of ground-water motion, J. Geol., 48, 785-944, 1940.

Nind, T.E.W., Influences of absolute and partial hydrologic barriers on pump test results, Can. J. Earth Sci., 2, 309-323, 1965.

Papadopulos, I.S., and H.H. Cooper, Jr., Drawdown in a well of large diameter, Water Resour. Res., 3(1), 241-244, 1967.

Stallman, R.W., Type curves for the solution of single boundary problems, Shortcuts and Special Problems in Aquifer Tests, compiled by Ray Bentall, U.S. Geol. Surv. Water Supply Pap., 1545-C, C45-C47, 1963.

Theis, C.V., The effect of a well on the flow of a nearby stream, EOS Trans. AGU, 22, 734-738, 1941.

Verruijt, A., Theory of Groundwater Flow, Gordon and Breach, New York, 1970.

Analysis of Constant Discharge Wells by Numerical Inversion of Laplace Transform Solutions

Allen Moench and Akio Ogata
U.S. Geological Survey, Menlo Park, California 94025

Introduction

Analytical methods are commonly used to analyze water level changes that occur in response to groundwater pumpage. These methods permit determination of aquifer properties and prediction of aquifer response which are needed in the evaluation of the groundwater resources of a given region. After the initial work of Theis [1935], which has found broad application in both groundwater hydraulics and petroleum engineering, it was necessary to consider different geometries in order to reconcile deviations, noted in field tests, from this highly idealized response. Refinements have been made in the Theis theory by including such effects as leakage from confining layers, fissured formations, anisotropic conditions, partial penetration of wells, well bore storage, and free-surface conditions. Hantush [1964] gives solutions to many groundwater flow problems that he and others developed up to that time. A recent review of the state of the art of the theory of well tests is given by Weeks [1977]. He points out that although many analytical solutions are available to groundwater hydrologists, not all are in a form that can be readily evaluated.

There exists extensive literature [e.g., Hantush, 1964] on the methodology of developing solutions of various boundary value problems for which the diffusion equation (groundwater equation) is the controlling equation. One commonly used method that is ideally suited for the analysis of transient problems in diffusion involves an integral transform known as the Laplace transform. The method allows the independent time variable of the diffusion equation to be eliminated so that the transformed equation and boundary conditions simplify the problem and solution in the Laplace plane is usually obtained without difficulty. The real-plane solution, or solution to the original problem, can be obtained by analytically inverting the Laplace transform solution; however, the resulting expressions are often such that they are difficult to evaluate numerically. Even with large, high-speed digital computers some evaluations are unsatisfactory, and in many instances [e.g. Hantush, 1960; Cooley and Case, 1973], analysis had to be confined to evaluation at short and long times.

Doetsch [1961] points out that Laplace transform solutions are simpler in appearance than the corresponding real-plane solutions. Because the original function is damped by multiplication with an exponential function, solutions in the Laplace plane decay rapidly and lend themselves readily to numerical evaluation. This is one reason why it is advantageous to work in the Laplace plane. Many of the problems inherent in evaluation of complicated real-plane solutions can be avoided when numerical inversion techniques are applied. Although other methods are available, in this paper a numerical inversion algorithm described by Stehfest [1970] is used. The method has been used extensively in petroleum engineering but has seldom been seen in groundwater literature. In many instances the simplicity of the inversion algorithm makes the approach readily applicable to solution on hand-held calculators (H. Ramey, oral communication, 1978).

The purpose of this paper is to demonstrate the advantage of working in the Laplace plane. The Stehfest algorithm of the Laplace transform inversion is applied here to several problems pertinent to the analysis of well test data. The study is limited to wells that discharge at a constant rate. It is shown that previously intractable mathematical solutions can be easily evaluated by this method. In addition, it is shown that new Laplace transform solutions to relevant well test problems can be obtained simply by extending existing solutions.

Theory

The Laplace transform method for solution of the diffusion or heat flow equation has been used extensively in the analysis of various boundary value problems [e.g., Carslaw and Jaeger, 1959]. The transform is defined as

$$L\{h(t)\} = \bar{h}(p) = \int_0^\infty e^{-pt}\, h(t)\, dt \tag{1}$$

where h(t) is the original function and $\bar{h}(p)$ the image function or Laplace transform of h(t). The real-plane solution or solution to the original problem must be obtained either by analytical or numerical inversion. In the radial flow system that is of interest in this paper, analytical inversion, using the methods of operational calculus, results in solutions expressed as integral equations. These integrals are often composed of the oscillating Bessel functions J_0 and Y_0, which converge slowly. The solutions may be difficult to evaluate numerically, and in some instances, satisfactory convergence may not be possible at all. Thus it is often more efficient and accurate to obtain the real-plane solution through numerical inversion of the image function.

In applying numerical inversion of Laplace transform solutions, two-dimensional flow to a pumping well is considered. The medium is assumed to be homogeneous with principal components of hydraulic conductivity in the radial and vertical directions. Both line source wells and large-diameter wells are considered, including fully and partially penetrating wells. The equation describing the flow is the cylindrically symmetrical, diffusion equation with a line source function

$$K \frac{1}{r}\frac{\partial}{\partial r}\left(r\frac{\partial h}{\partial r}\right) + K_z \frac{\partial^2 h}{\partial z^2} = S_s \frac{\partial h}{\partial t} + q_b \tag{2}$$

where q_b is the source function located at the boundary separating the aquifer and a leaky confining bed. The form q_b takes depends upon the confining-bed geometry and boundary conditions. In this paper it is assumed that flow within the aquitard is strictly vertical.

The initial condition for all problems in this paper is specified as h(r,z,o)=0. To apply the Laplace transform, (2) is multiplied by e^{-pt} and integrated term by term from zero to infinity. In all instances considered in this paper, q_b is linearly related to h. Thus when q_b=G(r,z,t)h, evaluated at z=b, this process leads to the subsidiary equation

$$K \frac{1}{r} \frac{\partial}{\partial r}\left(r \frac{\partial \bar{h}}{\partial r}\right) + K_z \frac{\partial^2 \bar{h}}{\partial z^2} = S_s p\bar{h} + \bar{q}_b \tag{3}$$

where $\bar{h}$ is the Laplace transform of the function h(r,z,t) and $\bar{q}_b$=G(r,b,p)$\bar{h}$.

For fully penetrating wells the second term on the left-hand side of (3) is zero. In this instance, (3) becomes an ordinary differential equation, recognized as the modified Bessel equation, whose solution is given by

$$\bar{h} = AI_o (rk^{\frac{1}{2}}) + BK_o (rk^{\frac{1}{2}}) \tag{4}$$

where $rk^{\frac{1}{2}} = r[(p/\alpha) + (G/K)]^{\frac{1}{2}}$. The functions I_o and K_o are the modified Bessel functions of the first and second kind, respectively. The coefficients A and B are determined from the boundary conditions for the particular problem. In this paper the aquifer systems are considered infinite in radial extent; hence it can be shown that A is zero in all instances.

Hantush [1964, pp. 304-308] describes the procedure that may be used to solve (2) for partially penetrating wells. The method involves the reduction of (3) to an ordinary differential equation by applying finite Fourier transforms. The resulting equation appears as

$$K \frac{1}{r} \frac{d}{dr}\left(r \frac{d\bar{h}_f}{dr}\right) - K_z \left(\frac{n\pi}{b}\right)^2 \bar{h}_f = S_s p \bar{h}_f + G\bar{h}_f \qquad (5)$$

where $\bar{h}_f$ is the finite Fourier transform of the variable $\bar{h}(p)$ and $n=0,1,2,3,\cdots$. The solution to (5) is the same as (4) but with the arguments of the Bessel functions redefined as

$$rk^{\frac{1}{2}} = r\left[\frac{p}{\alpha} + \frac{G}{K} + \frac{K_z}{K}\left(\frac{n\pi}{b}\right)^2\right]^{\frac{1}{2}}$$

The Laplace transform solution is obtained by using the inverse finite Fourier transform [Hantush, 1964].

Numerical Inverter

Many numerical methods of Laplace transform inversion are given in the literature (see, for example, the review by Davies and Martin [1979]). Most are in the form of a polynomial approximation of which the Fourier series is commonly used because of calculation ease [Durbin, 1973]. Other investigators have used the Gaussian formula for integration of the inversion integral [Piessens, 1971]. Here a numerical method developed by Gaver [1966], revised and presented in the form of an algorithm by Stehfest [1970], is used because it is extremely simple and requires relatively little computation. In this approach the weighting coefficients depend only on the number of terms used in the computation.

The Stehfest [1970] algorithm is written as follows:

$$F_a \approx \left[\frac{(\ln 2)}{t}\right]\sum_{i=1}^{N} V_i P\left[\frac{i(\ln 2)}{t}\right] \qquad (6)$$

where F_a is the approximate value of the inverse $F(t)$ at t and $P[p]$ is the Laplace-transformed function to be inverted. The coefficients V_i are given by

$$V_i = (-1)^{(N/2)+i} \sum_{k=(i+1)/2}^{\min(i,N/2)} \frac{k^{N/2}\,(2k)!}{(N/2-k)!\,k!\,(k-1)!\,(i-k)!\,(2k-i)!} \qquad (7)$$

where N is an even number and k is computed using integer arithmetic. Two properties of (7) are that for a given N the V_i sum to zero, and as N increases the V_i tend to increase in absolute value.

As the V_i depend only on N, they need to be calculated only once for any chosen value of N. According to Stehfest, F_a becomes more accurate with increasing values of N; however, after a point, rounding errors worsen the result. The value of N to use for maximum accuracy is approximately proportional to the number of significant figures possessed by the computer in use.

Requirements for use of (6) and (7) are that the function F(t) have no discontinuities or rapid oscillations. Stehfest recommends that the accuracy be checked by employing different values of N and other inversion techniques. Unless otherwise indicated the results that follow were obtained using N=18 in the Stehfest algorithm and double precision on an IBM 3033 computer. (The use of the brand name in this report is for identification purposes only and does not imply endorsement by the U.S. Geological Survey.) The results were checked, wherever possible, against published figures and tables.

Application

In this section, solutions of the subsidiary or transformed differential equation and boundary conditions to several well test problems are inverted numerically by using the Stehfest algorithm. The utility of the method of numerical inversion increases with the complexity of the problem; therefore many of the simpler solutions to which the technique could be applied with ease are not considered. However, the Theis [1935] nonequilibrium flow problem is included because it is instructive to note the form of the Laplace transform solution, a modified Bessel function of the second kind and order zero. Solutions to many of the more complex line source problems are of the same form and require only that changes be made in the argument of the Bessel function.

Most of the examples involve an aquifer with an overlying aquitard containing storage. Flow in the aquitard is assumed to be vertical, and flow in the aquifer is assumed to be horizontal. For this to be the case, horizontal hydraulic conductivity in the aquitard must be small compared with horizontal hydraulic conductivity in the aquifer. These examples are essentially variations of the modified leaky aquifer theory presented by Hantush [1960]. Solutions vary depending upon manner in which the upper boundary of the aquitard is characterized.

A constant head at the upper boundary was assumed by Hantush [1960, case 1] in the development of the modified leaky aquifer theory. Neuman and Witherspoon [1969a] modified the problem by assuming that drawdown occurs in an unpumped aquifer overlying the aquitard. The upper boundary was also modified by Cooley and Case [1973], who assumed that the aquifer was overlain by a water table aquitard. Boulton and Streltsova [1977] in their study of flow in a fissured water-bearing formation, which was idealized as an aquifer-aquitard combination, assumed an impermeable boundary at the top of the aquitard.

In these examples the well was considered to penetrate fully the aquifer; however, as indicated by (5), further variation is obtained with partial penetration of the well. An analytical solution for transient flow appears to have been first given by Hantush [1957]. As with fully penetrating wells, additional complication is readily added to the problem by considering a leaky aquifer with storage in the confining layer. Vertical anisotropy [Hantush, 1964] is included in this example because of its importance in partial penetration; however, horizontal anisotropy is omitted for simplicity in illustrating the results. Although not considered here, partial penetration can be handled equally well when an aquifer is overlain by a water table aquitard.

Another important consideration in well test analysis is the effect that a large-diameter well with well bore storage has upon the drawdown within and around the well. The solution for this case involves modified Bessel functions of the second kind of order zero and unity. Two examples are considered in this study, one involves a nonleaky aquifer [Papadopulos and Cooper, 1967] and the other involves a leaky aquifer with storage in the confining bed. Lai and Su [1974] and Abdul Khader and Ramadurgaiah [1977] also present solutions for a large-diameter well in a leaky aquifer, but storage in the confining bed is not included. Again, although not illustrated, the large-diameter well solution could be applied with equal ease to the problem of an aquifer overlain by a water table aquitard.

Expressions in the examples to follow are solutions of the subsidiary equation. Expressions obtained by analytical inversions are not shown. The reader, if interested, should refer to the original publication cited.

Example 1: Line Source in a Nonleaky Aquifer [Theis, 1935]

The solution to the Theis nonequilibrium flow problem is

$$\bar{h}_D = \frac{K_o(r\,k^{\frac{1}{2}})}{p} \tag{8}$$

where $rk^{\frac{1}{2}} = \eta$. Numerical inversion reproduced precisely the tabulation of the well function W(u) given by Ferris et al. [1962] for all values of u less than 5. To obtain the same agreement for values of u up to 9.9, it was necessary to have N=26 in the inversion algorithm.

Example 2: Line Source in a Leaky Aquifer With Storage in the Confining Layer [Hantush, 1960]

The solution for a line source in a leaky aquifer with storage in an overlying confining layer is

$$\bar{h}_D = \frac{K_o(r\,k^{\frac{1}{2}})}{p} \tag{9}$$

where

$$r\,k^{\frac{1}{2}} = \left\{\eta^2 + 4\eta\beta \coth \frac{4\eta\beta}{(r/B)^2}\right\}^{\frac{1}{2}}$$

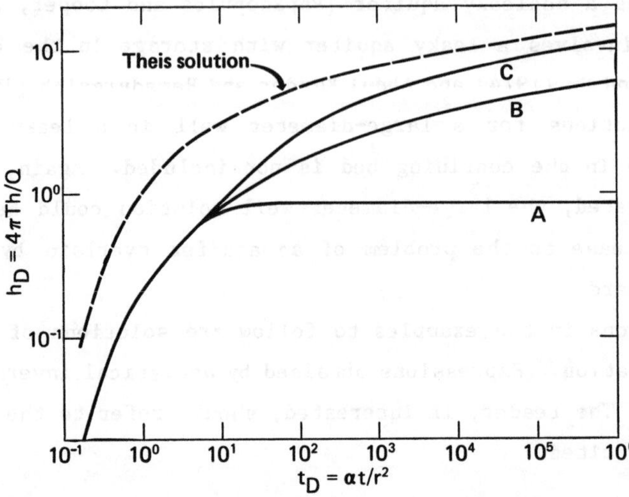

Fig. 1. Dimensionless drawdown in the pumped aquifer versus dimensionless time in the two-aquifer system of Neuman and Witherspoon [1969a] for values of parameters given in Table 1.

$$\beta = \frac{K'r}{4Kb} \left(\frac{\alpha}{\alpha'}\right)^{\frac{1}{2}}$$

$$\frac{r}{B} = \frac{r}{b} \left(\frac{K'b}{Kb'}\right)^{\frac{1}{2}}$$

The parameters β and r/B are dimensionless groups attributed to Hantush [1960]. In this example the system was assumed to have only one leaky confining layer. The lower aquitard was assumed to be impermeable.

Results obtained by numerical inversion appear to be identical with the type curves presented by Neuman and Witherspoon [1969b, 1971] over the entire time domain. They also reproduce precisely the table of the function $H(u,\beta)$ given by Hantush [1960] for the short time asymptotic solution for all values of u less than 5. As with example 1, in order to obtain the same agreement for values of u up to 10.0 it was necessary to use a larger value of N in the inversion algorithm. A typical type curve is given by curve A in Figure 1.

Example 3: Line Source in a Two-Aquifer System [Neuman and Witherspoon, 1969a]

The solution for drawdown in the pumped aquifer of a two-aquifer system is:

$$\bar{h}_D = \frac{1}{p}\,\frac{(A_2-k_1)}{D}\;K_o(rk_1^{\frac{1}{2}}) - \frac{1}{p}\,\frac{(A_2-k_2)}{D}\;K_o(r\;k_2^{\frac{1}{2}}) \tag{10}$$

where

$$k_1 = \frac{1}{2}\left[A_1+A_2-D\right]$$

$$k_2 = \frac{1}{2}\left[A_1+A_2+D\right]$$

$$D = \left[4B_1B_2+(A_1-A_2)^2\right]^{\frac{1}{2}}$$

$$A_1 = r^{-2}\left[\eta^2+4\eta\beta_{11}\;\coth\;(\psi)\right]$$

$$A_2 = r^{-2}\left[\eta^2\;\frac{\alpha_1}{\alpha_2} + 4\eta\beta_{21}\left(\frac{\alpha_1}{\alpha_2}\right)^{\frac{1}{2}}\;\coth\;(\psi)\right]$$

$$B_1 = r^{-2}\;\;4\eta\beta_{11}\left[\frac{1}{\sinh\;(\psi)}\right]$$

$$B_2 = r^{-2}\;\;4\eta\beta_{21}\left(\frac{\alpha_1}{\alpha_2}\right)^{\frac{1}{2}}\left[\frac{1}{\sinh\;(\psi)}\right]$$

$$\psi = 4\eta\beta_{11}\left(\frac{r}{B_{11}}\right)^{-2}$$

$$\beta_{11} = \frac{1}{4}\;\;\frac{K'}{K_1}\,\frac{r}{b_1}\left(\frac{\alpha_1}{\alpha'}\right)^{\frac{1}{2}}$$

$$\frac{r}{B_{11}} = \frac{r}{b_1}\;\left(\frac{K'}{K_1}\;\;\frac{b_1}{b'}\right)^{\frac{1}{2}}$$

$$\beta_{21} = \beta_{11} \frac{T_1}{T_2} \left(\frac{\alpha_2}{\alpha_1} \right)^{\frac{1}{2}}$$

$$\frac{r}{B_{21}} = \frac{r}{B_{11}} \left(\frac{T_1}{T_2} \right)^{\frac{1}{2}}$$

The subscript 1 refers to the pumped aquifer, the subscript 2 refers to the unpumped aquifer, and the primed parameters refer to the aquitard. In terms of the last four parameters the ratio of aquifer diffusivities which appears above in A_2 and B_2 can be expressed as

$$\frac{\alpha_1}{\alpha_2} = \left(\frac{\beta_{11}}{\beta_{21}} \right)^2 \left(\frac{r}{B_{21}} \right)^4 \left(\frac{r}{B_{11}} \right)^{-4}$$

For the unpumped aquifer the expression obtained is

$$\bar{h}_{D2} = \frac{B_2}{pD} \left[K_o(r\,k_1^{\frac{1}{2}}) - K_o(r\,k_2^{\frac{1}{2}}) \right] \qquad (11)$$

and for the aquitard

$$\bar{h}'_D = \frac{\sinh(\psi z/b')}{\sinh(\psi)}\, \bar{h}_{D2} + \frac{\sinh[\psi(1-z/b')]}{\sinh(\psi)}\, \bar{h}_{D1} \qquad (12)$$

where z is the distance within the aquitard from the interface between the aquitard and the pumped aquifer.

For two aquifers with identical properties, (10) becomes

$$\bar{h}_{D1} = \frac{1}{2p} \left[K_o\,(r\,k_1^{\frac{1}{2}}) + K_o\,(r\,k_2^{\frac{1}{2}}) \right] \qquad (13)$$

and the coefficient B_2/D in (11) becomes 1/2.

Results obtained by numerical inversion, of (10)-(13) appear identical with the type curves given by Neuman and Witherspoon [1969a, 1971]. Figure 1 shows a comparison of the results obtained for the two-aquifer problem with the Theis solution and the Hantush modified leaky-aquifer problem. The parameters used are shown in Table 1. Curve A represents the drawdown that occurs in the pumped aquifer when the head in the source bed (unpumped aquifer) is held

TABLE 1. Values of Parameters for Figure 1
(Any System of Consistent Units)

	Curve A	Curve B	Curve C
b_2	1	1	0.01
b'	1	1	1
b_1	1	1	1
K_2	∞	1	1
K'	0.01	0.01	0.01
K_1	1	1	1
α_2	∞	1	1
α'	1/1600	1/1600	1/1600
α_1	1	1	1
β_{11}	1	1	1
r/B_{11}	1	1	1
β_{21}	0	1	100
r/B_{21}	0	1	10

constant. Curves B and C represent drawdown that occurs in the pumped aquifer when head in the unpumped aquifer declines in response to pumpage. In the case of curve C the drawdown at large time is enhanced because the unpumped aquifer is quite thin compared with the unpumped aquifer represented by curve B.

Example 4: Line Source in a Fissured Formation [Boulton and Streltsova, 1977]

The solution for this problem is

$$\bar{h}_D = \frac{1}{p}\int_0^\infty \frac{aJ_o(ra)}{(a^2+k)}\,da = \frac{K_o(r\,k^{\frac{1}{2}})}{p} \tag{14}$$

where

$$r\,k^{\frac{1}{2}} = \left\{\eta^2 + \eta\left(\frac{b'K'}{bK}\right)\left(\frac{r}{b'}\right)\left(\frac{\alpha}{\alpha'}\right)^{\frac{1}{2}}\tanh\left[\eta\left(\frac{b'}{r}\right)\left(\frac{\alpha}{\alpha'}\right)^{\frac{1}{2}}\right]\right\}^{\frac{1}{2}}$$

The solution for the reservoir block is (14) multiplied by the expression

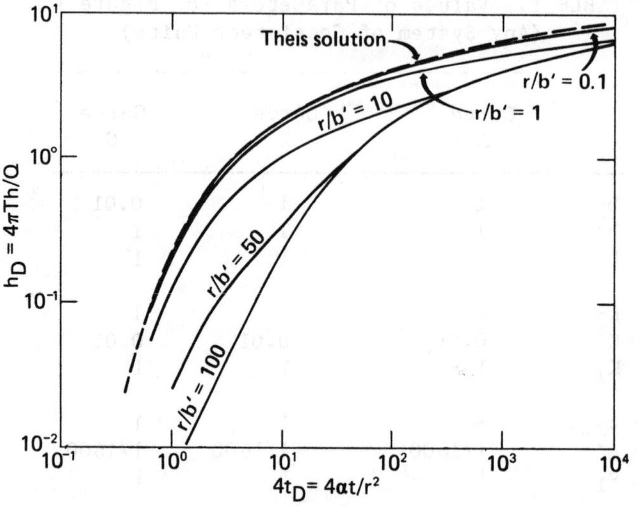

Fig. 2. Dimensionless drawdown versus 4 times dimensionless time in the fissured formation of Boulton and Streltsova [1977] for the parameters $\alpha'/\alpha = 10^{-4}$ and $K'b'/Kb = 10^{-3}$.

$$\cosh\left[\eta\left(\frac{b'}{r}\right)\frac{z}{b'}\left(\frac{\alpha}{\alpha'}\right)^{\frac{1}{2}}\right] - \tanh\left[\eta\left(\frac{b'}{r}\right)\left(\frac{\alpha}{\alpha'}\right)^{\frac{1}{2}}\right]\sinh\left[\eta\left(\frac{b'}{r}\right)\frac{z}{b'}\left(\frac{\alpha}{\alpha'}\right)^{\frac{1}{2}}\right] \quad (15)$$

Figure 2 shows results obtained by numerical inversion using parameters given by Boulton and Streltsova [1977, Figure 8]. The abscissa in Figure 2 has been multiplied by 4 to correspond with the dimensionless time defined by Boulton and Streltsova. The results differ markedly from those presented by Boulton and Streltsova, especially for small values of the ratio r/b' and small values of dimensionless time, and will be discussed later in this paper. Numerical inversions of (14) and (15) are corroborated by results of another study using a finite-difference method [Moench and Denlinger, 1980].

Example 5. Line Source in an Aquifer Overlain by a Water Table Aquitard [Cooley and Case, 1973]

The solution for this problem is

$$\bar{h}_D = \frac{K_o(r\,k^{\frac{1}{2}})}{p} \quad (16)$$

where

$$r\,k^{\frac{1}{2}} = \left\{ \eta^2 + 4\eta\beta \tanh\left[\frac{4\eta\beta}{(r/B)^2}\right] + \left[\eta^2 \operatorname{sech}^2 (4\eta\beta/(r/B)^2\right] / \right.$$

$$\left. \left[\eta^2 \frac{L/b'}{(r/B)^2} + \frac{(r/B)^2 S'}{16\beta^2 S_y} + \frac{\eta}{4\beta} \tanh\left(\frac{4\eta\beta}{(r/B)^2}\right)\right]^{-1} \right\}^{\frac{1}{2}}$$

The parameter r/B is modified slightly by Cooley and Case [1973] as

$$\frac{r}{B} = \frac{r}{b} \frac{K'b}{K(b'+L)}$$

where L is the thickness of the capillary fringe, a nearly saturated zone above the water table.

Curve A in Figure 3 shows that the results of numerical inversion are in good agreement with a finite-difference solution developed by Cooley [1971]. The parameters used are those given by Cooley and Case [1973, Figure 4].

Curves B and C in Figure 3 demonstrate the effects of a constant-head boundary and an impermeable boundary, respectively, in the aquitard. These correspond to Hantush's [1960] case 1 and case 2. The solution for Hantush's case 2 is identical to (14). At large time, drawdown in the presence of a free surface in the aquitard is intermediate between case 1 and 2. At early time, before the effect of the upper boundary of the aquitard is felt in the aquifer, the three solutions yield identical results.

Example 6: Partial Penetration of a Line Source in an Anisotropic Leaky Aquifer With Storage in the Confining Layer--A Combination of Solutions by Hantush [1960, 1964]

The solution for this problem is

$$\bar{h}_D = \frac{1}{p} [K_o(r\,k^{\frac{1}{2}}) + F] \tag{17}$$

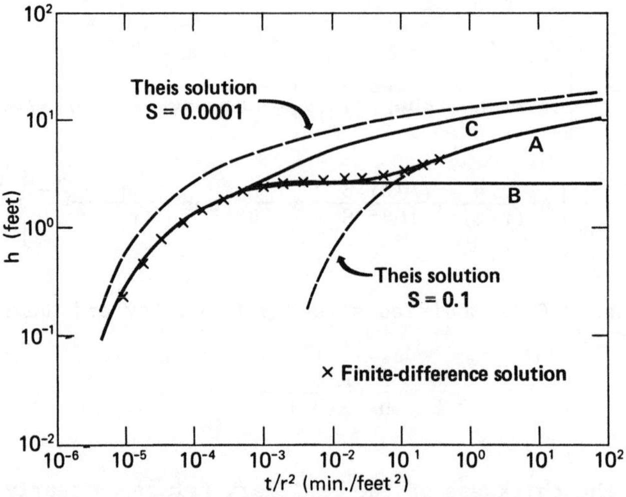

Fig. 3. Drawdown versus t/r^2 in an aquifer overlain by a water table aquitard (curve A), an aquitard with constant head at the top boundary (curve B), and an aquitard with no flow at the top boundary (curve C). The parameters used are $r/B=0.39$, $\beta=0.32$, $L/b'=0.06$, $S'/S_y=0.01$, $S=10^{-4}$, $T=4.64$ ft^2/min, $r=200$ ft, and $Q=66.8$ ft^3/min [after Cooley and Case, 1973].

where

$$F = \frac{2}{(x_\ell - x_d)(x'_\ell - x'_d)} \sum_{n=1}^{\infty} \frac{1}{n^2} \left[\sin(x_\ell n) - \sin(x_d n)\right]$$

$$\left[\sin(x'_\ell n) - \sin(x'_d n)\right] K_o \left[r^2 k + \gamma \left(\frac{n\pi r}{b}\right)^2\right]^{\frac{1}{2}}$$

$$r\,k^{\frac{1}{2}} = \left\{\eta^2 + 4\eta\beta \coth\left[\frac{4\eta\beta}{(r/B)^2}\right]\right\}^{\frac{1}{2}}$$

$$x_\ell = \pi\ell/b$$

$$x_d = \pi d/b$$

$$x'_\ell = \pi\ell'/b$$

$$x'_d = \pi d'/b$$

$$\gamma = K_z/K$$

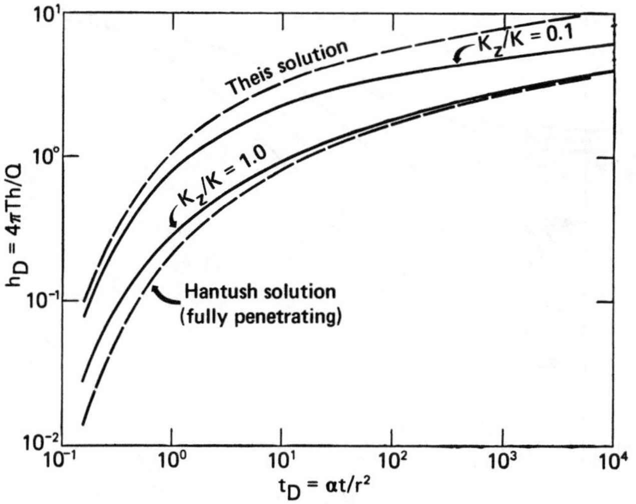

Fig. 4. Dimensionless drawdown versus dimensionless time in a well partially penetrating a leaky, vertically anisotropic aquifer compared with the fully penetrating solution for the parameters $r/B=0.1$, $\beta=1.0$, $x_\ell = x_\ell' = 0.1$, and $x_d = x_d' = 0$.

Equation (17) represents the average drawdown in a partially penetrating observation well. It derives from the solution presented by Hantush [1964, p. 350], for drawdown in a piezometer located at any point within the pumped aquifer, by integrating in the vertical direction over the screened interval of the observation well [Hantush, 1961].

Figure 4 presents example-type curves obtained by numerical inversion. Results are compared with the fully penetrating solutions of Theis [1935] and Hantush [1960]. Figure 4 illustrates that partial penetration of the pumping and observation well not only increases drawdown but also enhances effects of vertical anisotropy.

Example 7: Large-Diameter Well With Well Bore Storage [Papadopulos and Cooper, 1967]

The solution for drawdown around the pumped well is

$$\bar{h}_D = \frac{K_o(r\,k^{1/2})}{p\left[W_D(\eta^2/r_D^2)\,K_o\,(r\,k^{1/2}/r_D) + (r\,k^{1/2}/r_D)\,K_1\,(r\,k^{1/2}/r_D)\right]} \qquad (18)$$

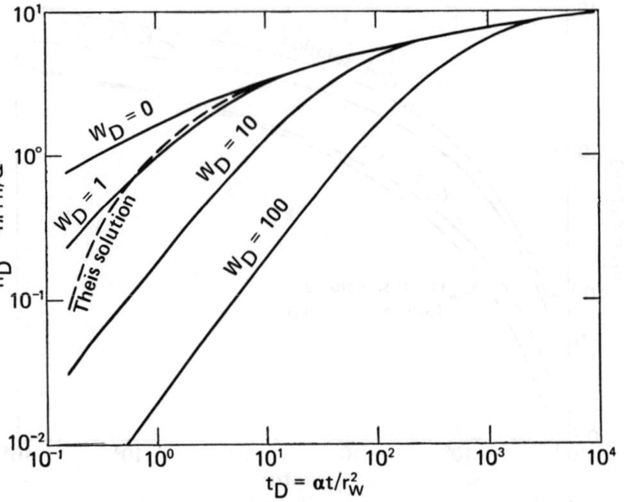

Fig. 5. Dimensionless drawdown versus dimensionless time around a well of large diameter in a nonleaky aquifer.

where

$$W_D = \frac{r_c^2}{2r_w^2 S}$$

$$r_D = \frac{r}{r_w}$$

$$r\,k^{\frac{1}{2}} = \eta$$

Drawdown in the pumped well is obtained when $r=r_w$. The parameter W_D is the one dimensionless well bore storage factor. Papadopulos and Cooper [1967] define a similar factor, α, as $r_w^2\,S/r_c^2$.

Results obtained by numerical inversion are in agreement with tabulations of Papadopulos [1967] and Papadopulos and Cooper [1967] to within as many significant figures as given in their tables. Figure 5 illustrates typical type curves compared with the Theis [1935] line source solution. When W_D=0, the solution is that of Hantush [1964, p. 340] for drawdown due to a large-diameter well without well bore storage.

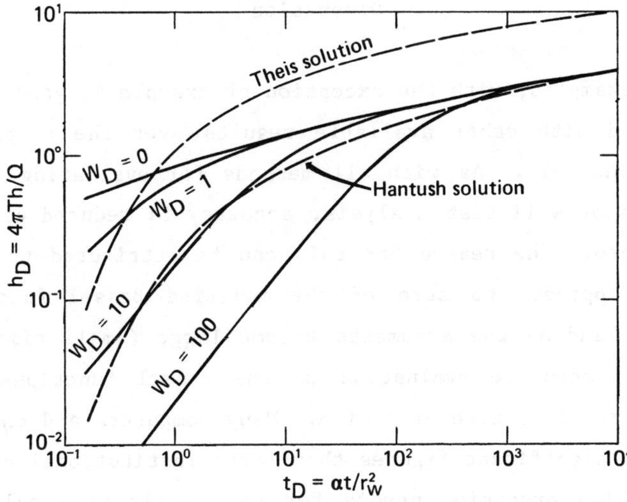

Fig. 6. Dimensionless drawdown versus dimensionless time around a well of large diameter in a leaky aquifer compared with the line source solution for the parameters $r/B=0.1$ and $\beta=1.0$.

Example 8: Large-Diameter Well With Well Bore Storage in a Leaky Aquifer With Storage in the Confining Layer--A Combination of Solutions by Papadopulos and Cooper [1967] and Hantush [1960]

The solution for drawdown around the pumped well is

$$\bar{h}_D = \frac{K_o(r\,k^{\frac{1}{2}})}{p\,\{W_D(\eta^2/r_D^2)\,K_o\,(r\,k^{\frac{1}{2}}/r_D) + (r\,k^{\frac{1}{2}}/r_D)\,K_1\,(r\,k^{\frac{1}{2}}/r_D)\}} \tag{19}$$

where

$$r\,k^{\frac{1}{2}} = \{\eta^2 + 4\eta\beta\,\coth[\frac{4\eta\beta}{(r/B)^2}]\}^{\frac{1}{2}}$$

As with (17), drawdown in the pumped well is obtained when $r=r_w$.

Results obtained by numerical inversion are presented in Figure 6 for selected values of well bore storage. The line source solutions of Theis [1935] and Hantush [1960] are shown for comparison.

Discussion

In all examples, with the exception of example 4, good agreement was obtained with other available results over the complete time domain of interest. As with all methods for evaluating solutions to problems of well test analysis, accuracy is reduced as time approaches zero. The reason for this can be attributed to the slow asymptotic approach to zero of the modified Bessel functions of the second kind as the arguments become large (small time). As a consequence, accurate evaluation of the Bessel functions at very early time requires high precision. Many computers and calculators carry fewer significant figures than large institutional computers; hence the high precision needed for very early time calculations may not be available. Fortunately, the need for such early time analysis rarely arises in real problems; however, if need be and if precision of the computer permits, accuracy can usually be increased by increasing the value of N in the inversion algorithm, as discussed in examples 1 and 2.

The complete real-time solution to the Hantush modified leaky aquifer problem (example 2) can be obtained by numerical inversion of (9) without resorting to the short-time or large-time asymptotic solutions developed by Hantush [1960]. Neuman and Witherspoon [1969a, b] obtained an analytical solution in the real plane for the problem, but because the solution is in the form of an integral equation of an oscillating function, its evaluation can incur large computational expense.

The solutions to the two-aquifer problem (example 3) of Neuman and Witherspoon [1969a] are easily obtained by numerical inversion of (10)-(13). The analytical solutions to the two-aquifer problem in the real plane are considerably more complex than the solution to the Hantush problem; hence they are also expensive to evaluate numerically. Neuman and Witherspoon in developing type curves used asymptotic approximations wherever possible to save computer time and evaluated their analytical expressions only for a few selected values of the parameters.

Boulton and Streltsova [1977] inverted (14) analytically to obtain a real-plane solution to the problem of a line source in a fissured formation (example 4). Unfortunately, evaluation of the resulting equation may lead to erroneous results. This is no better exemplified than in Boulton and Streltsova's Figures 7 and 8, which show, for small values of r/b', that there is increased departure from the Theis curve as r/b' decreases. Comparison of (8) and (14) clearly shows that as r/b' decreases, the Boulton and Streltsova solution approaches the Theis solution. The results presented by Boulton and Streltsova [1977], seen also in the word by Streltsova-Adams [1978, Figure 25], demonstrate the difficulties sometimes encountered in evaluating real-plane solutions to boundary value problems of this type. In attempting to evaluate the real-plane solution given by Boulton and Streltsova, we encountered these same difficulties and were able to obtain satisfactory convergence only for selected values of the parameters.

The solution to the problem of an aquifer overlain by a water table aquitard (example 5) by Cooley and Case [1973] is likewise easily obtained by numerical inversion. Because of analytical complexity, Cooley and Case, like Hantush, developed short- and long-time asymptotic solutions which are not needed when the numerical inversion is used.

Hantush's modified theory of leaky aquifers is a significant improvement over the earlier theory of Hantush and Jacob [1955]. Neuman and Witherspoon's theory for the two-aquifer system is, in turn, a significant improvement over the Hantush modified leaky aquifer theory; however, possibly because of its complexity and because the range of published type curves may be inadequate, it has been rarely applied. Likewise, Cooley and Case's theory of an aquifer overlain by a water table aquitard and Boulton and Streltsova's theory for a fissured formation appear to be applied seldom. With the Stehfest numerical Laplace transform inversion algorithm, evaluation of solutions such as these should no longer be a barrier to their application in well test analysis.

The effects of partial penetration in an anisotropic, leaky aquifer with storage in the confining layer given in example 6 illustrate how solutions can be obtained for more complicated systems by combining Laplace transform solutions for different problems and inverting numerically the equation obtained. Type curves for this case can be generated for use in well test analysis; however, in view of the number of parameters, the approach may prove more useful for predicitng aquifer response using known or estimated parameters.

Solutions for a large-diameter well in a nonleaky aquifer (example 7) and a large-diameter well in a leaky aquifer with storage in the confining layer (example 8) also illustrate the advantages of working in the Laplace plane. Additional factors could be added to these solutions by the interested user. For example, in the petroleum industry the so-called 'skin' factor is often used to account for decreased or increased resistance to flow in the immediate vicinity of the well bore. Sandal et al. [1978] discuss the use of the Stehfest numerical inversion scheme in this instance.

The computer time required in making the inversions for examples 1-8 is difficult to estimate accurately. As an example, however, the calculation of 35 values of drawdown in both a pumped and unpumped aquifer using the Neuman and Witherspoon two-aquifer equations took 1.3 s of CPU time on an IBM 3033.

Conclusions

Analytical solutions to many boundary value problems for wells discharging at a constant rate have not been frequently applied by groundwater hydrologists because they are difficult to evaluate numerically. In this paper the authors have shown that these solutions can be accurately evaluated using a simple numerical Laplace transform inverter. In addition, an attempt is made to demonstrate the advantage of working in the Laplace plane by combining solutions of the subsidiary differential equation for various boundary value problems to form new solutions for well test problems. Type curves for such extended problems can be generated without difficulty with the numerical Laplace transform inverter. Such solutions obtained analytically might otherwise be mathematically intractable.

Notation

- a variable of integration.
- b aquifer thickness, L.
- b' aquitard thickness, L.
- d depth of penetration of unscreened portion of pumping well, L.
- d' depth of penetration of unscreened portion of observation well, L.
- G relating function for leakage, 1/LT.
- h pumped aquifer drawdown, L.
- h_D dimensionless drawdown, equal to $4\pi Th/Q$.
- h' aquitard drawdown, L.
- $\bar{h}$ Laplace transform of h.
- $\bar{h}'$ Laplace transform of h'.
- $\bar{h}_D = 2\pi T\bar{h}/Q$.
- $\bar{h}_D' = 2\pi T\bar{h}'/Q$.
- $\bar{h}_f$ finite Fourier transform of $\bar{h}$.
- $I_0(x)$ modified Bessel function of the first kind and order zero.
- $J_0(x)$ Bessel function of first kind and order zero.
- $K_0(x)$ modified Bessel function of second kind and order zero.
- $K_1(x)$ modified Bessel function of second kind and order unity.
- K aquifer horizontal hydraulic conductivity, L/T.
- K' aquitard vertical hydraulic conductivity, L/T.
- K_z aquifer vertical hydraulic conductivity, L/T.
- L thickness of capillary fringe, L.
- ℓ depth of penetration of pumping well, L.
- ℓ' depth of penetration of observation well, L.
- p Laplace transform variable.
- Q well discharge rate L^3/T.
- q_b source function for leakage, 1/T.
- r radial coordinate, L.

r_c radius of well casing, L.

r_w effective radius of well, L.

S aquifer coefficient of storage.

S_s aquifer coefficient of specific storage, equal to S/b, 1/L.

S' aquitard coefficient of storage.

S_y aquitard coefficient of specific yield.

T aquifer transmissivity, equal to Kb, L^2/T.

t time, T.

t_D dimensionless time, equal to $\alpha t/r^2$.

$u = r^2S/4Tt$.

x argument of Bessel functions.

z vertical coordinate L.

α aquifer diffusivity ($=T/S=K/S_s$), L^2/T.

$\alpha' = K'b'/S'$, L^2/T.

$\eta = (r^2 p/\alpha)^{\frac{1}{2}}$.

Subscripts

1 pumped aquifer.

2 unpumped aquifer.

References

Abdul Khader, M.H., and D. Ramadurgaiah, Flow towards a well of large diameter in a leaky confined aquifer, paper presented at the Sixth Australasian Hydraulics and Fluid Mechanics Conference, Adelaide, Australia, Dec. 5-9, 1977.

Boulton, N.S., and T.D. Streltsova, Unsteady flow to a pumped well in a fissured water-bearing formation, J. Hydrol., 35, 257-269, 1977.

Carslaw, H.S., and J.C. Jaeger, Conduction of Heat in Solids, 2nd ed., 510 pp., Oxford University Press, New York, 1959.

Cooley, R.L., A finite-difference method for unsteady flow in variably saturated porous media: Application to a single pumping well, Water Resour. Res., 7(6), 1607-1625, 1971.

Cooley, R.L., and C.M. Case, Effect of a water table aquitard on drawdown in an underlying pumped aquifer, Water Resour. Res., 9(2), 434-447, 1973.

Davies, B., and B. Martin, Numerical inversion of the Laplace transform: A survey and comparison of methods, J. Comput. Phys., 33, 1-32, 1979.

Doetsch, G., Guide to the Applications of Laplace Transform, 255 pp., D. Van Nostrand, New York, 1961.

Durbin, R., Numerical inversion of Laplace transforms: An efficient improvement to Durbin and Abate's method, Comput. J., 17(4), 371-376, 1973.

Ferris, J.G., D.B. Knowles, R.H. Brown, and R.W. Stallman, Theory of aquifer tests, U.S. Geol. Surv. Water Supply Pap., 1536-E, 69-174, 1962.

Gaver, G.P., Jr., Observing stochastic processes, and approximate transform inversions, Oper. Res., 14(3), 444-459, 1966.

Hantush, M.S., Non-steady flow to a well partially penetrating an infinite leaky aquifer, Proc. Iraqi Sci. Soc., 1, 10-19, 1957.

Hantush, M.S., Modification of the theory of leaky aquifers, J. Geophys. Res., 65(11), 3713-3725, 1960.

Hantush, M.S., Drawdown around a partially penetrating well, J. Hydraul. Div. Am. Soc. Civ. Eng., 87(HY4), 83-98, 1961.

Hantush, M.S., Hydraulics of wells, Adv. Hydrosci., 1, 281-432, 1964.

Hantush, M.S., and C.E. Jacob, Nonsteady radial flow in an infinite leaky aquifer, EOS Trans. AGU, 36(1), 95-100, 1955.

Lai, R.Y., and C.W. Su, Nonsteady flow to a large well in a leaky aquifer, J. Hydrol., 22, 333-345, 1974.

Moench, A.F., and R. Denlinger, Fissure-block model for transient pressure analysis in geothermal steam reservoirs, paper presented at the Sixth Workshop on Geothermal Reservoir Engineering, Stanford Univ., Stanford, Calif., Dec. 16-18, 1980.

Neuman, S.P., and P.A. Witherspoon, Theory of flow in a two-aquifer system, Water Resour. Res., 5(4), 803-816, 1969a.

Neuman, S.P., and P.A. Witherspoon, Applicability of current theories of flow in leaky aquifers, Water Resour. Res., 5(4), 817-829, 1969b.

Neuman, S.P., and P.A. Witherspoon, Transient flow of ground water to wells in multiple-aquifer systems, Sea-Water Intrusion: Aquitards in the Coastal Ground Water Basin of Oxnard Plain, Ventura County, Calif. Bull. 63-4, Appendix A, pp. 159-360, Calif. Dep. of Water Resour., Sacramento, 1971.

Papadopulos, I.S., Drawdown distribution around a large-diameter well, paper presented at the National Symposium on Groundwater Hydrology, Am. Water Resour. Assoc., San Francisco, Calif., 1967.

Papadopulos, I.S., and H.H. Copper, Jr., Drawdown in a well of large diameter, Water Resour. Res., 3(1), 241-244, 1967.

Piessens, R., Gaussian quadrature formulas for the numerical integration of Bromwich's integral and the inversion of Laplace transforms, J. Eng. Math., 5(1), 1-9, 1971.

Sandal, H.M., R.N. Horne, H.J. Ramey, Jr., and J.W. Williamson, Interference testing with wellbore storage and skin effect at the producing well, 12pp., paper presented at the 53rd Annual Fall Technical Conference and Exhibition, Soc. of Pet. Eng. of AIME, Houston, Tex., Oct. 1-3, 1978.

Stehfest, H., Numerical inversion of Laplace transforms, Commun. ACM, 13(1), 47-49, 1970.

Streltsova-Adams, T.D., Well hydraulics in heterogeneous aquifer formations, Adv. Hydrosci., 11, 357-423, 1978.

Theis, C.V., The relation between the lowering of the piezometric surface and the rate and duration of discharge of a well using ground-water storge, EOS Trans. AGU, 16, 519-524, 1935.

Weeks, E.P., Aquifer, tests - state of the art in hydrology, paper presented at the Invitational Well-Testing Symposium, sponsor, U.S. Dep. Energy, Berkeley, Calif., Oct. 19-21, 1977.

Pumping Test Analysis in Fractured Aquifer Formations: State of the Art and Some Perspectives

C. Sauveplane
Alberta Research Council, Edmonton, Alberta T6H SR7, Canada

1. General Considerations About Aquifer Test Analysis

The purpose of aquifer test interpretation is to identify an unknown system. The solution to this problem, the 'inverse problem,' involves the search of a well-defined model to simulate the behavior of the actual system and to reproduce, as closely as possible, its observed response(s).

The matching process between the model and the unknown system enables one to calculate the aquifer characteristics, provided that a set of initial and boundary conditions similar to the known or supposed ones of the real system has been previously introduced into the model. The selection of a wrong model will lead to incorrect results. Also, the solution of the inverse problem is usually not unique; different models can give a response similar to the observed one. Fortunately, the degree of nonuniqueness is reduced when the number and the range of observed responses increase.

Unsteady state flow models are represented by a family of type curve solutions graphed on a log-log paper as dimensionless drawdown s_D versus dimensionless time t_D, and a specific dimensionless number characterizes each type curve. These dimensionless parameters are introduced for convenience when solving the partial differential equation and associated initial and boundary conditions that govern the flow toward the well. The function $s_D(t_D)$ represented by a given type curve is independent of the specific conditions of a test (pumping rate, for instance), and the type curves represent a typical response of the simulated system.

Qualitative information may be deduced by comparing the plotted real responses to the theoretical responses; however, this information should be used carefully and in conjunction with other available hydrogeological arguments to select the most appropriate model for interpretation.

In the last decade, type curve solutions have been derived for an impressive number of porous media flow problems; the analytical treatment of a well-defined problem (partial differential equation and associated initial and boundary conditions) generally involves Laplace, Hankel, and Fourier transforms or the use of Green's function. The complexity of the derived functions $s_D(t_D)$ often requires the use of numerical techniques to compute s_D. It is then possible to match the totality of observed responses, including the early time behavior that may help considerably for a satisfactory determination of aquifer parameters.

2. Groundwater Flow Models in Fractured Formations

Specific models are based on the following two major sets of conceptual ideas:

1. Starting from the assumption that flow through intergranular pores of the rock matrix makes no significant contribution to the total flow, various authors have been interested in characterizing the flow exclusively through the fractures simulated by a system of pipes or horizontal plates. Individual fractures are described by parameters such as aperture size, roughness, friction coefficient, and hydraulic radius. The flow is either laminar or turbulent according to the Reynolds number; influence of stress conditions also is considered. Emphasis is given to laboratory experiments, core analysis, geophysical logging, and in situ geotechnical tests that are more related to rock hydraulics than to conventional hydrogeology. (Rock hydraulics is a branch of hydrogeotechnics that focuses on water-related stress conditions of essentially discontinuous massifs of consolidated rocks. The fron-

tier of this discipline with the hydrogeology of fractured aquifers is obviously unclear; however, state-of-the-art techniques have been, so far, developed separately.) Resulting models simulate the productivity of a single fracture of limited extent; a fractured aquifer is viewed as the result of a certain pattern of fractures or the conjunction of a few extended fractures, each of them being characterized by a specific hydraulic behavior. The purely fractured medium considered by this approach cannot be treated as a continuum. Type curve solutions are not therefore available.

2. Another conceptual approach is to regard a network of fractures as a continuum where a fictitious velocity can be defined as a continuous function throughout the entire medium (rock matrix plus fractures). In porous media, detailed analysis leads to the concept of a statistically homogeneous medium and to Darcy's macroscopic velocity single valued at any point of this medium. This analysis is not applicable to fractured media because of the strong mechnical discontinuity represented by the fractures. A distinction must be made between fracture and rock matrix characteristics. The difficulty can be surmounted by assuming that the rock matrix 'feeds its fluid into the highly permeable fractures' [Muskat, 1937]. Two overlapping continuous media, each of them having different hydraulic conductivities and storativities, can be imagined to represent an actual fractured and porous aquifer. This concept is usually known as the 'double-porosity medium' theory.

The main feature of the flow mechanism is a process of pressure equalization between the blocks of rock material (low K, high pore volume) and the fractures (high K, low pore volume). Depending on the model, this process is either described as steady state [Barenblatt et al., 1960; Warren and Root, 1963] or time dependent [Streltsova, 1976; Duguid and Lee, 1977]. With the double-porosity models, type curve solutions have been derived for various initial and boundary conditions.

3. Double-Porosity Models

3.1. Barenblatt et al. Model

The rock mass is broken into blocks of irregular size by the fractures. The elementary volume of the model has to include a large number of porous blocks (large in comparison with the real size of the blocks) but has to remain small as compared with the total volume of the aquifer.

From a dimensional analysis, Barenblatt et al. [1960] proposed that the drainage rate from blocks to fractures per unit volume of rock is proportional to the pressure differential between the two components of the model. Additional assumptions are as follows:

1. The flow from blocks to fractures is steady state.
2. The change in volume of liquid resulting from the compressibility of the fractures is negligible when compared with the change of volume of liquid caused by the flow from the blocks.
3. No flow is entering the blocks.
4. In the blocks the change in volume due to the liquid moving out of the blocks is neglected when compared to the change of volume due to liquid expansion.
5. The blocks are isotropic, and the aquifer is confined and of lateral extent.

With these assumptions, drawdown is the fracture s_1 at a point located at a distance r from a fully penetrating well discharging at a constant rate Q from time t = 0 is described by the equation

$$K_1 \Delta s_1 = Ss_2 \left[\frac{\partial s_1}{\partial t} - B_1{}^2 \frac{\partial}{\partial t} (\Delta s_1)\right] \qquad (1)$$

where

$$\Delta s_1 = \frac{\partial^2 s_1}{\partial r^2} + \frac{1}{r} \frac{\partial s_1}{\partial r}$$

is the Laplacian of s_1

$$B_1{}^2 = \frac{K_1}{\alpha} \frac{\mu}{\rho g}$$

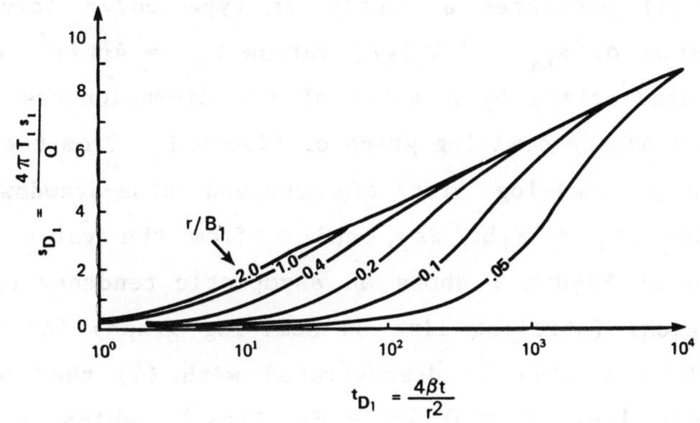

Fig. 1. Type curves for Barenblatt et al [1960] model.

with $[B_1] = [L]$ and where α is a dimensionless characteristic of the fractured rock defined by

$$q = \frac{\alpha}{\mu}(p_2 - p_1) = \alpha \frac{\rho g}{\mu}(s_1 - s_2) \quad (2)$$

where q is the interporosity flow rate per unit volume $[T^{-1}]$. The following initial and boundary conditions for the variable s_1 (r, t) are associated to (1):

$$s_1(r, 0) = 0 \quad \text{for all } r \quad (3)$$

$$s_1(\infty, t) = 0 \quad \text{for all } t$$

$$\lim_{r_w \to 0} \left(r_w \frac{\partial s_1}{\partial r}\right) = 0$$

when $t \to 0^+$, initial flow comes entirely from the blocks

$$\lim_{r_w \to 0} \left(r_w \frac{\partial s_1}{\partial r}\right) = \frac{Q}{2\pi T_1}\left(1 - e^{-\beta t/B_1^2}\right)$$

where $\beta = K_1/Ss_2$.

Solution of (1) and (3) is obtained by means of successive Laplace and Hankel transforms, yielding for s_1

$$s_1 = \frac{Q}{4\pi T_1}\int_0^\infty Jo(xr)\left[1 - \exp\left(\frac{-\beta t x^2}{1 + B_1^2 x^2}\right)\right]\frac{dx}{x} \quad (4)$$

where x is a dummy variable of integration.

Equation (4) generates a family of type curve solutions by plotting values of $s_{D_1} = 4\pi T_1 s_1/Q$ versus $t_{D_1} = 4\beta t/r^2$; each type curve is characterized by a value of the dimensionless parameter r/B_1 as shown on the semi-log graph of Figure 1. From the matching process with a semi-log plot of observed time-drawdown curve, one can deduce $T_1 = K_1 b$, Ss_2, and α from the value of r/B_1.

The graph of Figure 1 shows an asymptotic tendency toward the Theis conditions (straight line on semi-log graph) for long time behavior. It can also be demonstrated with (4) that when r/B_1 becomes large, i.e., $B_1 \to 0$, $s_{D_1} \to E_1\ (1/t_{D_1})$, which is the Theis solution.

At early times, the slope of the time-drawdown curves is finite, since from (4)

$$\lim_{t \to 0^+} \left(\frac{\partial s_1}{\partial t}\right) = \frac{Q}{2\pi T_1} \frac{\beta}{B_1^2} \mathrm{K0} \left(\frac{r}{B_1}\right)$$

which is an interesting departure from the radial flow curves (infinite slope at early times).

3.2. Warren-Root Model

Warren and Root [1963] keep the same basic assumptions as Barenblatt et al. [1960] but allow fracture compressibility ($S_1 \neq 0$) and solve the problem for an idealized fracture network consisting of three orthogonal, continuous, and uniform sets of fractures delimiting a systematic array of identical, rectangular parallelepipeds; each fracture is parallel to one of the principal axis of hydraulic conductivity. The aquifer is thus anisotropic with respect to the fracture hydraulic conductivity. Flow between blocks and fractures is steady state, and the flow to the well through the fractures only is unsteady state. Kazemi et al. [1969] have proposed a solution of the Warren-Root model for observation well responses.

Kazemi et al. [1969] write the system of partial differential equations in a dimensionless form as follows:

$$\left\{\begin{aligned} \frac{\partial^2 s_{D_2}}{\partial r_D^2} + \frac{1}{r_D}\frac{\partial s_{D_2}}{\partial r_D} - (1-\omega)\frac{\partial s_{D_1}}{\partial t_D} &= \omega \frac{\partial s_{D_2}}{\partial t_D} \\ (1-\omega)\frac{\partial s_{D_1}}{\partial t_D} &= \lambda\,(s_{D_2} - s_{D_1}) \end{aligned}\right\} \quad r_D \text{ and } t_D > 0 \tag{5}$$

where

$$s_{D_1} = \frac{2\pi T s_1}{Q} \qquad s_{D_2} = \frac{2\pi T s_2}{Q} \qquad t_D = \frac{T\ t}{(S_1 + S_2)\ r_w^2}$$

$$r_D = \frac{r}{r_w} \qquad \omega = \frac{S_1}{S_1 + S_2} \qquad \lambda = k\,\frac{T_2}{T}\,r_w^2$$

The k is a shape factor $[L^{-2}]$, and λ is often called 'interporosity flow' parameter (dimensionless). T is defined as the effective transmissivity of the fractures expressed by

$$T = \sqrt{K_{1x}\ K_{1y}}\ \times b$$

where K_{1x} and K_{1y} are the principal hydraulic conductivities of the fracture network and b is the aquifer thickness.

Initial and boundary conditions associated with (5) are

$$\begin{aligned} &s_{D_1}(0, r_D) = s_{D_2}(0, r_D) = 0 \quad \text{for all } r_D \\ &\lim_{r_D \to 0} \left(r_D \frac{\partial s_{D_1}}{\partial r_D}\right) = -1 \text{ for all } t_D \\ &\lim_{r_D \to \infty} (s_{D_1}) = 0 \text{ for all } t_D \end{aligned} \tag{6}$$

It is shown that the solutions of (5) and (6) in the Laplace plane are

$$\overline{s_{D_1}} = \frac{K_o\,[r_D\sqrt{p f(p)}]}{p} \qquad \overline{s_{D_2}} = \frac{\lambda\,K_o\,[r_D\sqrt{p f(p)}]}{p\,[(1-\omega)p + \lambda]}$$

with

$$f(p) = \frac{\omega\,(1-\omega)p + \lambda}{(1-\omega)\,p + \lambda}$$

and p is the parameter of Laplace transformation.

In the real plane, the solution for s_{D_1} is obtained by inversion of $\overline{s_{D_1}}$:

$$s_{D_1} = \frac{1}{2}\int_0^\infty \frac{Jo(r_D\sqrt{\omega x})}{\rho_1 - \rho_2}[(1 + \frac{\gamma}{\rho_1})(e^{\rho_1 t_D} - 1) - (1 + \frac{\gamma}{\rho_2})(e^{\rho_2 t_D} - 1)]dx \quad (7)$$

where ρ_1 and ρ_2 are the roots of the auxiliary equation

$$\rho^2 + (\frac{\gamma}{\omega + x})\ \rho + \gamma x = 0$$

and $\gamma = \lambda/1\text{-}\omega$ (dimensionless) and x is a dummy variable of integration.

The value of s_{D_1} (r_D, t_D) can be obtained from (7) for various couples of (λ, ω), using a numerical technique of evaluation of the integral; this calculation is done with a computer; the tabulated results have not been published so far. The inversion of $\overline{s_{D_2}}$ to arrive at an expression of s_{D_2} (r_D, t_D) is judged of limited interest by Kazemi et al. [1969], since it is assumed in this model that the flow to the well bore occurs only through the fracture network.

Asymptotic solution: Considering the expression of $\overline{s_{D_1}}$, an approximative form of the Bessel function that will be valid for small values of p (or large values of t_D) can be used:

$$Ko\ [r_D\sqrt{pf(p)}] \simeq -\{0.57722 + \ln\ [\frac{r_D}{2}\sqrt{pf(p)}]\}$$

After inversion, the approximate solution for (s_{D_1}) becomes

$$s_{D_1}\ (r_D, t_D) \simeq \frac{1}{2}\ \{\ln(\frac{t_D}{r_D^{\ 2}}) + E_1(\frac{\lambda t_D}{1-\omega}) - E_1[\frac{\lambda t_D}{\omega(1-\omega)}] + 0.80908\} \quad (8)$$

for $t_D > 100\ r_D^{\ 2}$

E_1 ( ) = -Ei (-) is the tabulated exponential integral function. Long times dimensionless fracture drawdown at the well $(s_{D_{1_w}})$ can be calculated from (8) by setting $r_D = r/r_w = 1$.

Tables 1 and 2 give values of (s_{D_1}) and $(s_{D_{1_w}})$ as computed with (8), and Figure 2 is a semi-log plot of $(s_{D_{1_w}})$ from Table 2. The graph of Figure 2 shows clearly that for very large dimen-

TABLE 1. Long Time Values of s_{D_1} (Equation (8)) When (1) $\lambda = 10^{-6}$ and $\omega = 10^{-3}$ and (2) $\lambda = 5 \times 10^{-6}$ and $\omega = 10^{-2}$

t_D	r_d = 20		50		100		500
	1	2	1	2	1	2	
4×10^4	4.048	3.315					
6×10^4	4.058	3.359					
8×10^4	4.067	3.402					
2.5×10^5	4.145	3.695	3.229	2.779			
5×10^5	4.244	3.982	3.328	3.066			
8×10^5	4.358	4.207	3.441	3.290			
10^6	4.424	4.317	3.508	3.400	2.815	2.708	
4×10^6	5.012	5.010	4.095	4.093	3.402	3.400	
8×10^6	5.356		4.440		3.747		
10^7	5.468		4.551		3.858		
2.5×10^7	5.926		5.010		4.316		2.707
6×10^7	6.364		5.447		4.754		3.145
10^8	6.619		5.703		5.010		3.400

sionless times ($t_D > 10^6$), or when $\lambda \rightarrow \infty$, the solution obtained with the Warren-Root model is a straight line on a semi-log plot similar to the Jacob logarithmic approximation of the Theis model (homogeneous and isotropic behavior of the aquifer and radial flow toward the well).

3.3. Boulton-Streltsova Model

Boulton and Streltsova [1977] propose to replace the irregular network of blocks and fractures of Barenblatt et al. [1960] model by a regular pattern of horizontal strata. The block units representing the rock mass have a thickness (2H) equal to the average thickness of the actual blocks; the fracture units have a thickness (2h), which is the average thickness of the actual fractures and 2h << 2H.

Both (2h) and (2H) are constant over the infinite lateral exten-

TABLE 2. Long Time Values of $(s_{D_{1_w}})$
(From Equation (8))

t_D	$\lambda = 10^{-6}$ $\omega = 10^{-3}$	$\lambda = 5 \times 10^{-6}$ $\omega = 10^{-2}$
10^2	6.154	4.984
3×10^2	6.582	5.486
5×10^2	6.749	5.695
8×10^2	6.870	5.871
10^3	6.915	5.940
5×10^3	7.023	6.229
10^4	7.028	6.239
5×10^4	7.048	6.333
10^5	7.029	6.435
5×10^5	7.239	6.978
10^6	7.420	7.313
10^7	8.463	8.463
10^8	9.615	9.615

sion of the confined aquifer. Due to the vertical symmetry of this system, it is sufficient to study a representative element of dimensions H and h in the vertical plane of coordinates (z,r).

For this representative element, it is assumed that (1) the two elementary units (porous block and fracture) are compressible, (2) flow is vertical in the block and horizontal in the fracture; this implies that the well is screened only against the fracture, (3) there is no contact resistence to flow between block and fracture, (4) both flows in block and fracture are obeying Darcy's law, and

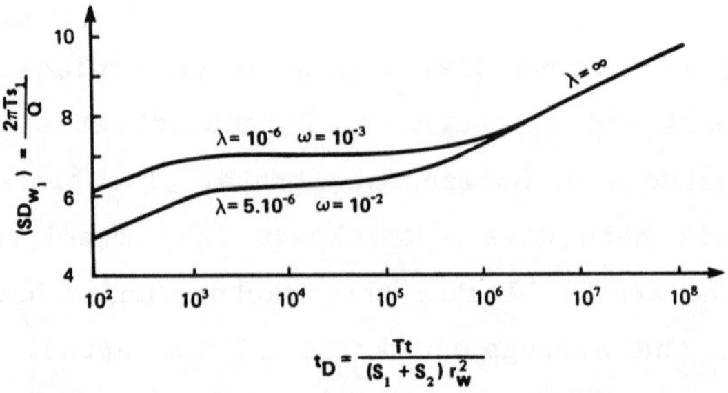

Fig. 2. Type curves for Warren and Root [1963] model (long time approximation at the pumping well).

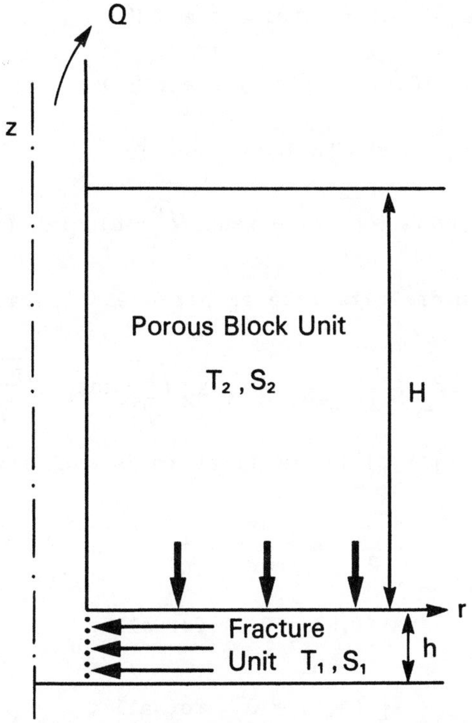

Fig. 3. Idealized representation Boulton-Streltsova model.

(5) the radius of the well is vanishingly small and the discharge in the well per unit length of fracture unit is constant. (The case when both block and fracture units are tapped by the well was also treated by the authors; very little difference with the case treated here is obtained for the type curve solutions representing the dimensionless fracture drawdown.)

In these conditions, the drawdown s_2 (z,t) resulting from vertical flow in the block is described by the equations (see Figure 3)

$$\frac{\partial^2 s_2}{\partial z^2} = \frac{1}{\eta_2} \frac{\partial s_2}{\partial t}$$

with $\eta_2 = T_2/S_2$ = block diffusivity

$$\frac{\partial s_2}{\partial z} = 0 \text{ for } z = H, \ t \geqslant 0 \qquad (9)$$

$$s_2\ (z,0) = 0 \text{ for } 0 \leqslant z \leqslant H$$

$$s_2\ (0,t) = s_1(r,t) \text{ for } t \geqslant 0$$

The solution of (9) in the Laplace plane is

$$\overline{s_2} = \overline{s_1}\ [\cosh\ (\sqrt{\frac{p}{\eta_2}}\ Z) - \tanh\ (\sqrt{\frac{p}{\eta_2}}\ H)\ \sinh\ (\sqrt{\frac{p}{\eta_2}}\ Z)]$$

In this plane, the drainage rate v_D per unit time and unit area at $z = 0$ is

$$\overline{v_D} = -K_2\ (\frac{\partial \overline{s_2}}{\partial z})_{z=0} = \overline{s_1}\ K_2 \sqrt{\frac{p}{\eta_2}}\ \tanh\ (\sqrt{\frac{p}{\eta_2}}\ H)$$

Now, the drawdown $s_1(r,t)$ in the fracture is such that

$$\frac{\partial^2 s_1}{\partial r^2} + \frac{1}{r}\frac{\partial s_1}{\partial r} = \frac{1}{\eta_1}\frac{\partial s_1}{\partial t} + \frac{v_D}{T_1} \qquad (10)$$

$$s_1(r,0) = 0 \quad \text{for all } r$$

$$s_1\ (\infty,t) = 0 \quad \text{for all } t$$

$$\lim_{r\to 0}\ (r\ \frac{\partial s_1}{\partial r}) = \frac{-Q}{2\pi T_1}$$

(assumption 5 above). The solution of (10) after Laplace and zero-Hankel transformations is obtained after replacing $\overline{v_D}$ by its expression derived from (9):

$$s_1^* = +\frac{Q}{2\pi T_1}\ \frac{1/p}{a^2 + (p/\eta_1) + (K_2/T_1)\ \sqrt{(p/\eta_2)}\ \tanh\ (\sqrt{(p/\eta_2)}\ H)}$$

where a is the parameter of the zero-Hankel transformation.

Defining the function g(p) as

$$g(p) = \frac{1}{\eta_1} + \frac{K_2}{T_1\sqrt{\eta_2}}\ \frac{1}{\sqrt{p}}\ \tanh\ (\sqrt{\frac{p}{\eta_2}}\ H)$$

one can express s_1^* as

$$s_1^* = \frac{Q}{2\pi T_1}\ \frac{1/p}{a^2 + pg(p)}$$

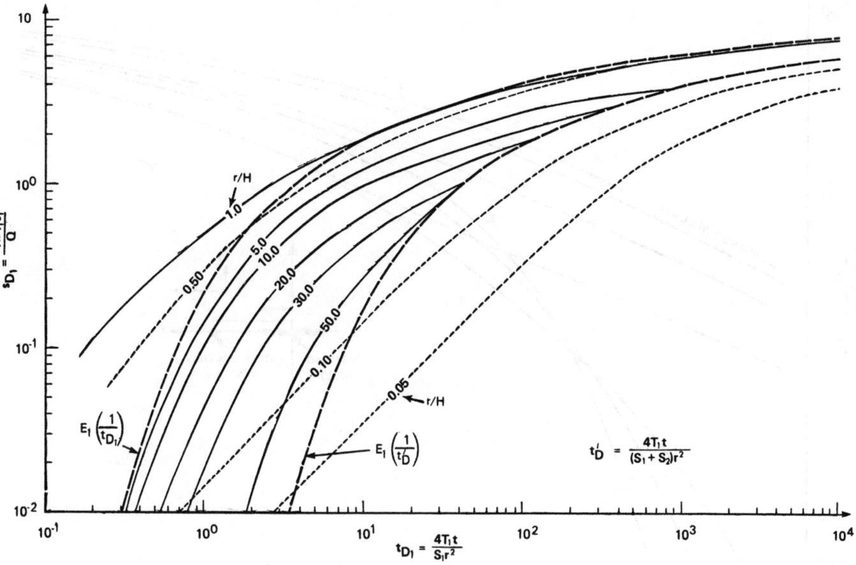

Fig. 4. Type curves for Boulton and Streltsova model for $\eta_2/\eta_1 = 10^{-4}$ and $T_2/T_1 = 10^{-3}$.

The inverse zero-Hankel transform of $s_1{*}$ yields $\overline{s_1}$ as

$$\overline{s_1} = \frac{Q}{2\pi T_1} \frac{1}{p} \int_0^\infty a \, Jo(ra) \frac{da}{a^2 + pg(p)}$$

and since

$$\int_0^\infty a \, Jo(ra) \frac{da}{a^2 + \eta^2} = Ko(\eta r)$$

one can express $\overline{s_1}$ in a somehow different form than Boulton and Streltsova:

$$\overline{s_1} = \frac{Q}{2\pi T_1} \frac{Ko\left[r\sqrt{pg(p)}\right]}{p} \tag{11}$$

Exact Mellin inversion of (11) was derived by Boulton and Streltsova [1977]: a complex analytical solution is obtained from which type curve solutions are calculated and characterized by r/H for various values of the coupled parameters (η_2/η_1, T_2/T_1), as shown, for example, by Figure 4.

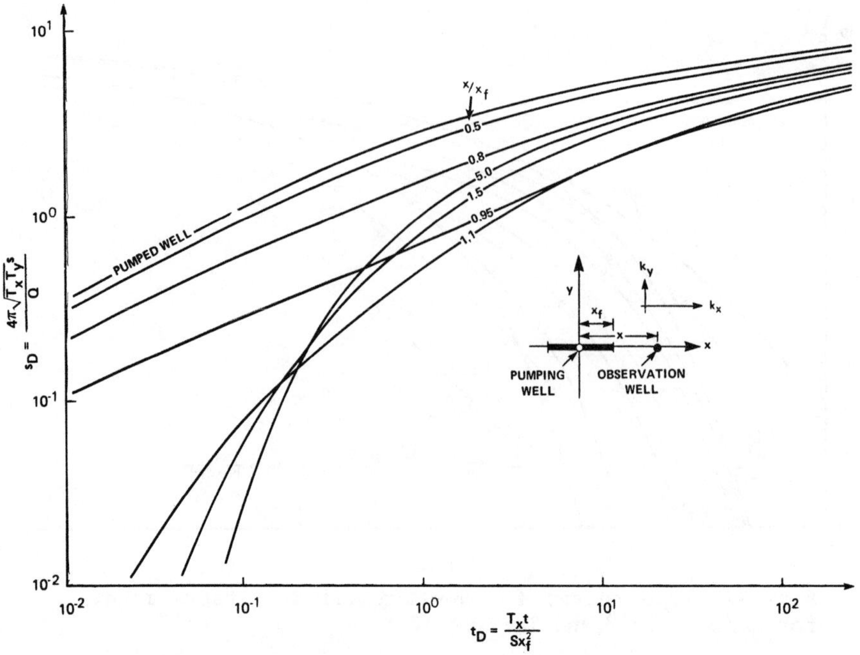

Fig. 5a. Type curves for Gringarten and Witherspoon [1972] vertical fracture model with an observation well located on x axis.

Using $\overline{s_1}$ obtained from (11) and the solution of (9) for $\overline{s_2}$, s_2 can be, in turn, expressed in an analytical form, and type curves have been derived for the mean drawdown in the block.

A log-log plot of measured drawdowns in observation wells versus pumping time is matched as usual with one of the type curves to determine T_1 and S_1 and to estimate T_2, S_2, and H from η_2/η_1, T_2/T_1, and r/H ratios. However, sets of type curves other than the published ones may be necessary to obtain a satisfactory match with the field data. This requires new computations of s_{D_1} with the Boulton-Streltsova solution, which means considerable amount of computer time, above all when r/H or t_D are small.

4. Single-Fracture Models

The naturally fractured as well as porous medium is now simulated by an homogeneous medium with an usually strong anisotropy.

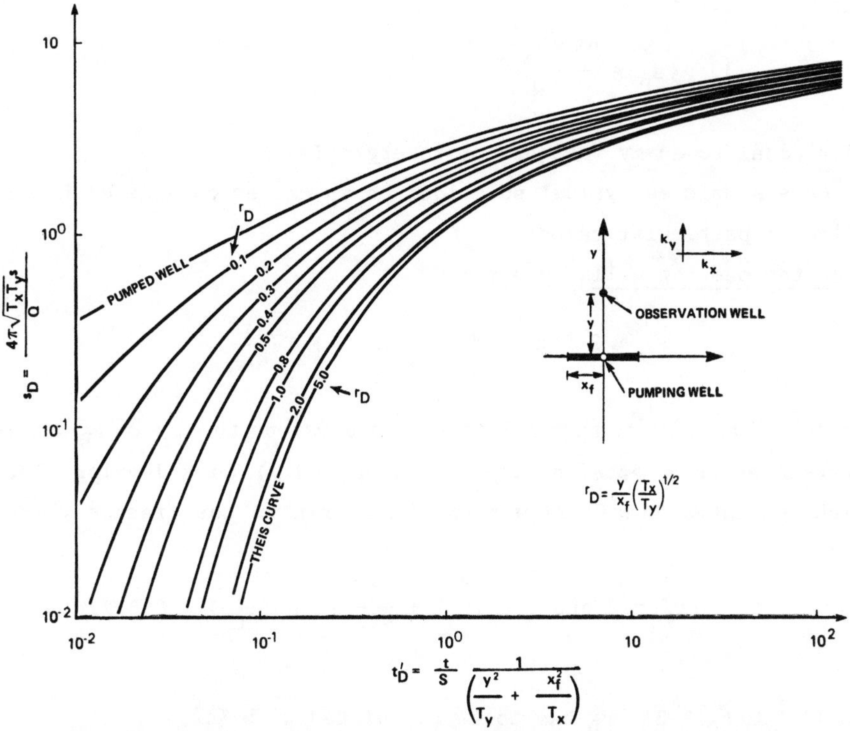

Fig. 5b. Type curves for Gringarten and Witherspoon [1972] vertical fracture model with an observation well located on y axis.

These models study the cases when a single vertical or horizontal fracture with finite extent and uniform flux (water enters the fracture at the same rate per unit area) is intersected by a pumping well. The model and solutions presented here follow the work of Gringarten and Ramey [1973, 1974] and Gringarten and Witherspoon [1972].

In Figures 5a and 5b, a plane view of the vertical fracture model is represented. The fracture aperture is assumed to be infinitely small. By product of Green's functions, the dimensionless drawdown at a point M (x,y) of the aquifer is obtained as follows:

$$s_D = \frac{\sqrt{\pi}}{2}\int_0^{t_D} \left[\text{erf}\left(\frac{1 - \frac{x}{x_f}}{2\sqrt{\tau}}\right) + \text{erf}\left(\frac{1 - \frac{x}{x_f}}{2\sqrt{\tau}}\right)\right] \exp\left[-\frac{1}{4\tau}\left(\frac{y^2}{x_f}\right)\frac{T_x}{T_y}\right]\frac{d\tau}{\sqrt{\tau}} \quad (12)$$

with

$$s_D = \frac{4\pi \sqrt{T_x T_y}\ s}{Q} \qquad t_D = \frac{T_x t}{S x_f^2}$$

and τ equal to dummy variable of intergration.

More simple analytical solutions of s_D can be considered in the following particular cases:

At the pumping well: $x = y = 0$

$$s_D = 2\sqrt{\pi t_D}\ \mathrm{erf}\ (\frac{1}{2\sqrt{t_D}}) + E_1\ (\frac{1}{4t_D})$$

for $10^{-2} \leqslant t_D < 10^{-1}$, this solution is equivalent to $s_D = 2\sqrt{\pi t_D}$ and is represented by a straight line of slope (0.5) on a log-log plot, which indicates linear flow from the matrix to the fracture. For $t_D \geqslant 25$,

$$s_D = 2\ \exp\ ({}^{-1}/{4t_D}) + \frac{1}{4t_D} - \ln\ (\frac{1}{4t_D}) - 0.57722$$

For $10^{-1} \leqslant t_D < 25$, s_D has to be calculated with (12).

At an observation well located on x axis (Figure 5a): $y = 0$

$$s_D = \frac{\sqrt{\pi}}{2} \int_0^{t_D} [\mathrm{erf}\ \frac{1 - (x/x_f)}{2\ \sqrt{\tau}} + \mathrm{erf}\ \frac{1 + (x/x_f)}{2\ \sqrt{\tau}}]\ \frac{d\tau}{\sqrt{\tau}}$$

Resulting type curves for (x/x_f) ratios are shown in Figure 5a. Note that $x/x_f = 1$ is not included in this solution and that the shapes of type curves are different for $x/x_f < 1$ and $x/x_f > 1$. The Theis solution is obtained for $x/x_f \geqslant 5$.

At an observation well located on y axis: $x = 0$

$$s_D = \sqrt{\pi} \int_0^{t_D} \mathrm{erf}\ (\frac{1}{2\sqrt{\tau}})\ \exp\ (\frac{-r_D^2}{4\tau})\ \frac{d\tau}{\sqrt{\tau}}$$

where

$$r_D = \frac{y}{x_f} \sqrt{\frac{T_x}{T_y}}$$

Figure 5b shows the type curves solutions as $s_D = f(t'_D)$, where

$$t'_D = \frac{t_D}{1 + r_D^2}$$

5. Approximate Solutions for Double-Porosity Models

The inconvenience with Kazemi et al. [1969] (section 3.2) and Boulton and Streltsova [1977] (section 3.3) solutions is that they depend on more than one parameter, and thus type curve solutions have to be calculated from complex analytical forms whenever the field data do not correspond to already tabulated type curves (i.e., for other values of the dependent parameters). Also, many pumping test data do not clearly show the tendency to a pseudoradial flow, and late time Theis-type solutions are either too imprecise or inappropriate to interpret these data.

To circumvent these difficulties, it is proposed to arrive at simpler analytical solutions by using Schapery's [1961] technique of approximate inversion of the derived functions in the Laplace plane.

According to this technique

$$F(t) = L^{-1}\,[f(p)] \simeq [pf(p)]_{\,p = \frac{1}{2t}}$$

where F(t) is the unknown function in the real plane and f(p) is its Laplace transform. The condition to apply this approximation is that the derivative of f(p) with respect to (ln p) approaches a straight line.

5.1. Application to Boulton-Streltsova Model

Equation (11) can be rewritten in terms of dimensionless fracture drawdown as

$$\overline{s_{D_1}} = \frac{2\,K0\,(r\sqrt{pg(p)}}{p} \qquad g(p) = \frac{1}{\eta_1} + \frac{k_2}{T_1\sqrt{\eta_2}}\,\frac{1}{\sqrt{p}}\,\tanh(\sqrt{\frac{p}{\eta_2}}\,H)$$

Doing $s_{D_1} \simeq [p\,\overline{s_{D_1}}(p)]\;p=1/2t$ gives $s_{D_1} \simeq 2\,K0\,[r\sqrt{g(1/2t)/2t}]$ and using the dimensionless time, $t_{D_1} = 4\eta_1 t/r^2$:

$$s_{D_1} \simeq 2Ko(v) \quad v = [\frac{2}{t_{D_1}} + \frac{T_2}{T_1}\frac{r}{H}\sqrt{\frac{\eta_1}{\eta_2}}\sqrt{\frac{2}{t_{D_1}}}\tanh(\frac{H}{r}\sqrt{\frac{\eta_1}{\eta_2}}\sqrt{\frac{2}{t_{D_1}}})]^{1/2} \quad (13a)$$

The function (v) can be expressed by the equivalent forms

1. With Boulton-Streltsova's notations, $b = \eta_2/\eta_1$, $c = T_2/T_1$

$$v = [\frac{2}{t_{D_1}} + \frac{r}{H}\frac{c}{\sqrt{b}}\sqrt{\frac{2}{t_{D_1}}}\tanh(\frac{1}{r/_H}\frac{1}{\sqrt{b}}\sqrt{\frac{2}{t_{D_1}}})]^{1/2} \quad (13b)$$

2. With Streltsova-Adams' [1978] notations, $B = H\sqrt{T_1/T_2}$ and expressing in function of parameters r/B and S_1/S_2:

$$v = [\frac{2}{t_{D_1}} + \frac{1}{\sqrt{S_1/_{S_2}}}\sqrt{\frac{2}{t_{D_1}}}\tanh(\frac{B}{r}\frac{1}{\sqrt{S_1/_{S_2}}}\sqrt{\frac{2}{t_{D_1}}})]^{1/2} \quad (13c)$$

3. With the author's notations, $r_D = r/\sqrt{Hh}$, $m = \sqrt{K_2/K_1}$, $n = \sqrt{S_2/S_1}$:

$$v = [\frac{2}{t_{D_1}} + r_D mn\sqrt{\frac{2}{t_{D_1}}}\tanh(\frac{n}{r_{D}m}\sqrt{\frac{2}{t_{D_1}}})]^{1/2} \quad (13d)$$

This last formulation uses the definitions: $T_1 = K_1h$ and $T_2 = K_2H$.

A direct comparison of Streltsova-Adams [1978, Table VI, p. 393] published values of s_{D_1} with this approximate solution as calculated with (13c) for v and (13a) for s_{D_1} is done on Table 3 for $S_1/S_2 = 0.10$ and some values of r/B.

In Table 3 the three calculations used are (1) approximate Schapery's [1961] technique of inversion (equation (13a)), (2) Boulton and Streltsova [1977] results as obtained from their equations (63), (65), and (67) and published by Streltsova-Adams [1978], and (3) our results from the same equations (63), (65), and (67). (These results are obtained with a double-precision algorithm based on an adaptive Simpson's rule technique to evaluate the integral of equation (63). Relative error $\simeq 10^{-3}$. Procedure and results of this work are the object of D. Cuthiell and C. Sauveplane (unpublished manuscript, 1983). The equations (63),

TABLE 3. Results for s_{D_1} Using Three Different Calculations

t_{D_1}	r/B 0.1			0.10			2.0		
	1	2	3	1	2	3	1	2	3
1	0.468	0.121	0.215	0.393	0.197	0.184	0.050	-	-
2	0.824	0.235	-	0.685	0.478	0.465	0.109	0.022	0.050
5	1.436	0.534	1.197	1.176	1.011	0.998	0.272	0.078	0.143
10	1.979	0.928	1.779	1.597	1.476	-	0.496	0.221	0.286
20	2.563	1.470	-	2.036	1.953	1.940	0.828	0.493	0.558
50	3.366	2.296	3.243	2.620	2.571	-	1.424	1.093	1.158
100	3.979	2.930	-	3.127	3.015	3.002	1.968	1.675	1.740
1,000	5.923	4.922	5.869	4.485	4.361	4.347	4.084	3.878	3.944
10,000	7.617	6.600	7.438	6.426	6.250	6.237	6.366	6.178	6.236
100,000	9.155	7.984	-	8.778	8.551	-	8.772	8.540	-

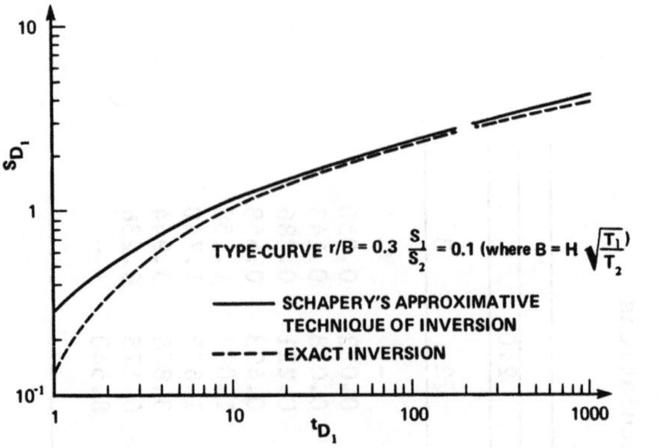

Fig. 6. Comparison between exact and approximate Boulton and Streltsova [1977] model.

(65), and (67) discussed here can be found in the work by Streltsova-Adams [1978, p. 393].) Figure 6 is a log-log plot of results obtained by calculations 1 and 2 for r/B = 0.30. The following comments can be made concerning these results:

1. The complexity of equations (63), (65), and (67) derived by Boulton and Streltsova [1977] is such that different techniques of evaluation of s_{D_1} may yield substantially different results, especially for low values of r/B. Comparing columns numbered 2 and 3 of Table 3, one can notice that for a given r/B the difference between s_{D_1} values is nearly constant for most of t_{D_1} values. Also, our results are systematically above Boulton and Streltsova's for low and high values of r/B.

2. Comparing columns numbered 1 with 2 and 3 of Table 3, it is clear that the approximate inversion technique overestimates s_{D_1}. However, absolute departures tend to gradually decrease when t_{D_1} increases, although for r/B = 0.1, a minimum is obtained for $t_{D_1} \simeq$ 50.

3. As shown on the log-log plot of Figure 5, the type curve obtained for r/B = 0.3 with the approximate technique of inversion is in good agreement for $t_{D_1} > 10$ with Boulton-Streltsova type curve. The error introduced in the determination of the aquifer parameters (matching process) by using the approximate solution becomes unsignificant beyond $t_{D_1} = 10$.

Log-log plots for other values of r/B show the same agreement except for r/B = 0.01, where this is only true when comparing columns numbered 1 and 3 of Table 3. Also, when r/B = 2, the agreement is satisfactory for $t_{D_1} > 100$ only.

Evaluation of s_{D_1} with the proposed approximate inversion technique (equation (13a)) is reliable for intermediate and long dimensionless times only. It generally overestimates s_{D_1}, and this should result in a pessimistic determination of the aquifer parameters (when any appreciable error is committed). Its interest lays in its simplicity and rapidity to compute s_{D_1} for a wide range of the three depending parameters η_2/η_1, T_2/T_1, and r/H (or r/B or $r/\sqrt{Hh}$).

As an illustration, values of s_{D_1} calculated with (13a) and (13d) are given in Table 4 and plotted in figures 7a, 7b, and 7c.

Considering (13a) and (13d), one can deduce the following asymptotic solutions:

1. If no contribution to the flow is coming from the blocks $K_2 = 0$ and $m \to 0$, expression (13d) becomes $v = \sqrt{2/t_{D_1}}$ and (13a) will be

$$s_{D_1} = 2 \text{ Ko } \left(\sqrt{\frac{2}{t_{D_1}}}\right) = E_1 \left(\frac{1}{t_{D_1}}\right)$$

which is the Theis solution for the fracture.

2. When r_D is large ($H \to 0$ or $h \to 0$) with (Hh) finite),

$$v \sim \left[(1 + n^2) \frac{2}{t_{D_1}}\right]^{\frac{1}{2}}$$

$$s_{D_1} \sim E_1 \left(\frac{1 + n^2}{t_{D_1}}\right)$$

which corresponds to a shift to the right of the Theis solution for fractures $E_1(1/t_{D_1})$ of the quantity $1 + n^2 = [1 + (S_2/S_1)]$ on a log-log plot. These two limiting cases are represented by dashed lines on Figures 7a, 7b, and 7c.

TABLE 4. Values of S_{D_1} Computed From

t_{D_1}	r_D							
	0.1			1			5	
	1	2	3	1	2	3	1	2
0.1	0.013	0.013	0.013	0.0111	0.0111	0.013	0.0060	0.0060
0.2	0.057	0.057	0.057	0.0483	0.0483	0.055	0.026	0.026
0.5	0.223	0.223	0.2264	0.189	0.189	0.215	0.1013	0.1013
1.0	0.469	0.469	0.475	0.394	0.394	0.449	0.212	0.212
2.0	0.824	0.824	0.836	0.685	0.685	0.786	0.373	0.373
5.0	1.439	1.439	1.464	1.177	1.177	1.364	0.652	0.652
10.0	1.984	1.984	2.024	1.598	1.598	1.868	0.906	0.902
20.0	2.573	2.573	2.629	2.043	2.043	2.412	1.198	1.177
50.0	3.382	3.382	3.465	2.628	2.620	3.149	1.678	1.564
100.0	4.065	4.065	4.201	3.077	3.051	3.696	2.134	1.872
1,000.0	5.923	5.923	---	4.556	4.386	---	4.198	---
10,000.0	7.617	7.617	---	6.426	5.611	---	6.369	---

Sets (1) $m = 0.1$, $n = \sqrt{10}$ or $K_1 = 100K_2$, $S_2 = 10S_1$; (2) $m = n = 100$ or $K_1 = 10^6K_2$, $S_2 = 10^4S_1$.

5.2. Application to a Hybrid Model (Warren-Root/Boulton-Streltsova)

When the fractured and porous aquifer is idealized through an array of orthogonal cubes, a first approximation is to replace the matrix cubes by spheres of equivalent radius R (equal volume). Assumptions made by Boulton-Streltsova [1977] are kept, but the matrix representative element of thickness H is now divided into cubes by an array of orthogonal fractures, and the cubes have an equivalent radius of R, as shown on Figure 8. Note that a finer partition of the matrix unit can be imagined, so that, for instance, H = 10R or even H = 20R. The drainage rate per unit area from the infinite matrix strata into the horizontal fracture (aperture h) is described as for the Boulton-Streltsova model with a convolution. In these conditions, DeSwaan [1976] and later Najurieta [1976] give the Laplace transform of the time-dependent diffusivity function:

$$h(p) = \frac{1}{\eta_1} + \frac{2K_2}{T_1R}\ \frac{1}{p}\ [R\sqrt{\frac{p}{\eta_2}}\ \coth\ (R\sqrt{\frac{p}{\eta_2}}) - 1]$$

Equations (13a) and (13d) for Three Sets

	10			20			50	
3	1	2	3	1	2	3	2	3
0.010	0.003	0.003	0.008	0.0009	0.0009	0.005	---	0.001
0.044	0.0132	0.013	0.034	0.0044	0.0044	0.021	---	0.007
0.171	0.0544	0.054	0.132	0.0205	0.0204	0.085	---	0.03
0.355	0.118	0.118	0.275	0.0505	0.0488	0.179	7.92×10^{-3}	0.068
0.616	0.218	0.217	0.478	0.1095	0.098	0.318	9.98×10^{-3}	0.132
1.055	0.414	0.401	0.823	0.272	0.202	0.564	0.564	0.262
1.432	0.631	0.579	1.123	0.500	0.315	0.789	0.100	0.395
1.824	0.935	0.786	1.445	0.828	0.457	1.040	0.175	0.559
2.355	1.487	1.098	1.885	1.426	0.692	1.403	0.351	0.819
2.738	2.010	1.358	2.222	1.974	0.911	1.696	0.579	1.045
4.075	4.180	2.434	3.374	4.175	2.159	2.732	2.056	1.939
---	6.367	4.318	4.569	6.366	4.274	3.902	4.261	2.988

$^1/\sqrt{1000}$, n = 10 or $K_1 = 1000K_2$, $S_2 = 100S_1$; and (3) $m = 10^{-3}$,

and show that

$$\overline{s_{D_1}} = \frac{2\ \text{Ko}\ [r\sqrt{ph\ (p)}]}{p}$$

Applying Schapery's [1961] technique of inversion to $\overline{s_{D_1}}$, the following expressions are obtained:

$$w = [2/t_{D_1} + r'_D\ c/\sqrt{b}\ \sqrt{2/t_{D_1}}\ \coth\left(\frac{1}{r'_D\sqrt{b}}\sqrt{2/t_{D_1}}\right) - c\ r'_D{}^2]^{1/2} \qquad (14)$$

$$s_{D_1} \simeq 2\ \text{Ko}\ (w)$$

where $c = T_2/T_1$, $b = \eta_2/\eta_1$, and $r'_D = r/R$. Equation (14) has been derived for the case when H = 2R. Table 5 compares the DeSwaan model as approximated by (14) and the Boulton-Streltsova model (exact and approximate solutions) for the case when $\eta_2/\eta_1 = 10^{-4}$, $T_2/T_1 = 10^{-3}$, and $r'_D = r/R = 10$ (which is equivalent to r/H = 5). An approximate solution of the Boulton-Streltsova model is computed with (13a) and (13b), that is, with the same

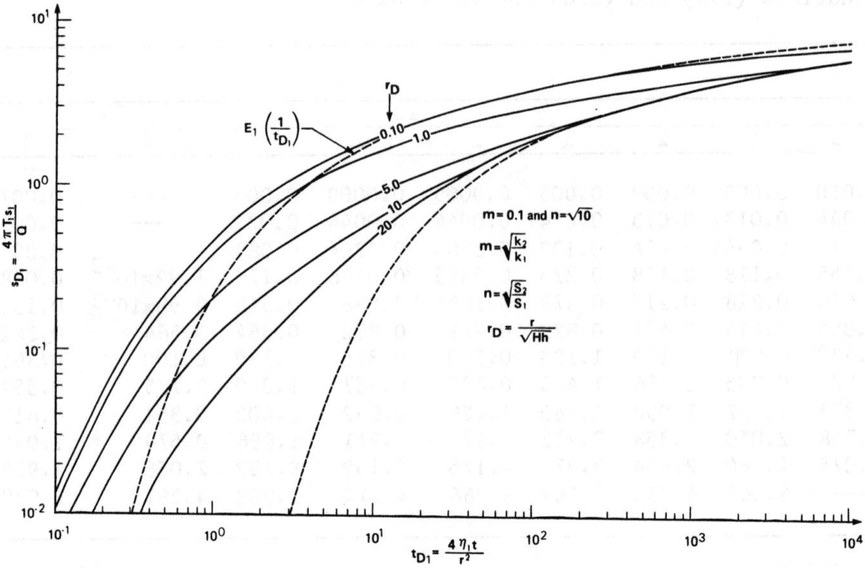

Fig. 7a. Type curves for approximate Boulton and Streltsova [1977] model when m = 0.1, n = $\sqrt{10}$.

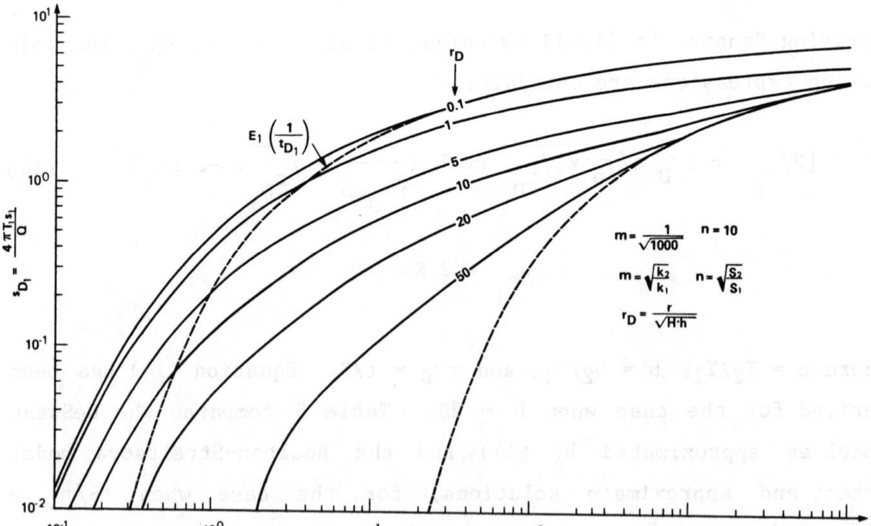

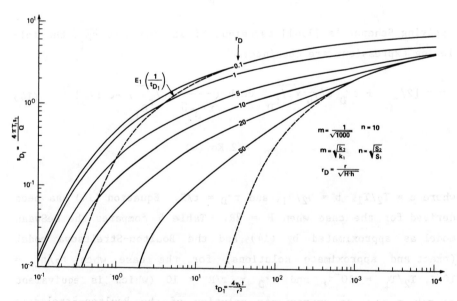

Fig. 7b. Type curves for approximate Boulton and Streltsova [1977] model when m = $1/\sqrt{1000}$, n = 10.

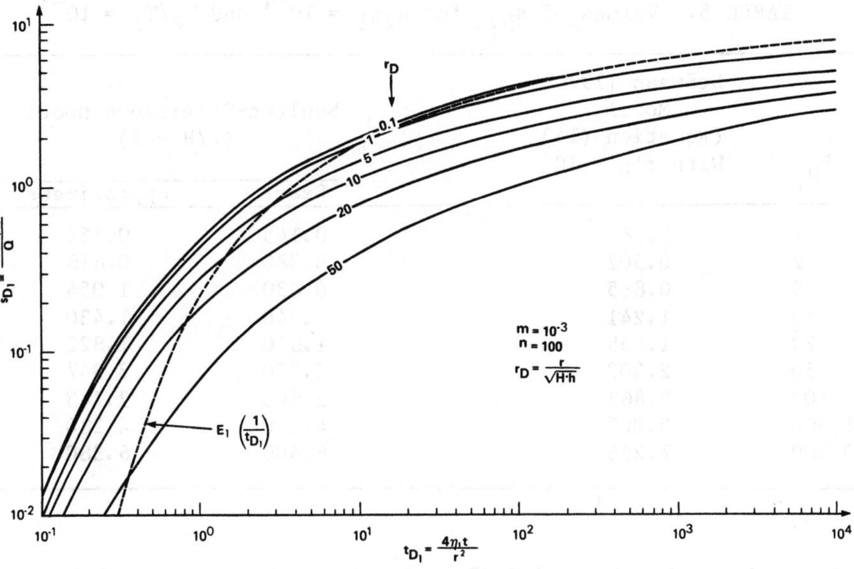

Fig. 7c. Type curves for approximate Boulton and Streltsova [1977] model when $m = 10^{-3}$, $n = 10^2$.

Schapery technique of inversion that was used to derive (14).

When these results are plotted on a log-log scale, it becomes apparent that for $t_{D_1} > 5$, the two models are nearly identical for the determination of the aquifer parameters. However, it should be noted that the asymptotic Theis-type behavior for very large t_{D_1} is obtained earlier with the Boulton-Streltsova model ($t_{D_1} > 10^3$) than with the DeSwaan [1976] model (where at $t_{D_1} = 10^4$, approximate type curve is still above the shifted Theis type curve). Of course, to be strictly valid, these conclusions should be verified for

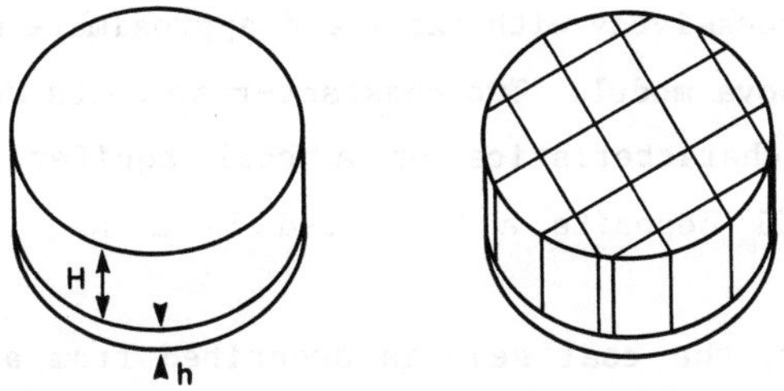

Fig. 8. Representation of DeSwaan's model, H = 2R.

TABLE 5. Values of s_{D_1} for $\eta_2\eta_1 = 10^{-4}$ and $T_2/T_1 = 10^{-3}$

t_{D_1}	DeSwaan [1976] Model (Equation (14) With $r'_D = 10$	Boulton-Streltsova Model (r/H = 5)	
		Exact	Approximate
1	0.284	0.145	0.355
2	0.502	0.380	0.616
5	0.885	0.830	1.054
10	1.241	1.240	1.430
20	1.655	1.650	1.822
50	2.302	2.300	3.347
100	2.863	2.800	2.743
1,000	5.007	4.100	4.280
10,000	7.295	6.400	6.380

other values of η_2/η_1 and T_2/T_1 and values of s_{D_1} computed from the correct inversion of the DeSwaan solution in the Laplace plane.

6. Examples of Pumping Test Data Interpretation

Often in Alberta, observation well time-drawdown curves for coal sandstone fractured aquifers plot as quasi-straight lines on a log-log paper. Interpretation with the Theis model is either impossible or very imprecise, and the use of specific type curves (as described in sections 3 and 4) will only give satisfactory results.

6.1. Halkirk Coal Aquifer

On this example, two different sets of observation well data are interpreted successively with exact and approximate solutions of the Boulton-Streltsova model. Two constant-rate tests were conducted to determine the characteristics of a coal aquifer, 3.05 m thick, confined by till deposits and located 19 m deep in central east Alberta.

Geologically, the coal seam is described from an area where it was strip mined as a flat continuous layer with extensive series of

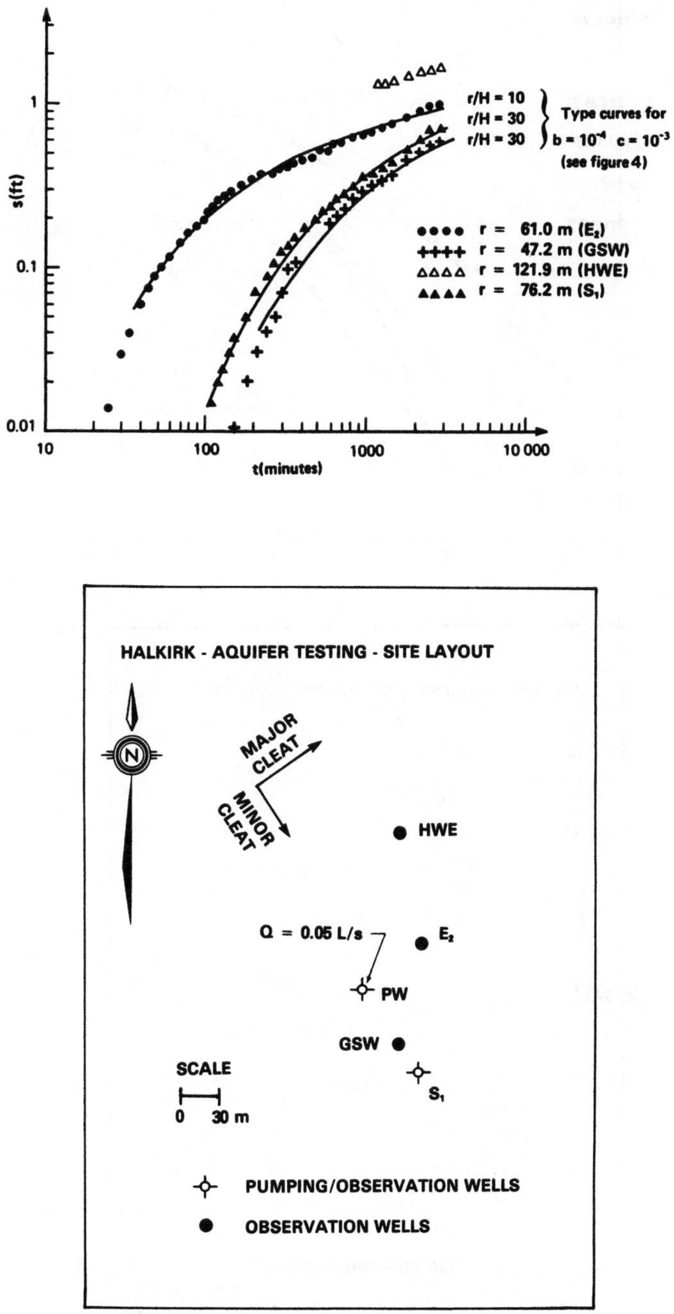

Fig. 9a. Test 1, Halkirk coal aquifer interpreted with Boulton-Streltsova model: exact solution.

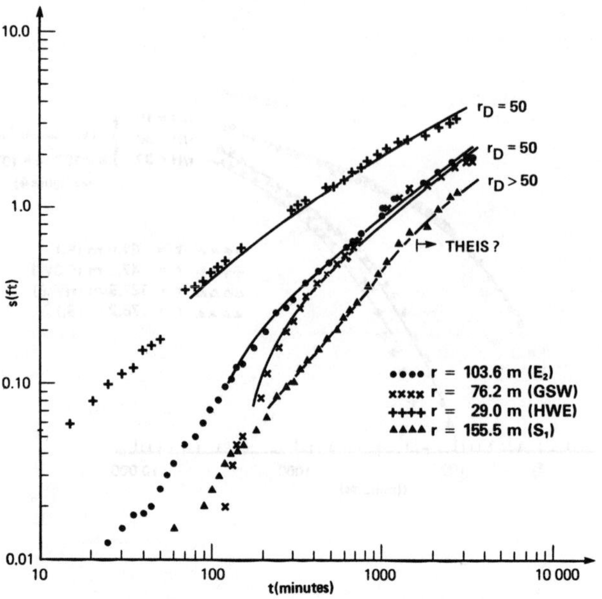

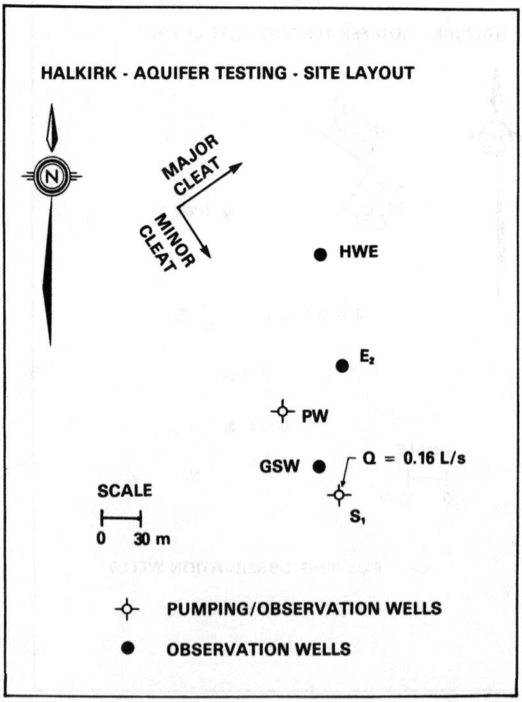

Fig. 9b. Test 2, Halkirk coal aquifer interpreted with Boulton-Streltsova model: approximate solution.

vertical fractures (see major and minor cleats on diagram of Figure 9a) and horizontal partings associated with chert bands. The basal parting located about 0.5 m above the base of the coal seam is an important aquifer zone.

Confined conditions are present throughout the coal seam but mainly over the bottom 1.5 m, as testified to by an excellent correlation between piezometric levels and barometric pressure. Groundwater flow occurs along vertical and horizontal fractures.

These hydrogeological conditions are supporting the choice of Boulton-Streltsova model for interpretation with possible $H \simeq 2.5$ m and $h \simeq 0.10$ m. The interpreted results of each test are identical with both exact and approximate solutions of this model. Figure 9a shows how adequately the observed data of test 1 are fitted with the type curves of the exact solution for $b = 10^{-4}$ and $c = 10^{-3}$. The data of test 2 (Figure 9b) are interpreted with type curves of Figure 7b (where $m = 1/\sqrt{1000}$ and $n = 10$), and the agreement is quite satisfactory.

Interpreted results of Figures 9a and 9b are condensed on Table 6. Results for S_1 may be significantly different with the two solutions (compare observation wells GSW and HWE for tests 1 and 2), but the agreement for S_2 is much better. Results for T_1 and T_2 are satisfactorily coherent. An important variation is noticed on H values; assuming that $H = 2.5$ m is the most realistic one and since $\sqrt{Hh} \simeq 0.6$, then $h \simeq 0.14$ m.

6.2. Barrhead Coal Aquifer

For this coal seam (1.22 m thick, 34 m deep), three pumping tests have been done using successively each well of the diagram of Figure 10 as a pumping well and the two other wells as observation wells. The coal aquifer is confined. Unfortunately, geological information, specially that related to the fracture network, is not available for this example, and both the Boulton-Streltsova and the vertical fracture models have been used for interpretation of the measured data.

TABLE 6. Halkirk Coal Aquifer: Comparison Between Approximate and Exact Solutions of Boulton and Streltsova Model

Observation well	Test No.	Distance to Pumping Well, m	T_1,[a] $\times10^{-5}$ m^2/s	S_1, $\times10^{-4}$	T_2,[b] $\times10^{-8}$ m^2/s	S_2, $\times10^{-4}$	H, m	$\sqrt{Hh}$, m	k_2, $\times10^{-8}$ m/s	k_1/k_2
GSW	1	47.2	1.4	2.4	1.4	24	1.6	–	0.9	–
	2	29.0	1.8	0.3	≃1.8	28	–	0.6	–	1000
PW	2	76.2	0.9	0.3	≃0.9	30	–	0.6	–	1000
S_1	1	76.2	1.6	0.7	1.6	7	2.5	–	0.6	–
E_2	1	61.0	3	0.6	3.0	6	6.1	–	0.5	–
	2	103.6	1	0.1	≃1	9	–	0.6	–	1000
HWE	1	121.9	1.3	0.13	1.3	1.3	–	–	–	–
	2	155.5	2.5	0.03	≃2.5	3.3	–	>0.6	–	1000
Mean values			1.7	0.6	1.7	13	–	–	0.7	1000

Test 1, pumping well PW Q = 0.05 l/s, exact solution; test 2, pumping well S_1 Q = 0.16 l/s, approximate solution (see Figures 9a and 9b).

[a] $k_1 \simeq 6\times10^{-6}$ m/s.

[b] $k_2 \simeq 6\times10^{-9}$ m/s.

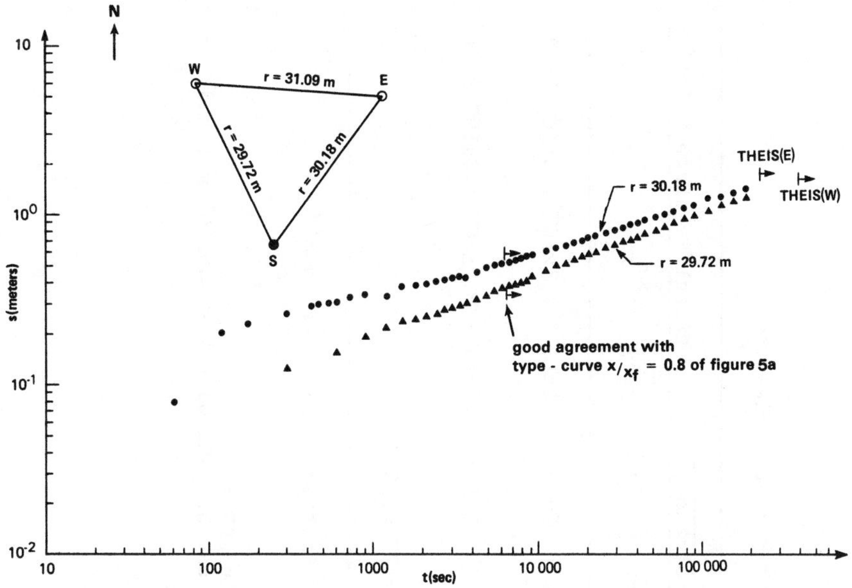

Fig. 10. Barrhead coal aquifer interpretation with Gringarten-Witherspoon model (pumped well S - Q = 0.54 L/S).

Agreement with the vertical fracture model is shown on Figure 10 where responses of observation wells E and W are interpreted for $t > 7000$ s with type curve ($x/x_f = 0.8$) of Figure 5a.

Table 7 summarizes the results obtained with the three sets of type curves: first, for the case of Figure 10 and then when the interpreted results of the three tests are averaged for each of the interpretative models.

Considering these averaged parameters, it can be noted that approximate solutions of the Boulton-Streltsova and vertical fracture models are in good agreement for the transmissivities (T_1 and $\sqrt{T_x T_y}$); the exact solution of the Boulton-Streltsova model gives higher results (by a factor 10).

A conjunctive use of the two models is done since the vertical fracture model needs prior evaluation of the storage coefficient for complete interpretation; values are taken as $S = \sqrt{S_1 S_2}$, where S_1 and S_2 are the storage coefficients determined with the Boulton-

TABLE 7. Barrhead Coal Aquifer: Interpretation Results

	Boulton and Streltsova Model								Vertical Fracture Model[d]			
	Exact Solution				Approximate Solution							
	T_1	T_2	S_1	S_2	T_1	T_2	S_1	S_2	$\sqrt{T_xT_y}$	T_x	T_y	x_f
Well												
W[a]	$8.8x10^{-5}$	$8.8x10^{-8}$	$4.2x10^{-4}$	$4.2x10^{-3}$	$1x10^{-4}$	$2.9x10^{-7}$	$2x10^{-5}$	$2x10^{-3}$	$1.1x10^{-4}$	$7.8x10^{-5}$	$1.5x10^{-4}$	37.2
E[b]	id	id	$1.9x10^{-4}$	$1.9x10^{-3}$	$9.5x10^{-5}$	$2.9x10^{-7}$	$3x10^{-5}$	$3x10^{-3}$	$1.3x10^{-4}$	$9.2x10^{-5}$	$1.8x10^{-4}$	37.7
Averaging values[c]	$7x10^{-4}$	$7x10^{-7}$	$7.9x10^{-4}$	$7.9x10^{-3}$	$8x10^{-5}$	$1.4x10^{-7}$	$4.2x10^{-5}$	$4x10^{-3}$				

Test 1, pumping well S at 0.54 l/s; responses of wells E and W; transmissivities in m^2/s.

[a] r = 29.72 m

[b] r = 30.18 m

[c] Of tests 1, 2, and 3 (where successively, wells SE and W were pumped).

[d] Vertical fracture model: $T = \sqrt{T_xT_y} = 1.2x10^{-4} m^2/s$; major cleat SSW-NNE with $k \simeq 4.5x10^{-4}$ m/s and minor cleat NNW-ESE with $k \simeq 1.5x10^{-5}$ m/s.

Streltsova model from the same time-drawdown curve. The directions of major and minor cleats are presumed from the results of the vertical fracture model, and hydraulic conductivities can be assigned to these two directions. This information should be controlled by another aquifer test, where the observation wells would be located in the directions of the major (x axis) and minor (y axis) supposed cleats.

7. Conclusions

This paper presents an overview of available specific models than can be used to analyze constant-rate test data for fractured and confined aquifers. Basic assumptions and equations are given together with analytical and graphical solutions. Groundwater flow toward the pumped well takes into account idealized representations of the fracture network based on double-porosity medium theory or on an equivalent single fracture configuration. In both types of specific models, the Theis type curve derived for homogeneous and porous media is obtained as a long time or a large geometric parameter asymptotic solution. As a consequence, the Theis model is usually inapplicable to the interpretation of tests of economical duration, and even though it may be, the matching process is not accurate.

An approximate solution to the Boulton-Streltsova 'layered' model is derived and tested on field data; its accuracy, compared to the exact solution, is judged sufficient for interpretative purposes; its advantage is to allow hand calculations (or very fast and stable computer ones) to generate type curves that will allow better match of time-drawdown curves than the already tabulated and exact solutions. Based on the same approximative technique of inversion of Laplace transforms, a solution to the Warren and Root model as modified by DeSwaan [1976] and Najurieta [1976] is also proposed. Type curves calculated with this solution do not seem to be significantly different from those of the Boulton-Streltsova model.

The selection of the most appropriate specific model should be based primarily on available hydrogeological and geological information; however, the very early time response may be helpful and should be carefully monitored. Also, the location of observation wells with respect to the pumping well is not indifferent, especially for an optimum interpretation with the single-fracture model.

Field boundary conditions are often more complex than those assumed by the described specific models. Additional work is needed to incorporate, for instance, finite lateral boundaries to these models, as was done in the case of homogeneous/porous media. Inversion of solutions from the Laplace plane to the real plane may be performed with analytical (as done here) or numerical approximate techniques (such as Stehfest [1970]).

Notation

s_D drawdown (usually equal to $4\pi Ts/Q$), dimensionless.

t_D time (usually equal to $4Tt/Sr^2$), dimensionless.

s drawdown (difference in hydraulic head), L.

T transmissivity, $L^2\ T^{-1}$.

K hydraulic conductivity, $L\ T^{-1}$.

Q pumping rate, $L^3\ T^{-1}$.

S storativity or storage coefficient.

r Distance observation well/pumping well, L.

t pumping time, T.

ρ water density, $M\ L^{-3}$.

μ water dynamic viscosity, $M\ L^{-1}\ T^{-1}$.

r_w radius of the pumping well, L.

S_s specific storage, L^{-1}.

$Jo(x)$ Bessel function of first kind and zero order (of variable x).

$E_1(x)$ exponential integral function of the real variable x (noted here as $E_1(x) = -Ei(-x)$), equal to $\int_x^\infty e^{-u} du/u$.

$Ko(x)$ modified Bessel function of zero order of the real variable x.

p, a	Laplace and zero-Hankel transformations parameters.
x, y, z	Cartesian coordinates, L.
H	equivalent thickness of porous block unit, L.
h	equivalent thickness of fracture unit, L.
η	diffusivity, equal to T/S, $L^2\ T^{-1}$.
erf(x)	error function of the variable x, equal to $\frac{2}{\sqrt{\pi}} \int_0^x e^{-u^2} du$.
T_x, T_y	principal transmissivities in x and y directions, $L^2\ T^{-1}$.
b	diffusivity ratio, equal to η_2/η_1, dimensionless.
c	transmissivity ratio, equal to T_2/T_1, dimensionless.
m	parameter (so that $m = (K_2/K_1)^{1/2}$), dimensionless.
n	parameter (so that $n = (S_2/S_1)^{1/2}$), dimensionless.
x_f	half length of vertical fracture, L.

Subscripts

1 fracture characteristics.

2 rock matrix characteristics.

References

Barenblatt, G. E., I. P. Zheltov, and I. N. Kochina, Basic concepts in the theory of seepage of homogeneous liquids in fissured rocks, J. Appl. Math. Mech. Engl. Transl., 24(5), 1286-1303, 1960.

Boulton, N. S., and T. D. Streltsova, Unsteady flow to a pumped well in a fissured water-bearing formation, J. Hydrol., 35, 257-269, 1977.

DeSwaan, A., Analytic solutions for determining naturally fractured reservoir properties by well testing, Soc. Pet. Eng. J., 117-122, 1976.

Duguid, J. O., and P. C. Y. Lee, Flow in fractured porous media, Water Resour. Res., 13(3), 558-566, 1977.

Gringarten, A. C., and H. J. Ramey, The use of source and Green's functions in solving unsteady flow problems in reservoirs, Soc. Pet. Eng. J., 285-296, 1973.

Gringarten, A. C., and H. J. Ramey, Unsteady state pressure distributions created by a well with a single horizontal fracture, partial penetration, or restricted entry, Soc. Pet. Eng. J., 413-426, 1974.

Gringarten, A. C., and P. A. Witherspoon, A method of analyzing pump test data from fractured aquifers, paper presented at the Symposium on Percolation Through Fissured Rock, Int. Soc. of Rock Mech./Int. Assoc. Eng. Geol., Stuttgart, 1972.

Kazemi, H., M. S. Seth, and G. W. Thomas, The interpretation of interference tests in naturally fractured reservoirs with uniform fracture distribution, Soc. Pet. Eng. J., 463-472, 1969.

Muskat, M., The Flow of Homogeneous Fluids Through Porous Media, McGraw-Hill, New York, 1937.

Najurieta, H. L., A theory for the pressure transient analysis in naturally fractured reservoirs, Spec. Pap. SPE6017, Soc. of Pet. Eng. of AIME, Oct., Dallas, Tex., 1976.

Schapery, R. A., Approximate methods of transform inversion for viscoelastic stress analysis, Proc. U.S. Natl. Congr. Appl. Mech., 4th, 1075-1085, 1961.

Stehfest, H., Numerical inversion of Laplace transforms, Commun. ACM, 13(1), 47-49, 1970.

Streltsova, T. D., Hydrodynamics of groundwater flow in a fractured formation, Water Resour. Res., 12(3), 405-414, 1976.

Streltsova-Adams, T. D., Well hydraulics in heterogeneous aquifer formations, Adv. Hydrosci., 11, 357-423, 1978.

Warren, J. E., and P. J. Root, The behaviour of naturally fractured reservoirs, Soc. Pet. Eng. J., 9, 245-255, 1963.

Field Test for Effective Porosity and Dispersivity in Fractured Dolomite, the WIPP, Southeastern New Mexico

D. D. Gonzalez
Sandia National Laboratories, Albuquerque, New Mexico 87185

H. W. Bentley
Hydro Geo Chem, Inc., Tucson, Arizona 85716

Introduction

The Waste Isolation Pilot Plant (WIPP), a demonstration facility 26 mi (41.6 km) east of Carlsbad, New Mexico, used to store transuranic waste in Permian-bedded salts, has been under field investigation since 1975. Hydrologically, the area is characterized by a typical semiarid environment underlaid by four confined aquifer systems whose transmissivities range from 10 to 10^{-5} ft^2/d (1 to 10^{-6} m^2/d).

Previous local hydrogeologic investigations show a fractured dolomite, whose thickness and depth of burial range from 22 to 24 ft (6.7 to 7.3 m) and 498 to 897 ft (152 to 274 m), respectively, as the most likely groundwater vehicle to transport waste to the biosphere in the event such a repository is breached. To describe adequately and to predict solute transport, certain hydraulic characteristics of the transporting medium need to be estimated along a hypothetical flow path. Effective porosity and dispersivity are two parameters which are most difficult to predict, particularly in fractured rock exhibiting low transmissivities. This paper describes the results of the first of a series of two-well recirculation tracing tests to be performed. The recirculation tests were performed using an extraction-injection well couplet similar to that described by Grove and Beetem [1971]. Test duration was

270 days and resulted in a well-defined breakthrough curve. Specific test configurations are described in detail in a later section of this paper.

H-2 Site Description

The regional hydrogeology of the WIPP area has been described by Hiss [1976], Powers et al. [1978], and Mercer and Orr [1979]. Hydrologic testing at the H-2 site has concentrated on the three liquid-bearing zones above the proposed waste repository horizons. These are the Permian Rustler-Salado contact and the two beds within the Rustler Formation: the Culebra and the Magenta Dolomite members.

The H-2 wells were drilled by the U.S. Geological Survey (USGS) and Sandia National Laboratories in 1977. H-2A was completed to the Magenta at a depth of 563 ft (172 m). H-2B was drilled to the Culebra at 661 ft (202 m), then perforated in the Magenta, and completed as a dual-observation well. H-2C was drilled to the Rustler-Salado contact at 743 ft (227 m), perforated in the Culebra, and also completed as a dual-observation well [Mercer and Orr, 1979]. Figure 1 shows the three-well configuration, the zones each is open to, and their general orientation.

The three liquid-bearing zones were cored and analyzed by the USGS. A description of these cores, taken from Mercer and Orr [1979], is given in Tables 1a - 1c.

Hydrologic testing of the Rustler-Salado contact and of the Magenta Dolomite Member consisted of bailing each well dry and observing the recovery response. Estimates of transmissivity in the Rustler-Salado contact were between 10^{-1} and 10^{-4} ft^2/d (1 and 10^{-5} m^2/d). The transmissivity in the Magenta averaged about 0.1 ft^2/d (0.01 m^2/d) [Mercer and Orr, 1979].

Three aquifer tests with observation wells were made to determine hydraulic properties of the Culebra. The average transmissivity from these tests is 0.6 ft^2/d (0.06 m^2/d); storage coefficient is about 1.3×10^{-5} [Bentley and Walter, 1983].

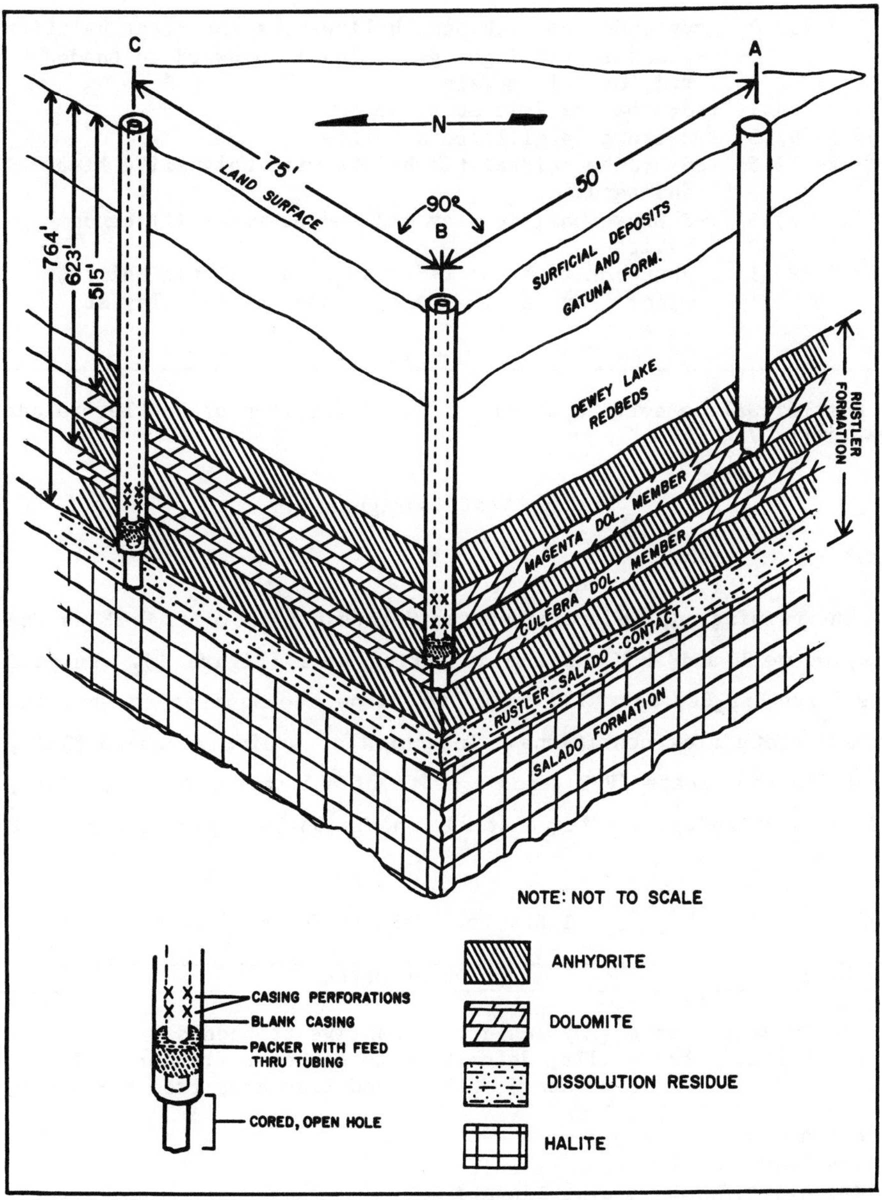

Fig. 1. Cross section through wells H-2A, H-2B, and H-2C.

TABLE 1a. Well H-2C

Depth, ft	Description
743-762.2	gray mudstone with pink halite vugs and clear halitic fracture fillings, gradational downward to banded, red, halitic mudstone
762.2-764.1	red-brown halitic mudstone
764.1-767.3	red-brown, argillaceous halite
767.3-772.5	red-orange polyhalitic halite with polyhalite blebs and bands
772.5-773.5	red-brown halitic clay and red-brown argillaceous halite
773.5-795.6	light pink to light red-orange polyhalitic halite, minor clay partins, with brown halitic clay at base

Rustler-Salado cored interval: 743-795 ft; top of Rustler-Salado contact: 764 ft.

Tracing Test Configuration

Test 1

The pumping and injecting system for test 1 included H-2B as the pumping well and H-2C as the injecting well (Figure 1). Pumping was begun on February 13, 1980, to allow the aquifer to approach steady state flow conditions. Injection of sodium benzoate (SBA), pentafluorobenzoate (PFB), and a suite of halocarbon tracers (CCl_4, $CFCl_3$, and CF_2Cl_2) was begun at 2330 hours on February 22, 1980, at

TABLE 1b. Well H-2B

Depth, ft	Description
611-624.2	dense gray anhydrite, massive to banded
624.2-642.0	brown silty dolomite with selentic fracture fillings and crystals, pitted and fractured from 629.5 to 642.0 ft
642.0-644.0	gray mudstone
644.0-652.0	red-brown selenitic siltstone
652.0-660.7	dense gray anhydrite

Culebra Dolomite cored interval: 611-661 ft; Culebra Dolomite thickness: 624-642 ft.

TABLE 1c. Well H-2A

Depth, ft	Description
511.7-513	cement
513.0-514.6	dense gray anhydrite
514.6-539.6	gray-brown silty dolomite with some fractures in the interval 537.5-539.6 ft
539.6-563	brown-gray banded anhydrite

Magenta Dolomite cored interval: 513-563 ft; Magenta Dolomite thickness: 515-540 ft.

which time the water flow rate through the system was 1140 ml/min. Injection continued until some time between 1200 hours on February 24 and 1200 hours on February 25, when the tracer injection line ruptured and the remainder of the tracers were lost.

Pumping continued in an attempt to complete the test successfully; however, sediment accumulation affected the injection system and the performance of the injection well, H-2C, until the test was terminated June 18, 1980.

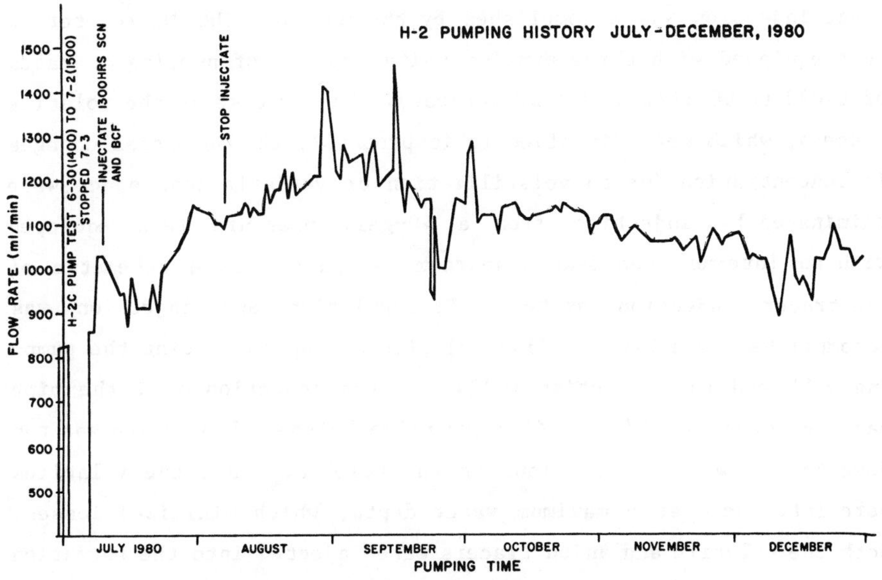

Fig. 2. H-2 pumping history July-December 1980.

Test 2

After the abortion of test 1 the site was reconfigured with H-2C as the pumped well and H-2B as the injection well. A combination of in-line filters and a settling tank alleviated almost all problems with sedimentation. After pumping to steady state, tracer injection was begun on July 10, 1980, with sodium thiocyanate (SCN) and bromochlorodifluoromethane (BCF) and continued for 28 days at rates of 1.9 ml/min for SCN and 0.75 ml/min for BCF. Pumping of H-2C continued for 274 days until the test was terminated when the data suite was completed on April 7, 1981. The discharge from H-2C varied from 930 to 1460 ml/min, as shown in Figure 2.

Tracer Injection

Analysis of the two-well recirculating test [Grove and Beetem, 1971] depends on a constant concentration of tracer having been injected for a significant fraction of the test period. This constant injection was accomplished by the use of a MPL Micrometering Pump equipped with three pumping units capable of pumping at rates of 0.092 to 60 ml/min with 2% accuracy. In the case of the volatile tracers, which were dissolved in isopropanol, the potential changes in concentration due to volatilization or variable head space were eliminated by injecting from a 40-gal. pressure tank equipped with an internal neoprene diaphragm. Figure 3 is a schematic of the tracer injection system. Recirculation and injection was accomplished by a 3/4-in. (1.9 cm) pipe string connecting the pumping well and the injection well. At the injection well the pipe was connected to a 1/2-in. (1.3 cm) polyethylene tubing which was run down to the well perforations at the Culebra. Thus the volatiles were introduced at a maximum water depth, which minimized losses. Both the volatile and anion tracers were injected into the formation with little dilution by the water stored in the well bore.

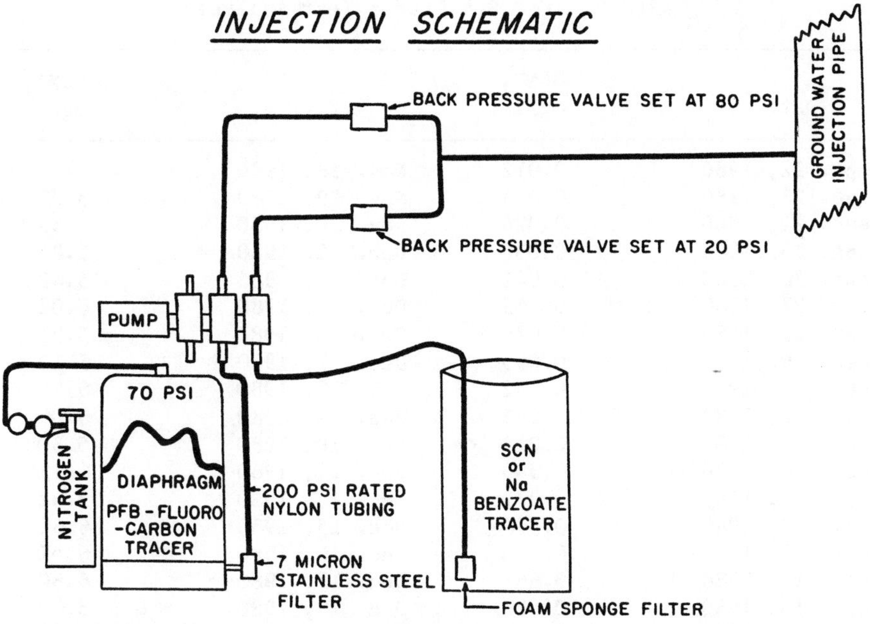

Fig. 3. Schematic of tracer injection circuit.

Tracer Collection

Anion tracers were collected from a faucet at the pumping head and stored in 250-ml polyethylene bottles and, later, in 30-ml plastic scintillation counting bottles. Volatiles were collected in 30-ml melt seal vials which were sealed within 5 min with a butane torch.

Tracer Analysis

Volatile tracer injectate concentrations were estimated at the site by direct injection into a Varian 3000 gas chromatograph equipped with a 6-ft (1.8 m) Carbopak 1/8-in. (0.3 cm) column and an electron capture detector. These analyses were qualitative, made only to establish that the tracers were in the proper range, approximately 1 mg/l. No subsequent analyses of these tracers were made. Anion tracer injectate concentrations were analyzed by

TABLE 2. H-2 SCN Tracer Test Analyses

Date	SCN^-, mg/l	Date	SCN^-, mg/l
Sept. 22, 1980	0.012	Nov. 28, 1980	4.61
Sept. 23, 1980	0.018	Nov. 29, 1980	3.79
Sept. 24, 1980	0.024	Nov. 30, 1980	4.59
Sept. 25, 1980	0.030	Dec. 2, 1980	5.84
Sept. 26, 1980	0.041	Dec. 3, 1980	5.62
Sept. 27, 1980	0.055	Dec. 5, 1980	6.02
Sept. 29, 1980	0.076	Dec. 6, 1980	5.96
Sept. 30, 1980	0.095	Dec. 7, 1980	6.12
Oct. 1, 1980	0.142	Dec. 8, 1980	6.26
Oct. 2, 1980	0.245	Dec. 9, 1980	6.06
Oct. 3, 1980	0.228	Dec. 10, 1980	5.80
Oct. 4, 1980	0.343	Dec. 11, 1980	6.12
Oct. 5, 1980	0.371	Dec. 14, 1980	5.42
Oct. 8, 1980	0.576	Dec. 15, 1980	6.60
Oct. 9, 1980	0.624	Dec. 16, 1980	6.57
Oct. 10, 1980	0.661	Dec. 17, 1980	6.80
Oct. 13, 1980	0.866	Dec. 19, 1980	5.31
Oct. 14, 1980	1.39	Dec. 20, 1980	6.65
Oct. 15, 1980	1.63	Dec. 22, 1980	6.76
Oct. 16, 1980	1.68	Dec. 23, 1980	7.38
Oct. 17, 1980	1.82	Dec. 24, 1980	7.53
Oct. 21, 1980	1.99	Dec. 26, 1980	7.54
Oct. 22, 1980	2.61	Dec. 28, 1980	7.49
Oct. 23, 1980	2.74	Dec. 29, 1980	7.72
Oct. 24, 1980	2.89	Dec. 31, 1980	8.07
Oct. 27, 1980	3.33	Jan. 4, 1981	7.94
Oct. 28, 1980	3.45	Jan. 5, 1981	8.23
Oct. 29, 1980	3.49	Jan. 7, 1981	7.75
Oct. 30, 1980	3.67	Jan. 8, 1981	7.98
Oct. 31, 1980	3.76	Jan. 11, 1981	6.94
Nov. 1, 1980	3.16	Jan. 12, 1981	7.89
Nov. 2, 1980	3.72	Jan. 14, 1981	7.42
Nov. 4, 1980	4.26	Jan. 16, 1981	6.66
Nov. 5, 1980	4.41	Jan. 20, 1981	10.31
Nov. 6, 1990	4.41	Jan. 21, 1981	10.28
Nov. 9, 1980	3.52	Jan. 22, 1981	10.31
Nov. 11, 1980	4.55	Jan. 23, 1981	10.36
Nov. 12, 1980	4.54	Jan. 27, 1981	10.26
Nov. 13, 1980	4.65	Jan. 28, 1981	10.49
Nov. 17, 1980	4.80	Jan. 29, 1981	10.71
Nov. 18, 1980	4.91	Jan. 30, 1981	11.14
Nov. 19, 1980	4.88	Jan. 31, 1981	7.09
Nov. 20, 1980	4.97	Feb. 2, 1981	12.11
Nov. 23, 1980	5.12	Feb. 3, 1981	9.04
Nov. 24, 1980	4.76	Feb. 4, 1981	11.89
Nov. 27, 1980	5.41	Feb. 5, 1981	12.24

TABLE 2. (continued)

Date	SCN^-, mg/l	Date	SCN^-, mg/l
Feb. 6, 1981	12.31	March 12, 1981	13.00
Feb. 12, 1981	12.47	March 13, 1981	12.88
Feb. 13, 1981	12.60	March 16, 1981	12.84
Feb. 15, 1981	12.60	March 17, 1981	7.81
Feb. 17, 1981	12.55	March 18, 1981	10.23
Feb. 18, 1981	12.51	March 22, 1981	12.39
Feb. 20, 1981	12.63	March 23, 1981	12.35
Feb. 22, 1981	12.25	March 24, 1981	12.33
Feb. 25, 1981	12.39	March 26, 1981	12.23
Feb. 26, 1981	12.63	March 27, 1981	12.20
Feb. 27, 1981	12.70	March 28, 1981	12.60
March 2, 1981	12.59	April 1, 1981	11.26
March 4, 1981	12.28	April 2, 1981	8.82
March 6, 1981	12.31	April 5, 1981	10.21
March 9, 1981	12.34	April 6, 1981	10.83
March 10, 1981	12.43	April 7, 1981	11.11

Injection July 10, 1980; average concentration 721 $\pm$ 27 mg/l.

high performance liquid chromatography (HPLC). HPLC was used onsite for the first tracing test for PFB and SBA tracer injection. Average injection well concentrations were 721 $\pm$ 27 mg/l for SCN^-, 650 mg/l for SBA, and 213 mg/l for PFB.

As the tracing test proceeded, tracer samples were sent to the Hydro Geo Chem laboratory in Tucson, Arizona, for further analysis. Laboratory analysis included HPLC measurement of SCN^- and PFB.

Table 2 shows the SCN^- results. Table 3 represents the PFB data. Standards were analyzed every fifth to tenth chromatogram, and two analyses were made for each sample. If the results did no agree within 2% or less, a third analysis was performed.

Tracer Stability

Both SCN^- and PFB appear to be refractory in the Culebra. Moreover, neither has shown any degradation at mg/l levels in Culebra H-2

TABLE 3. H-2 SCN Two-Well Pump Back Test

Date	PFB, mg/l	Date	PFB, mg/l
July 10, 1980	0.32	Aug. 24, 1980	8.85
July 15, 1980	0.21	Aug. 25, 1980	6.40
July 21, 1980	0.33	Aug. 30, 1980	8.19
July 25, 1980	1.26	Sept. 2, 1980	7.40
July 30, 1980	3.71	Sept. 5, 1980	5.64
Aug. 1, 1980	5.09	Sept. 10, 1980	3.92
Aug. 4, 1980	2.56	Sept. 15, 1980	3.28
Aug. 5, 1980	7.11	Sept. 18, 1980	2.23
Aug. 8, 1980	3.62	Oct. 5, 1980	0.70
Aug. 11, 1980	9.12	Oct. 10, 1980	0.40
Aug. 13, 1980	10.05	Oct. 15, 1980	0.50
Aug. 15, 1980	11.36	Oct. 21, 1980	0.29
Aug. 17, 1980	11.36	Oct. 24, 1980	0.24
Aug. 18, 1980	9.00	Oct. 30, 1980	0.19
Aug. 19, 1980	10.40	Nov. 5, 1980	0.20
Aug. 20, 1980	11.08	Nov. 20, 1980	0.13
Aug. 21, 1980	10.70	Dec. 10, 1980	0.30
Aug. 22, 1980	10.00	Dec. 20, 1980	0.20
Aug. 23, 1980	9.51		

water when stored in the laboratory for several months. PFB was first injected at H-2C on February 13, 1980, recovered in August 1980, and analyzed in September 1981, with no apparent losses occurring in the formation or in the laboratory. Studies with mixtures of barnyard soils and water yield similar results [H.W. Bentley, personal communication, 1983].

Results and Interpretation of H-2 Recirculation Test 2

A numerical analysis of the H-2 recirculation test was performed using the Grove and Beetem [1971] model. The model consists of a recharging-discharging well pair with a pattern of streamlines and treats the aquifer as uniform and isotropic and pumping as constant. To calculate the movement of the tracer, the model approximates the infinite number of streamlines by a finite number

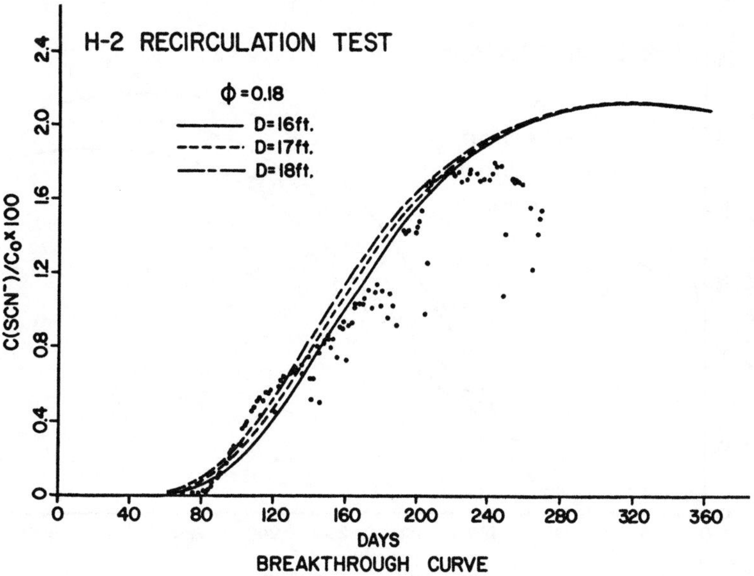

Fig. 4 Two-well recirculation SCN^- tracing test at the H-2 well pad, WIPP, SE New Mexico. Lines are output from the Grove and Beetem [1971] model with porosity ϕ fixed and varying dispersivities α.

of crescents. Each crescent is treated as a one-dimensional flow tube with only longitudinal dispersivity defined. This model uses the one-dimensional solution to the convective-dispersive equation for a finite column:

$$\frac{\partial c}{\partial t} = - v \frac{\partial c}{\partial x} + D \frac{\partial^2 c}{\partial_x^2}$$

as presented by Brenner [1962]. The boundary conditions used by Brenner are

$$qC_o = vc - D \frac{\partial c}{\partial x}$$

at x = 0 borehole surface of injection well and

$$\frac{\partial c}{\partial x} = 0$$

at x = L borehole surface of extraction well, where x = 0 and x = L are the coordinates of the injection and pumping wells, respectively, and

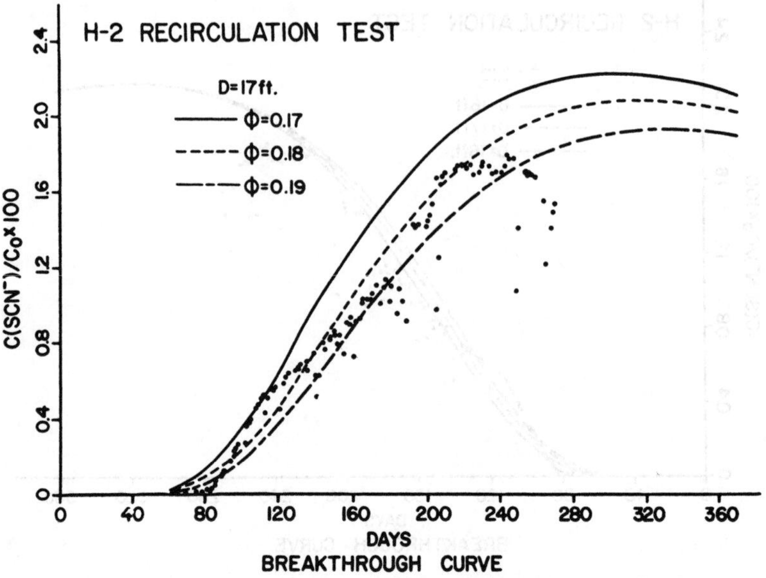

Fig. 5. Two-well recirculation SCN^- tracing test at the H-2 well pad, WIPP, SE New Mexico. Grove-Beetem curves show effects of changes in porosity ϕ by 0.01 with dispersivity α fixed at 17 ft.

q Darcy velocity, L/T;

C_o initial concentration, M/L^3;

v seepage velocity, L/T;

c observed concentration, M/L^3;

D coefficient of hydrodynamic dispersion, L^2/T;

x distance, L;

α dispersivity;

D αv.

The boundary condition used by Brenner at the extraction well implies no dispersive flux. Although this boundary condition has often been used to represent finite length columns, its physical validity is questionable.

An attempt was made to fit the observed data from the H-2 recirculation test with the Grove and Beetem model using a range of porosities, ϕ, and dispersivities, α. The result of variations in

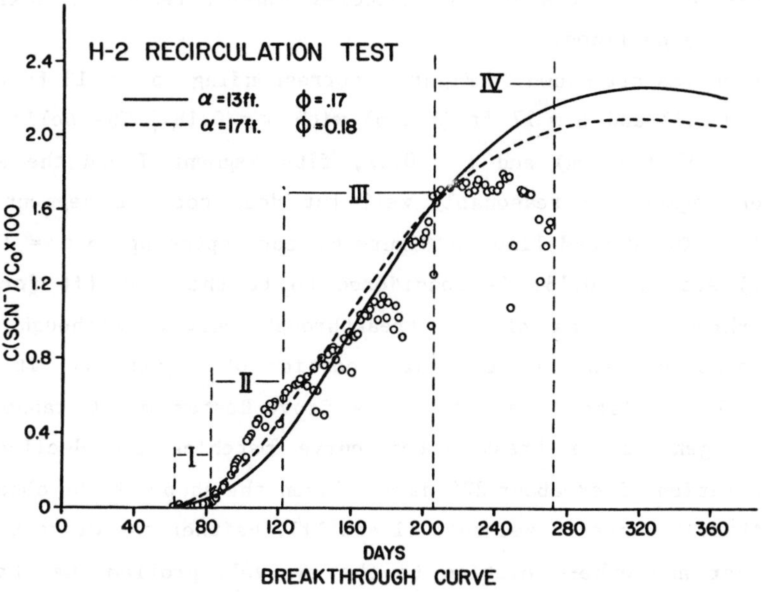

Fig. 6. Attempts to fit various parts of the SCN^- tracer breakthrough curve with varying dispersivity α and porosity ϕ.

α with a given ϕ is shown in Figure 4. Increases in α displace the curves toward earlier times. However, they have little effect on the curve slope or maximum.

Figure 5 demonstrates the variation in curves when α is fixed and ϕ is allowed to change. The porosity variations produce much larger effects in the first appearance time, slope of limb, peak time, and value of peak. For the values of α and ϕ shown, changes in porosity have a much larger effect on the breakthrough curve than changes in dispersivity.

Simplifying, the observed breakthrough curve was divided into four segments, as shown in Figure 6. Segment I extends from about 60 to 80 days after injection during which time the tracer first reached detectable concentrations. Segment II extends from 80 to 120 days during which time the concentration rose rapidly. Segment III includes that portion of the curve between 120 and 200 days where the curve is concave upward. Finally, segment IV begins at

200 days when the tracer curve becomes convex, reaches a maximum, and finally declines.

Two curves are shown in Figure 6 corresponding to α = 13 ft (4 m) with ϕ = 0.17 and α = 17 ft (5.2 m) with ϕ = 0.18. The solid line for α = 13 ft (4 m) and ϕ = 0.17, fits segment I and the early part of segment IV reasonably well but does not fit segments II and III. The dashed line in Figure 6, corresponding to α = 17 ft (5.2 m) and ϕ = 0.18, is considered to be the best fit for the first three segments of the breakthrough curve. Although this curve also matches the very early portion of segment IV, it does not match the late time data. The Grove-Beetem model cannot be used to generate a breakthrough curve matching the decline in concentration after about 220 days. Given the shape of the observed breakthrough curve in segments II and III, neither the Grove-Beetem model nor any other solution to the two-well problem that treats the formation as uniform and discharge as constant can be expected to represent accurately the observed breakthrough curve.

Conclusions

As a result of tracer performance and simplicity in handling and analysis, the anions PFB and SCN are the tracers of choice for use in evaluating aquifer parameters in dolomites at the WIPP.

Porosity and dispersivity at H-2 have been estimated at 18% and 17 ft (5.2 m), respectively. The data obtained from recirculating tests are valuable in establishing the flow and solute transport regime in the Culebra Dolomite of the Rustler Formation in the vicinity of the H-2 location. The type curves generated by the Grove-Beetem model fit the early portions of the data, however, do not match late-time data. The model cannot match the decline in concentrations without modification to include anisotropy, variable pumping rates and boundary conditions, and solute retardation in terms of matrix permeability.

A site specific numerical model is indicated, utilizing both recirculation and convergent flow tracer test for calibration. At

H-2, for instance, convergent flow tracer tests and aquifer tests for anisotropy should be performed. These results coupled with the H-2 recirculating tests will provide the basis to generate a model which describes these local phenomena and will provide guidance in establishing the field operations plan. Presently, anisotropy and tracer tests are planned at six hydropads along a hypothetical flow path originating at the center of the WIPP proposed facility and leading towards the discharge area near the Pecos River, 17 mi (27.4 km) away. Hydropad modeling will serve as the basis for a regional flow and solute transport model.

References

Bentley, H. W., and G. R. Walter, H-2 two-well recirculating tracer test, the proposed Waste Isolation Pilot Plant (WIPP), southeast New Mexico, Sandia Nat. Lab. Contract Report Sand 83-7014, Albuquerque, N. M., 1983.

Brenner, H., The diffusion model of longitudinal mixing in beds of finite length: Numerical values, Chem. Eng. Sci., 17, 229-243, 1962.

Grove, D. B., and W. A. Beetem, Porosity and dispersion constant calculations for a fractured carbonate aquifer using the two-well tracer method, Water Resour. Res., 7(1), 125-134, 1971.

Hiss, W. I., Structure of the Permian Guadalupian Capitan aquifer, southeast New Mexico and west Texas, Resour. Map 6, N. M. Bur. of Mines and Miner. Resour., Scorro, 1976.

Mercer, J. W., and B. R. Orr, Interim data report on geohydrology of the proposed water isolation pilot plant site, southeastern New Mexico, U. S. Geol. Surv. Water Resour. Invest., 79-98, 1979.

Powers, D. W., et al. (Eds.), Geological characterization report, Waste Isolation Pilot Plant (WIPP) site, southeastern New Mexico, Sandia Nat. Lab., Albuquerque, N. M., 1978.

Thompson, G. M., and J. M. Hayes, Trichlorofluoromethane in ground water--A possible tracer and indicator of groundwater age, Water Resour. Res., 15(3), 546, 1979.

Direct Calculation of Aquifer Parameters in Slug Test Analysis

V. Nguyen and G. F. Pinder
Water Resources Program, Princeton University,
Princeton, New Jersey 08544

Introduction

The objective of this paper is to present a conceptually and computationally simple methodology for determining aquifer parameters from water level observations in a single well. Because the mathematical manipulations necessary to arrive at practically useful formulae are rather tedious, we present the details of the development as an appendix to this paper. Let us begin the discussion with a mathematical description of our physical system.

Theoretical Development

We wish to consider problems which involve partially penetrating wells screened in aquifers where, at least in the short run, the effects of a water table or leakage from a confining bed can be disregarded. The analytical apparatus appears to be most appropriate in dealing with slug test analyses or short-term pumping tests in materials of moderate to low hydraulic conductivity. A schematic representation of the physical system we will consider is given in Figure 1. Because the well is of finite diameter, we must consider well bore storage effects. Moreover, to accommodate partial penetration, an axisymmetrical three-dimensional mathematical representation is needed.

The governing field equation for this class of problems is given by

$$D_{rr}s + \frac{1}{r} D_r s + D_{zz}s = \frac{S}{K} D_t s \qquad (1)$$

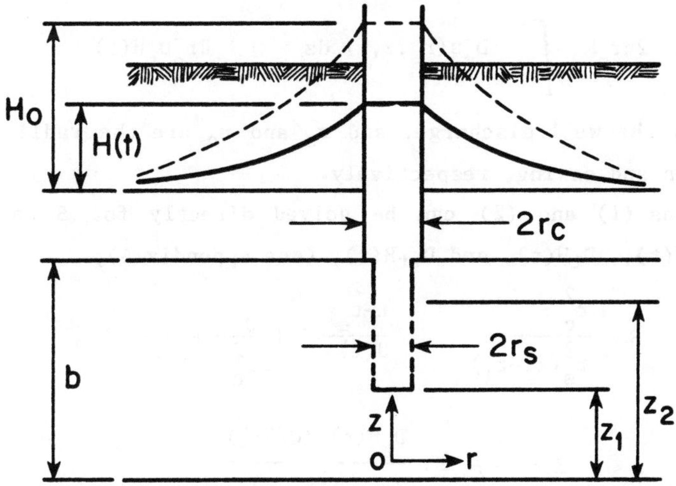

Fig. 1. Schematic representation of the artesian well.

where

s change in head relative to its initial state, L;

S specific storage, L^{-1};

K hydraulic conductivity, L T^{-1}.

Our task is to determine S and K, given observations in only one well.

Whether we employ slug test or pumping test methodology, the same initial and boundary conditions will be employed, namely,

$$s(r,z,0) = 0 \tag{2a}$$

$$D_z s(r,0,t) = D_z s(r,b,t) = 0 \tag{2b}$$

$$s(\infty,z,t) = 0 \tag{2c}$$

$$\frac{1}{z_2 - z_1} \int_{z_1}^{z_2} s(r_s,z,t)\, dz = H(t) \qquad t > 0 \tag{2d}$$

Equation (2d) states that the change in water level, H(t), observed in the pumping well is the average head measured along the well screen. We augment these equations with a description of the well bore behavior, namely,

$$2\pi r_s K \int_{z_1}^{z_2} D_r s(r_s, z, t)\, dz = Q + \Pi r_c^2 D_t H(t) \tag{2e}$$

where Q is the well discharge, and r_s and r_c are the radii of the well screen and casing, respectively.

Equations (1) and (2) can be solved directly for S and K in terms of H(t), $D_t H(t)$, and $D_{tt} H(t)$, (see appendix A),

$$S = \frac{r_c^2}{r_s^2 (z_2 - z_1)} \cdot \frac{t - t_e}{H(t)} \cdot \left[\frac{Q}{\Pi r_c^2} + D_t H(t)\right] \tag{3a}$$

$$K = \frac{-r_s^2}{4} S . \frac{D_t H(t) + (Q/\Pi r_c^2)}{(t - t_e)^2 D_{tt} H(t)} \tag{3b}$$

where t_e is the effective time, $t_e > t \geq 0$. The constant t_e can be determined from the formula

$$t_e = \left| \frac{k_1 t_1 - k_2 t_2}{k_1 - k_2} \right|$$

where

$$k_i = \frac{1}{H_i} \left[\frac{Q}{\pi r_c^2} + D_t H(t_i)\right] \qquad i = 1,2$$

and t_i are two time points.

For a discussion of the meaning of t_e, see appendix A.

Pumping Tests

Let us consider first the calculation of the specific storage S given (1) the water level in the well H(t) as a function of time, (2) the well discharge Q (constant), (3) the radii of the screen r_s and the casing r_c, and (4) the screen length $(z_2 - z_1)$. If S is a constant, as we assume it is in writing (1), then (3a) says there must be a linear relationship between the terms $t - t_e/H(t)$ and $[(Q/\Pi r_c^2) + D_t H(t)]$. In other words, if we plot on arithmetic paper $[(Q/\Pi r_c^2) + D_t H(t)]$ versus $H(t)/t - t_e$, the slope of the resulting line, C_1 say, is equal to $[r_s^2 (z_2 - z_1)/r_c^2]$ S, whereupon S can be easily obtained as

$$S = \frac{C_1 r_c^2}{r_s^2(z_2 - z_1)} \tag{4}$$

The determination of $D_t H(t)$ can be made directly from a continuous plot of $H(t)$ versus t. If only discrete measurements are available, one can employ a finite difference approximation such as

$$D_t H \Big|_t \approx \left(H \Big|_{t+\Delta t} - H \Big|_t \right) / \Delta t \tag{5}$$

where Δt is the time increment between two successive measurements of $H(t)$.

To determine the hydraulic conductivity K, one must eliminate S from (3b). Combination of (4) and (3b) yields the desired result,

$$K = \frac{r_c^2 C_1}{4(z_2 - z_1} \cdot \frac{D_t H(t) + (Q/\Pi r_c^2)}{(t - t_e)^2 D_{tt} H(t)} \tag{6}$$

Examination of (6) reveals that the slope of a line, C_2 say, relating $(t-t_e)^2 D_{tt}H(t)$ to $D_t H(t) + (Q/\Pi r_c^2)$ on an arithmetic plot will yield

$$\frac{r_c^2 C_1}{4K(z_2 - z_1)} = C_2 \tag{7}$$

From (7) we obtain directly

$$K = \frac{r_c^2 C_1}{4(z_2 - z_1) C_2} \tag{8}$$

In the calculation of K it is necessary to plot $D_{tt}H(t)$. Once again it is convenient to employ a finite difference representation of $D_{tt}H(t)$, i.e.,

$$D_{tt} H(t) \Big|_t \approx \left(H \Big|_{t+\Delta t} - 2H \Big|_t + H \Big|_{t-\Delta t} \right) / \Delta t^2 \tag{9a}$$

for a case of a constant Δt. When observations have been taken at irregular intervals, an appropriate approximation for the case of observations at times t_1, t_2, and t_3 would be

$$D_{tt}H(t)\Big|_{t_1} \approx [\Delta t_1 H\Big|_{t_3} - (\Delta t_1 + \Delta t_2)H\Big|_{t_2} + \Delta t_2 H\Big|_{t_1}] \frac{2}{\Delta t_1+\Delta t_2}\frac{1}{\Delta t_2 \Delta t_1} \quad (9b)$$

where $\Delta t_1 = t_2 - t_1$ and $\Delta t_2 = t_3 - t_2$. An example calculation for a pumping well is given in Table 1 and Figures 2a and 2b.

Slug Test

While (4) and (8) can be used directly for the analysis of a slug test by simply setting Q to zero, it is possible to develop an alternative formulation that is somewhat easier to implement. For the special case of zero discharge (Q = 0), the formula for specific storage is (see appendix A for development)

$$S = \frac{r_c^2 C_3}{r_s^2 (z_2 - z_1)} \quad (10)$$

where C_3 is the slope of the curve obtained by plotting ln $(H(t)/H(t_e - t_0))$ versus ln $[(t_e - t)/t_0]$ and $(t_e - t_0)$ is any suitable interval wherein the values of $H(t)/H(t_e - t_0)$ are considered meaningful in the sense described earlier. The hydraulic conductivity is obtained from the relationship

$$K = \frac{r_c^2 C_3}{4C_4 (z_2 - z_1)} \quad (11)$$

where C_4 is the slope of the line obtained by plotting ln $(D_t H(t))$ versus $[1/(t_e - t)]$. An example of a slug test analysis is provided in Table 2 and Figures 3a, 3b, and 3c.

If we want to include an energy loss factor F_L ($0 < F_L \leq 1$) due to the screen friction and other dissipation of energy mechanisms at the well, boundary condition (2d) may be rewritten as

Table 1. Pumping Test

t, min	H(t), ft	$\Delta H/\Delta t$, ft/min	$\overline{(\Delta H/\Delta t)}$, ft/min	$\Delta^2 H/\Delta t^2$, ft/min	$t_e - t$, min	$\frac{H(t)}{t-t_e}$, ft/min	$-(t_e-t)^2$ $\times\Delta^2 H/\Delta t^2$, ft	$\frac{Q}{\pi r_c^2}(\frac{\Delta H}{\Delta t})$, ft/min
1	-24				9			
		-9.0						
2	-33		-8.0	2.00	8	4.125	-128.	34.52
		-7.0						
3	-40		-6.5	1.00	7	5.714	-49.	36.02
		-6.0						
4	-46		-5.5	1.00	6	7.667	-36.	37.02
		-5.0						
5	-51		-4.5	1.00	5	10.200	-25.	38.02
		-4.0						
6	-55		-3.25	0.60	4	13.750	-9.6	39.27
		-2.5						
10	-65		-2.35	0.06				
		-2.2						
15	-76		-2.1					
		-2.0						
20	-86		-2.0					
		-2.0						
25	-94		-1.6					
		-1.2						
30	-100		-1.1					
		-1.0						
35	-105		-1.0					
		-1.0						
40	-110		-0.8					
		-0.6						
45	-113		-0.5					
		-0.4						
50	-117		-0.35					
		-0.3						
60	-120							

Pumping rate, $Q = 33.4$ ft^3/min, $Q/r_c^2 = 42.52$ ft/min; well radii, $r_s = r_c = 0.5$ ft; borehole opening length, $z_2 - z_1 = 247.$ ft; effective time, $t_e = 10$ min (when $\Delta H/\Delta t = -2.$ ft/min begins to fluctuate and loses accuracy or use formula (A19) with $t_1 = 4$ min and $t_2 = 5$ min); from Figure 2a, $S = 0.375/247 = 1.5 \times 10^{-3}$ ft^{-1} = 4.98×10^{-3} m^{-1}; and from Figure 2b, $K = (0.5)^2(0.375)/4(12)(247) = 7.9 \times 10^{-6}$ ft/min = 3.47×10^{-3}m/d.

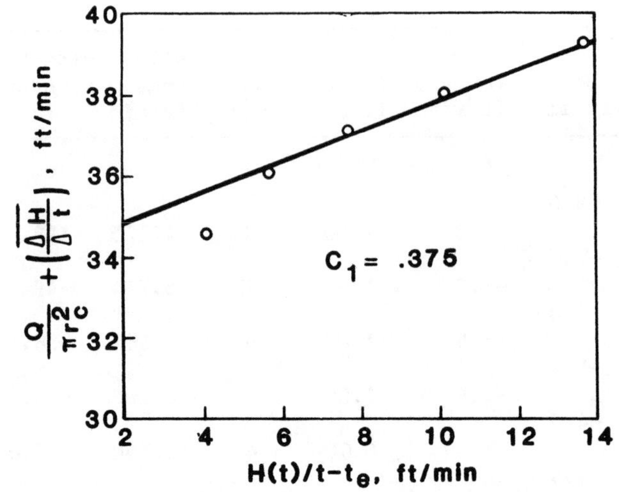

Fig. 2a. Graphic information for the identification of S and K.

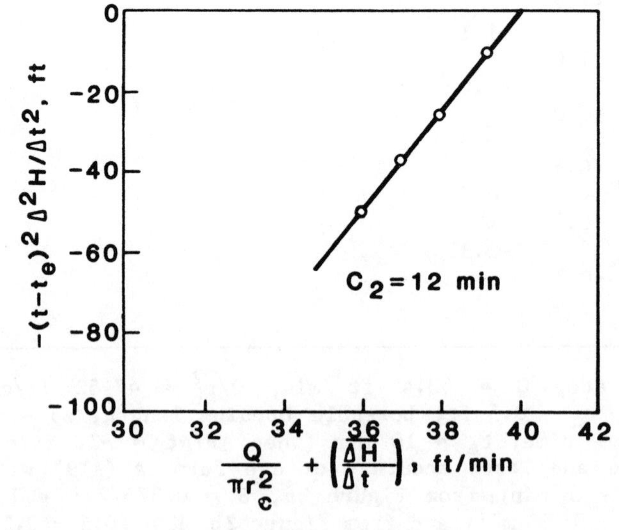

Fig. 2b. Graphic information for the identification of K.

Table 2. Slug Test

t, s	H(t) (Transducer Scale)	$\Delta H/\Delta t$ (Transducer Scale), s^{-1}	1/t, s^{-1}
1	13.0	*	1.000
2	11.0	-2.0	0.500
3	9.0	-2.0	0.330
4	6.5	-2.5	0.250
5	5.1	-1.4	0.200
6	4.5	-0.6	0.167
7	4.0	-0.5	0.143
8	3.6	-0.4	0.125
9	3.2	-0.3	0.111
10	3.0	-0.2	0.100
11	2.8	-0.2	0.091
12	2.5	*	0.083
13	2.4	*	0.077
14	2.2	*	0.071
15	2.1	*	0.066
16	2.0	*	0.063
17	2.0	*	0.058
18	2.0	*	0.055

Well radii, $r_s \cong 3.0$ cm, $r_c = 1.905$ cm; bore opening length, $z_2 - z_1 = 60.96$ cm; from Figure 3a, $S = (1.905)^2 (0.8)/(3.0)^2(60.96) = 5.3 \times 10^{-3}\ cm^{-1}$; from Figure 3c, $K = (1.905)^2(0.8)/4(7.05)(60.96) = 1.67 \times 10^{-3}$ cm/s = 1.44 m/d; note that the plot H(t) versus $(t_e - t)$ gives the slope C_3 (or the plot H(t) versus t gives the slope $-C_3$); similarly, the plot $\Delta H/\Delta t$ versus $1/(t_e - t)$ yields the scope C_4 (or the plot $-\Delta H/\Delta t$ versus 1/t also yields almost the same slope C_4; see Appendix A.

*Inaccurate measurements due to transducer sensitivity at early time and late time (see Figure 3c)

$$\frac{1}{z_2 - z_1} \int_{z_1}^{z_2} s(r_s, z, t)\,dz = F_L H(t) \qquad \text{for } t > 0 \tag{12}$$

subsequently, the modified estimates S* and K* for S and K would become

$$F_L S^* = S \qquad F_L K^* = K \tag{13}$$

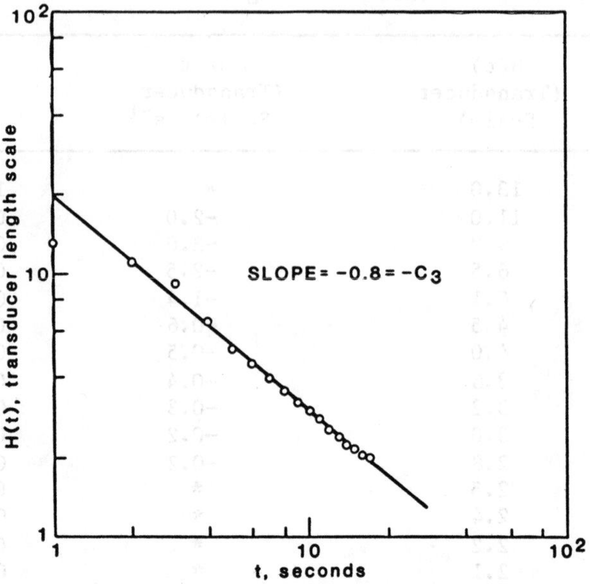

Fig. 3a. Slug test plot for the determination of S and K.

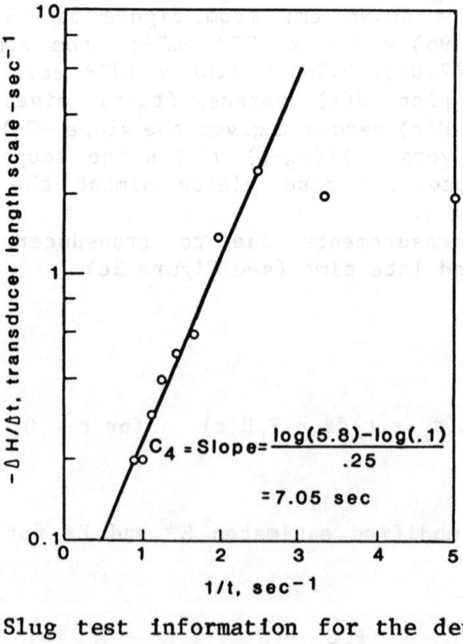

Fig. 3b. Slug test information for the determination of K.

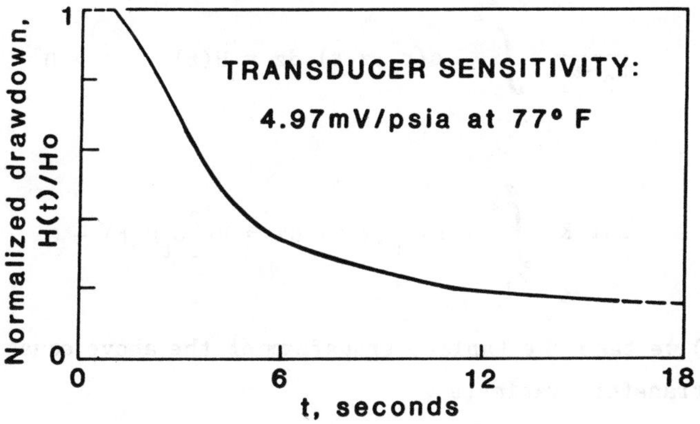

Fig. 3c. Sample of continuous transducer response in a slug test.

Discussion

The above analysis for S is devoid of mathematical approximation. Consequently, whenever a system satisfies the mathematical model proposed in (1) through (3a), the straight-line relationship described above should be obtained. Because a significant deviation from a straight-line plot indicates either inaccurate data or an inappropriate mathematical model, one is furnished with a useful check on the accuracy of the K and S determination.

Appendix A: Solution of the Drawdown Equation

We present here the method of solution to the field equation (1):

$$D_{rr}s + \frac{1}{r} D_r s + D_{zz}s = \frac{S}{K} D_t s$$

together with the initial and boundary conditions (2a)-(2d)

$$s(r,z,0) = 0$$

$$D_z s(r,0,t) = D_z s\,(r,b,t) = 0$$

$$s(\infty,z,t) = 0$$

$$\frac{1}{z_2-z_1}\int_{z_1}^{z_2} s(r_s,z,t)\,dz = H(t) \qquad t > 0$$

$$2\Pi r_s K \int_{z_1}^{z_2} D_r s(r_s,z,t)\,dz = \Pi r_c^2 D_t H(t) + Q$$

Using (2a) we take the Laplace transform of the above equations with p as the transform variable

$$D_{rr}\bar{s} + \frac{1}{r} D_r\bar{s} + D_{zz}\bar{s} = \frac{S}{K} p\bar{s} \tag{A1}$$

$$D_z\bar{s}(r,0,p) = D_z\bar{s}(r,b,p) = 0 \tag{A2}$$

$$\bar{s}(\infty,z,p) = 0 \tag{A3}$$

$$\frac{1}{z_2-z_1}\int_{z_1}^{z_2} \bar{s}(r_s,z,p)\,dz = \bar{H}(p) \tag{A4}$$

$$2\Pi r_s K\int_{z_1}^{z_2} D_r\bar{s}(r_s,z,p)\,dz = \Pi r_c^2(p\bar{H} - H_o) + \frac{Q}{p} \tag{A5}$$

Using (A2), the finite Fourier cosine transform of (A1) and (A3) with n as the transform variable would be

$$D_{rr}\bar{s}_c + \frac{1}{r} D_r\bar{s}_c - \left(\frac{S}{K} p + \left(\frac{\Pi n}{b}\right)^2\right)\bar{s}_c = 0 \tag{A6}$$

$$\bar{s}_c(\infty,n,p) = 0 \tag{A7}$$

It is well known that the solution of (A6) has the form

$$\bar{s}_c(r,n,p) = A_1\, I_o(\alpha_n r) + A_2 K_o(\alpha_n r)$$
$$\alpha_n^2 = \frac{S}{K}p + \left(\frac{\Pi n}{b}\right)^2 \qquad n = 0,1,\ldots \tag{A8}$$

A_1, A_2: generic constants to be specified,

where I_0 and K_0 are the zero-order modified Bessel function of the first and second kinds, respectively. Since $I_0\ (r\to\infty) = \infty$ and $K_0\ (r\to\infty) = 0$, (A7) reduces the solution to

$$\bar{s}_c(r,n,p) = A_2 K_0(\alpha_n r) \tag{A9}$$

By taking the inverse finite Fourier transform of (A9) we obtain

$$\bar{s}(r,z,p) = \frac{A_2}{b}\left[K_0(\alpha_0 r) + 2\sum_{n=1}^{\infty} K_0(\alpha_n r)\cos\left(\frac{n\Pi z}{b}\right)\right] \tag{A10}$$

Since $D_r K_0(\alpha_n r) = -\alpha_n K_1(\alpha_n r)$, the boundary condition (A5) gives

$$A_2 = -\frac{br_c^2}{2r_s K}\left(p\bar{H} - H_o + \frac{Q}{\Pi r_c^2 p}\right) \cdot \left[\alpha_0 K_1(\alpha_0 r_s)(z_2 - z_1)\right.$$

$$\left. + \frac{4b}{\Pi}\sum_{n=1}^{\infty}\frac{\alpha_n K_1(\alpha_n r_s)}{n}\cos\frac{n\Pi}{b}\left(\frac{z_2 + z_1}{2}\right)\sin\frac{n\Pi}{b}\left(\frac{z_2 - z_1}{2}\right)\right]^{-1} \tag{A11}$$

Let the Laplace transform of a function G(r,z,t) be

$$\bar{G}(r,z,p) = \frac{r_c^2}{2r_s K}\cdot\left\{\left[K_0(\alpha_0 r) + 2\sum_{n=1}^{\infty} K_0(\alpha_n r)\cos\left(\frac{n\pi z}{b}\right)\right] / \left[\alpha_0 K_1(\alpha_0 r_s)\right.\right.$$

$$\left.\left.(z_2 - z_1) + \frac{4b}{\pi}\sum_{n=1}^{\infty}\frac{\alpha_n K_1(\alpha_n r_s)}{n}\cos\frac{n\pi}{b}\left(\frac{z_2+z_1}{2}\right)\sin\frac{n\pi}{b}\left(\frac{z_2-z_1}{2}\right)\right]^{-1}\right\} \tag{A12}$$

Then the closed form solution of the field equation would be

$$s(r,z,t) = \int_{t_0}^{t}\left[\frac{Q}{\pi r_c^2} + D_t H(\tau)\right]G(r,z,t-\tau)\,d\tau \tag{A13}$$

Expression (A13) says that the solution of the forward problem composed of (1), (2a), (2b), (2c), and (2e) is uniquely determined. Due to the complexity of (A12), an exact form of the kernel function G(r,z,t) is not possible. Since our goal is to identify S and K, we must look for another approach equivalent to (A13). One approach is to rewrite (A10) and (A11) as follows:

$$\bar{s}(r,z,p)\left[(z_2 - z_1)\alpha_0 K_1(\alpha_0 r_s) + \frac{4b}{\pi}\sum_{n=1}^{\infty}\frac{\alpha_n K_1(\alpha_n r_s)}{n}\cos\frac{n\pi}{b}\right.$$

$$\cdot(\frac{z_2+z_1}{2}) \sin \frac{n\pi}{b} (\frac{z_2-z_1}{2})] = -\frac{r_c^2}{2r_s K} (p\bar{H} - H_0 + \frac{Q}{\pi r_c^2 p}) [K_0(\alpha_0 r) + 2 \sum_{n=1}^{\infty}$$

$$K_0 (\alpha_n r) \cos (\frac{n\pi z}{b})] \tag{A14}$$

Using suitable inverse Laplace transform formulae (see Appendix B) and the convolution theorem on (A14), we obtain

$$a_0 \int_{t_0}^{t} s(r,z,t)\{(z_2-z_1) \exp (\frac{a_0 r_s}{\tau-t}) + \frac{4b}{\pi} \sum_{n=1}^{\infty} \frac{1}{n} \exp [\frac{a_0 r_s}{\tau-t}$$

$$+ a_n (\tau-t)] \cdot \cos \frac{n\pi}{b} (\frac{z_2+z_1}{2}) \sin \frac{n\pi}{b} (\frac{z_2-z_1}{2})\} \frac{d\tau}{(\tau-t)^2}$$

$$= \frac{r_c^2}{4r_s K} \int_{t_0}^{t} [D_\tau H (\tau) + \frac{Q}{\pi r_c^2}] \{\exp \frac{a_0 r^2}{(\tau-t) r_s} + 2 \sum_{n=1}^{\infty} \exp [\frac{a_0 r^2}{(\tau-t) r_s}$$

$$+ a_n(\tau-t)] \cdot \cos (\frac{n\pi z}{b})\} \frac{d\tau}{(\tau-t)} \tag{A15}$$

where $a_0 = Sr_s/4K$ and $a_n = (K/S) (\pi n/b)^2$ for $n > 0$. Equation (A15) represents an equivalent alternative to the closed form solution (A13). The memory of the physical system is preserved in the convoluted integrals of (A15). By removing the integrals on both sides of (A15) and equating the integrands, we arrive at

$$s(r,s,t) \{(z_2-z_1) \exp \frac{a_0 r_s}{t-t_e} + \frac{4b}{\pi} \sum_{n=1}^{\infty} \frac{1}{n} \exp[\frac{a_0 r_s}{t-t_e} + a_n(t-t_e)] \cdot \cos$$

$$\frac{n\pi}{b} (\frac{z_2+z_1}{b}) \sin \frac{n\pi}{b} (\frac{z_2-z_1}{b})\} = \frac{r_c^2}{r_s S} [D_t H(t) + \frac{Q}{\pi r_c^2}]\{\exp \frac{a_0 r^2}{(t-t_e) r_s}$$

$$+ 2 \sum_{n=1}^{\infty} \exp [\frac{a_0 r^2}{(t-t_e) r_s} + a_n(t-t_e)] \cdot \cos (\frac{n\pi z}{b})\} (t-t_e) \tag{A16}$$

Here we have changed the variables τ to t and t to t_e and require that $t_0 \leq t < t_e$ so as to maintain the consistency of the convolution. However, the elimination of the physical memory has created a nonuniqueness in s(r,z,t) in the above relation through the appearance of a time parameter t_e, an unspecified upper bound for the time interval.

In order to regain the lost information and to close the system, we employ condition (2d). Combination of (2d) and (A16) leads to the expression

$$S = \frac{r_c^2}{r_s^2(z_2-z_1)} \cdot \frac{t-t_e}{H(t)} \cdot \left[\frac{Q}{\pi r_c^2} + D_t H(t)\right] \qquad \text{(A17)}$$

The compatability of the system and condition (2d) is implicitly satisfied throughout the derivation by the constancy of S. Setting $y(t) - y(t_0) = Q/\pi r_c^2 + D_t H(t)$ and $x(t) - x(t_0) = H(t)/t-t_e$, the slope of $y(t) - y(t_0)$ versus $x(t) - x(t_0)$ would, in principle, fulfill the compatibility constraint.

By virtue of constant positive S, t_e can be determined by selecting time points t_1 and t_2 and using (A17) to arrive at

$$\left|\frac{t_1 - t_e}{H_1}\left[\frac{Q}{\pi r_c^2} + D_t H(t_1)\right]\right| = \left|\frac{t_2 - t_e}{H_2}\left[\frac{Q}{\pi r_c^2} + D_t H(t_2)\right]\right| \qquad \text{(A18)}$$

where H_1 and $H(t_1)$ and $H_2 = H(t_2)$. As a result,

$$t_e = \left|\frac{k_1 t_1 - k_2 t_2}{k_1 - k_2}\right| \qquad \text{(A19)}$$

where

$$k_i = \frac{1}{H_i}\left[\frac{Q}{\pi r_c^2} + D_t H(t_i)\right] \qquad i = 1,2$$

Physically, t_e defines the time interval within which the identification problem is properly posed. We call t_e the effective time of the system.

In practical computation, the inequality $t_e > t \geq t_0$ must be enforced by having $t_e - t_{max} \geq \Delta t$, where t_{max} is the maximum time point within the time domain and Δt is the scale of accuracy from which $D_t H(t_{max})$ is estimated.

Computational experience indicates that the optimal t_e appears to coincide with the instant of questionable accuracy in the computed values of $D_tH(t)$ from the data. A sensitivity analysis can be performed on (A19) to see how much the measurement and numerical errors in k_1 can propagate through to the estimate of t_e. Let δk_1, δk_2, and δt_e be the error components of k_1, k_2, and t_e, respectively,

$$t_e + \delta t_e = \left| \frac{(k_1+\delta k_1)t_1 - (k_2+\delta k_2)t_2}{k_1+\delta k_1 - (k_2+\delta k_2)} \right| \tag{A20}$$

For the sake of simplicity, we assume $\delta k_1 - \delta k_2$ to be sufficiently small that

$$\delta t_e \cong \left| \frac{\delta k_1 t_1 - \delta k_2 t_2}{k_1 - k_2} \right| \tag{A21}$$

which implies that at large time, more accuracy on measurement is needed to damp out the error in t_e. Furthermore, the variations in t_e can be greatly reduced by selecting t_1 and t_2 as consecutive time points in the presence of k_1 errors.

Let us next integrate (1) from z_1 to z_2 and observe that

$$\int_{z_1}^{z_2} D_{zz} s dz = D_z s(t,z_2,t) - D_z s(r,z_1,t) \tag{A22}$$

where the right-hand side represents the difference between the fluxes across the planes $z = z_1$ and $z = z_2$. It is often that the casing is driven well below the upper confining bed, or well below the water table in the pheatic aquifer case, and consequently, we should have

$$\int_{z_2}^{z_1} D_{zz} s dz \simeq 0 \tag{A23}$$

If we now define

$$\langle s \rangle = \frac{1}{z_2 - z_1} \int_{z_1}^{z_2} s(r,z,t) dz$$

then from (A23), the governing field equation is averaged to

$$D_{rr} \langle s \rangle + \frac{1}{r} D_r \langle s \rangle = \frac{S}{K} D_t \langle s \rangle \tag{A24}$$

Near the well, (A24) may be written as

$$\lim_{r \to r_s} (D_{rr} \langle s \rangle + \frac{1}{r} D_r \langle s \rangle = \frac{S}{K} D_t \langle s \rangle) \tag{A25}$$

Substitute (A14) and (17) into (A25), we obtain the formula for the hydraulic conductivity:

$$K = \frac{-r_s^2}{4} \; S \cdot \frac{D_t H(t) + (Q/\pi r_c^2)}{(t-t_e)^2 D_{tt} H(t)} \tag{A26}$$

For the slug test analysis, (A17) and (26) can be put into simpler forms which are easier to implement. From (A17) with $Q = 0$, S can be viewed as a constant when the following relation is satisfied:

$$\frac{t-t_e}{H(t)} \cdot D_t H(t) = \text{const} \tag{A27}$$

or

$$\frac{H(t)}{H(t_e - t_o)} = (\frac{t_e - t}{t_o}) \; C_3 \qquad \text{for } 0 < t < t_e - t_o \tag{A28}$$

This suggests that

$$S = \frac{r_c^2 \, C_3}{r_s^2 (z_2 - z_1)} \tag{A29}$$

and C_3 can be obtained by plotting ln $[H(t)/H(t_e - t_o)]$ versus $\ln[(t_e^3 - t)/t_o]$; or by taking advantage of the fact that $\ln[H(t) / H(t_e - t_o)] = \ln H(t) - \ln H(t_e - t_o)$ and $\ln[t_e - t)/t_o] = \ln(t_e - t) - \ln t_o$, which implies a simple translation on the log-log plot, we may obtain C_3 by plotting $H(t)$ versus $(t_e - t)$ or versus t with a meaningful choice of t_e. The switching of the scale $(t_e - t)$ to t causes almost a reflection symmetry effect on the log-log plot and hence a change of sign in C_3. In any case, only the absolute value of C_3 is used in the formula.

Similarly, for the hydraulic conductivity, in a slug test formula (26) can be written as

$$K = \frac{r_c^2 C_3}{4(z_2 - z_1)} \cdot \frac{-D_t H(t)}{(t-t_e)^2 D_{tt} H(t)} \tag{A30}$$

K is regarded as a constant when

$$\frac{D_t H(t)}{(t-t_e)^2 D_{tt} H(t)} = \text{const} \tag{A31}$$

or

$$D_t H(t) = C_5 \exp\left(\frac{C_4}{t_e - t}\right) \tag{A32}$$

Consequently, a simplified formula for K is obtained

$$K = \frac{r_c^2 C_3}{4C_4(z_2 - z_1)} \tag{A33}$$

We may obtain C_4 by plotting $\ln D_t H(t)$ versus $1/(t_e - t)$, or versus $1/t$, which again causes almost a reflection symmetry effect on the semi-log plot and hence a change of sign in C_4. Again, only the absolute value of C_4 is meaningful for a suitable choice of t_e. These simplified formulae for S and K are independent of t_e and thus consistent with (A17), (A26), and (A19).

Appendix B: Some Useful Inverse Laplace Transform Formulae

$$L^{-1}[K_0(\alpha_0 r)] = L^{-1}[K_0(\sqrt{\frac{S}{K}}\, r \sqrt{p})]$$
$$= \frac{1}{2t} \exp\left(-\frac{Sr^2}{4Kt}\right) \tag{B1}$$

$$L^{-1}[K_0(\alpha_n r)] = L^{-1}[K_0(\sqrt{\frac{S}{K}}\, r \sqrt{p + \left(\frac{\Pi n}{b}\right)^2 \frac{K}{S}})]$$
$$= \frac{1}{2t} \exp\left[-\left(\frac{\Pi n}{b}\right)^2 \frac{K}{S} t - \frac{Sr^2}{4Kt}\right] \tag{B2}$$

Interchanging the order of D_r and L^{-1} operators gives the inverse transform of $\alpha_n K_1(\alpha_n r)$

$$L^{-1}[\alpha_o K_1(\alpha_o r_s)] = - \lim_{r \to r_s} D_r\{L^{-1}[K_o(\alpha_o r)]\}$$

$$= \frac{Sr_s}{4Kt^2} \exp\left(-\frac{Sr_s^2}{4Kt}\right) \qquad \text{(B3)}$$

$$L^{-1}[\alpha_n K_1(\alpha_n r_s)] = - \lim_{r \to r_s} D_r\{L^{-1}[K_0(\alpha_n r)]\}$$

$$= \frac{Sr_s}{4Kt^2} \exp\left[-\left(\frac{\Pi n}{b}\right)^2 \frac{K}{S} t - \frac{Sr_s^2}{4Kt}\right] \qquad \text{(B4)}$$

Acknowledgments. This work was supported in part by the National Science Foundation, grant CME-7920996 and the Department of Energy, contract DE-AC03-80SF11489.

3 HEAT TRANSPORT

Groundwater Hydraulics related to the problem of heat transport has received considerable attention by the hydrologist within the last decade. This attention has been focused primarily on problems associated with development of geothermal reservoirs and wells and thermal energy storage. The papers in this section address pertinent aspects of both problems including the relatively difficult problem of obtaining experimental and field data adequate for use for predictions.

Pressure Transient Analysis for Hot Water Geothermal Wells

S. K. Garg and J. W. Pritchett
Systems, Science and Software, La Jolla, California 92308

Introduction and Background

The line source solution to the linearized radial diffusivity equation has been traditionally employed in hydrology and petroleum engineering [see, e.g., Ferris et al., 1962; Matthews and Russell, 1967] to analyze pressure transient data from isothermal single-phase (water/oil/gas) reservoir systems. For constant rate of mass production M, the pressure at the bottom of the well $p_w(t)$ is given by [Matthews and Russell, 1967]

$$p_w(t) = p_i + \frac{M\nu}{4\pi Hk} \, Ei\left[-\frac{\phi r_w^2 \rho \nu C_T}{4kt}\right] \qquad (1)$$

where

p_i initial reservoir pressure;
ν kinematic fluid viscosity;
H formation thickness;
r_w well radius;
ρ fluid density;
k absolute formation permeability;
ϕ porosity;
t time;
C_T total formation compressibility, equal to $(1-\phi)/\phi \, C_m + C$;
C_m uniaxial formation compressibility;
C fluid compressibility.

The bottom-hole pressure $p_w(t)$ is thus principally a function of the kinematic mobility-thickness product kH/ν and the total formation compressibility C_T.

If $4kt/\phi r_w^2 \rho \nu C_T > 100$, (1) can be approximated as follows:

$$p_w(t) = p_i - \frac{1.15\ \nu M}{2\pi H k}\left[\log \frac{kt}{\phi r_w^2 \rho \nu C_T} + 0.351\right] \quad (2)$$

Equation (2) implies that a plot of p_w versus log t should be a straight line. If m denotes the slope of this straight line, then

$$k = \frac{1.15\ \nu M}{2\pi H m} \quad (3)$$

$$C_T = \frac{kt}{\text{antilog}\ [(p_i - p_w(t))/m - 0.351]} \cdot \frac{1}{\phi r_w^2 \rho \nu} \quad (4)$$

Superposition can be utilized to construct solutions for buildup (i.e., shutin after production for time t). The solution implies that a plot of p versus log $(t+\Delta t/\Delta t)$ (Δt = shutin time) should be a straight line. The slope of the straight line can be used, together with (3), to calculate formation permeability. (In the above, skin and well storage effects have been ignored. These effects, while important in practical well testing, are not germane to the present discussion.)

The fluid compressibility C can be defined in a number of ways (at constant internal energy C_E, at constant temperature C_t, and at constant enthalpy C_h):

$$C_E = \frac{1}{\rho}\left(\frac{\partial \rho}{\partial p}\right)_E \quad (5)$$

$$C_t = \frac{1}{\rho}\left(\frac{\partial \rho}{\partial p}\right)_T \quad (6)$$

$$C_h = \frac{1}{\rho}\left(\frac{\partial \rho}{\partial p}\right)_h \quad (7)$$

Table 1 gives liquid water compressibilities as a function of pressure and temperature. (Note that the data in Table 1 are based on the CHARGR equation of state for water [Pritchett, 1980] and may

TABLE 1. Constant Energy C_E, Isothermal C_t, and Isenthalpic C_h Compressibilities for Liquid Water as Given by the CHARGR Equation of State

	T, °C											
	0°			100°			200°			300°		
P, MPa	C_E	C_t	C_h	C_E	C_t	C_h	C_E	C_t	C_h	C_E	C_t	C_h
0.1	0.491	0.491	0.495	-	-	-	-	-	-	-	-	-
5.0	0.490	0.490	0.498	0.415	0.464	0.595	0.622	0.815	0.964	-	-	-
10.0	0.488	0.488	0.499	0.414	0.462	0.592	0.618	0.806	0.954	1.24	2.49	2.01
15.0	0.487	0.487	0.501	0.413	0.460	0.589	0.614	0.797	0.943	1.21	2.35	1.94

All compressibilities are in GPa^{-1} ($1\ GPa^{-1} = 10^{-9}\ Pa^{-1}$)

be in error by a few percent. Nevertheless, it is believed that the relative values of C_E, C_t, and C_h are approximately correct.) At 0°C, all three compressibilities are practically identical; thus for groundwater reservoirs it is of little concern which fluid compressibility is used. However, at elevated temperatures characteristic of geothermal systems, substantial differences do exist between the different compressibilities. In this paper we prove that for hot water reservoirs, isenthalpic fluid compressibility C_h is the appropriate one to use in the definition for total formation compressibility.

Since direct measurements by a downhole flow meter in a discharging geothermal well are usually not possible with available tools, cold water injection has been suggested by several investigators [see, e.g., Grant, 1979] for locating permeable horizons and for determining formation permeability. Application of the line source solution to analyze pressure injectivity and falloff data presents another problem area in so far as fluid kinematic viscosity ν (cf. equations (1)-(4)) is a strong function of temperature. In a subsequent section, we will show that the pressure buildup (i.e., injection) data are governed by the kinematic viscosity of the injected cold water. The pressure falloff data, on the other hand, asymptote to a straight line (p versus log $(t+\Delta t)/\Delta t$ plot) whose slope is determined by the kinematic viscosity of the hot reservoir fluid.

Hot Water Production and Radial Flow

Hot water flow in geothermal reservoirs follows a complex thermodynamic path. Although the variations in internal energy, temperature, and enthalpy are small, they do nevertheless occur. The fluid flow cannot be simply treated as isoenergetic, isothermal, or isenthalpic. The flowing enthalpy does, however, approach a definite limit as $r \to 0$ or $t \to \infty$. (It is worth emphasizing that the constancy of enthalpy means that internal energy and temperature must be varying.) The equations governing one-dimensional radial flow will now be used to prove that

$$\lim_{\substack{r \to 0 \\ \text{or } t \to \infty}} h \to h_o$$

Assuming that (1) the rock porosity depends only on the fluid pressure, (2) the rock matrix and the fluid are in local thermal equilibrium, (3) the global heat conduction is negligible, and (4) the fluid flow is governed by Darcy's law, the balance equations for mass and energy in radial geometry can be written as follows [see, e.g., Brownell et al., 1977; Garg and Pritchett, 1977]:

Mass (liquid)

$$\frac{\partial}{\partial t}\ (\phi\rho) - \frac{1}{r}\frac{\partial}{\partial r}\left[\frac{rk}{\nu}\ \frac{\partial p}{\partial r}\right] = 0 \tag{8}$$

Energy

$$\frac{\partial}{\partial t}\left[(1-\phi)\rho_r h_r + \phi\rho h - \phi p\right] - \frac{1}{r}\frac{\partial}{\partial r}\left[\frac{rk}{\nu}\ h\ \frac{\partial p}{\partial r}\right] = 0 \tag{9}$$

where ρ_r is the rock grain density and h_r is the rock enthalpy. The differential equations (8) and (9) are subject to the following boundary and initial conditions:

Boundary conditions

$$\lim_{r \to 0}\ r\ \frac{\partial p}{\partial r} = -\ \frac{M\nu}{2\pi Hk} \tag{10}$$

$$\lim_{r \to \infty}\ p = p_i,\ h = h_i \tag{11}$$

Initial conditions

$$t = 0:\ \ p = p_i,\ h = h_i \tag{12}$$

Following O'Sullivan [1981], the similarity variable η is introduced:

$$\eta = rt^{-0.5} \tag{13}$$

Substituting from (13) into (8)-(12), one obtains the following transformed equations:

Mass

$$\frac{d(\phi\rho)}{d\eta} + \frac{2}{\eta^2}\frac{d}{d\eta}\left[\frac{k}{\nu}\,\eta\,\frac{dp}{d\eta}\right] = 0 \tag{14}$$

Energy

$$\frac{d}{d\eta}\left[(1-\phi)\rho_r h_r + \phi\rho h - \phi p\right] + \frac{2}{\eta^2}\frac{d}{d\eta}\left[\frac{k}{\nu}\,h\eta\,\frac{dp}{d\eta}\right] = 0 \tag{15}$$

Boundary and initial conditions

$$\lim_{\eta\to 0} \eta\,\frac{dp}{d\eta} = -\frac{M\nu}{2\pi Hk} \tag{16}$$

$$\lim_{\eta\to\infty} p = p_i,\; h = h_i \tag{17}$$

Also, note that [Brownell et al., 1977]

$$\frac{d\phi}{dp} = (1-\phi)C_m \tag{18}$$

$$h = c_r T \tag{19}$$

where T is the common local temperature of the rock matrix and the pore fluids and c_r is the rock grain heat capacity.

Regarding p and h as independent thermodynamic variables and utilizing (18) and (19), (14) and (15) may be algebraically manipulated to yield

$$\phi\rho\left\{\frac{1-\phi}{\phi}C_m + C_h - \frac{1}{\rho}\left(\frac{\partial\rho}{\partial h}\right)_p \frac{(1-\phi)\rho_r c_r(\partial T/\partial p)_h - (1-\phi)\,C_m(\rho_r h_r + p) - \phi}{(1-\phi)\rho_r c_r(\partial T/\partial h)_p + \phi\rho + \frac{2}{\eta}(k/\nu)(dp/d\eta)}\right\}\cdot\frac{dp}{d\eta} + \frac{2}{\eta^2}\frac{d}{d\eta}\left[\left(\frac{k}{\nu}\right)\eta\,\frac{dp}{d\eta}\right] = 0 \tag{20}$$

$$\frac{dh}{d\eta} = -\frac{\left\{(1-\phi)\rho_r c_r\;(\partial T/\partial p)_h - (1-\phi)\;C_m(\rho_r h_r + p) - \phi\right\}}{(1-\phi)\rho_r c_r\;(\partial T/\partial h)_p + \phi\rho + \frac{2}{\eta}(k/\nu)(dp/d\eta)}\cdot\frac{dp}{d\eta} \tag{21}$$

Equations (16) and (21) can be combined to give

$$\lim_{\eta \to 0} \frac{dh}{d\eta} = 0 \tag{22}$$

Equation (22) demonstrates that the fluid enthalpy approaches a constant value in the limit $\eta \to 0$ (i.e., in the limit $r \to 0$ or $t \to \infty$). Since

$$\lim_{\eta \to 0} \eta \frac{dp}{d\eta} \neq 0$$

Equation (22) also implies that internal energy E and temperature T are varying as $t \to \infty$.

Although (22) is only valid as $\eta \to 0$, we now assert that $dh/d\eta = 0$ is a reasonable approximation to use in interpreting pressure transient data. With (22) and taking k/ν to be constant, (20) leads to the usual diffusivity equation:

$$\phi \rho C_T \frac{dp}{d\eta} + \frac{2}{\eta^2} \left(\frac{k}{\nu}\right) \frac{d}{d\eta} \left(\eta \frac{dp}{d\eta}\right) = 0 \tag{23}$$

with

$$C_T = \frac{1-\phi}{\phi} C_m + C_h \tag{24}$$

The solution of (23) subject to the boundary conditions (16) and (17) is identical with the classical line source solution (1).

Hot Water Reservoir Production Behavior

To test the validity of the preceding theory, the CHARGR reservoir simulator was exercised in its one-dimensional radial mode. The radially infinite reservoir was simulated using a 60-zone [$\Delta r_1 = 0.11$ m; $\Delta r_2 = 1.2\ \Delta r_1$; $\Delta r_3 = 1.2\ \Delta r_2$, ..., $\Delta r_{60} = 1.2\ \Delta r_{59}$] radial grid. The outer radius of the grid is 25,825 m and is sufficiently large such that no signal reaches this boundary for the production/shut-in periods considered. The formation thickness is

TABLE 2. Rock Properties Employed in Numerical Simulation

	Well Block (i = 1)	Rock Matrix ($2 \leq i \leq 60$)
Porosity ϕ	0.9999	0.1
Permeability k, m^2	50×10^{-12}	5×10^{-14}
Uniaxial formation compressibility C_m, MPa^{-1}	0	0
Rock grain density ρ_r, kg/m^3	1	2650
Grain thermal conductivity K_r, W/m °C	0	5.25
Heat capacity c_r, kJ/kg°C	0.001	1

H = 250 m. The well is assumed to be coincident with zone 1. (In the CHARGR code, a well can be represented as an integral part of the grid by assigning to the well block sufficiently high permeability and porosity.) The reservoir rock is assumed to be a typical sandstone. The relevant rock properties are given in Table 2. The mixture (rock/fluid) thermal conductivity is approximated by Budiansky's formula [Pritchett, 1980].

The initial formation pressure and temperature are 9.3917 MPa and 300°C, respectively. The reservoir is produced at a constant rate

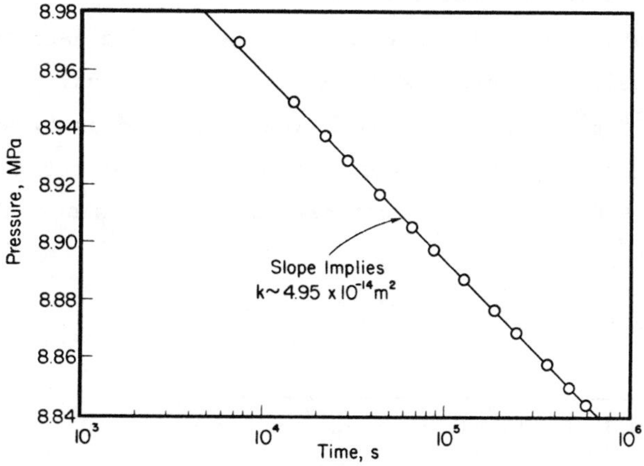

Fig. 1. Drawdown data for hot water geothermal reservoir (p_i = 9.3917 MPa, T_i = 300°C).

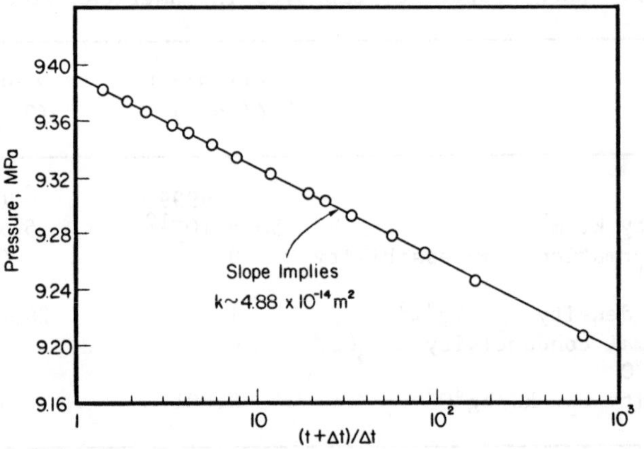

Fig. 2. Buildup data for hot water geothermal reservoir (p_i = 9.3917 MPa, T_i = 300°C).

of 35 kg/s for t = 5.868 x 10^5 s, and is then shut-in for Δt = 1.3932 x 10^6 s. Figures 1 and 2 show the calculated drawdown and buildup response of the well. The drawdown curve has a slope m of 0.0644 x 10^6 Pa/cycle. With ν = 1.244 x 10^{-7} m^2/s (= kinematic viscosity of fluid at initial reservoir pressure and temperature), we obtain for formation permeability k

$$k = \frac{1.15\ \nu M}{2\pi\ H\ m} = \frac{1.15 \times 1.244 \times 35 \times 10^{-7}}{2\pi \times 250 \times 0.0644 \times 10^{6}} \sim 4.95 \times 10^{-14}\ m^2$$

The slope of the buildup curve (~ 0.0653 x 10^6 Pa/cycle) yields k = 4.88 x 10^{-14} m^2. Both the drawdown and the buildup data thus yield permeability values in close agreement with the actual permeability of 5.00 x 10^{-14} m^2. With ρ = 713.9 kg/m^3 (fluid density at initial reservoir conditions), the following is obtained for total formation compressibility C_T (= fluid compressibility C since C_m = 0):

$$C_T = \frac{1}{\phi r_w^2 \rho \nu}\ \frac{kt}{\text{antilog}[p_i - p_w(t)/m] - 0.351}$$

$$= \frac{1}{0.1\ (0.11)^2\ 1.244 \times 10^{-7} \times 713.9}$$

$$\frac{4.95 \times 10^{-14} \times 5 \times 10^{5}}{\text{antilog}[(9.3917 - 8.8490/0.0644) - 0.351]} = 1.93\ GPa^{-1}$$

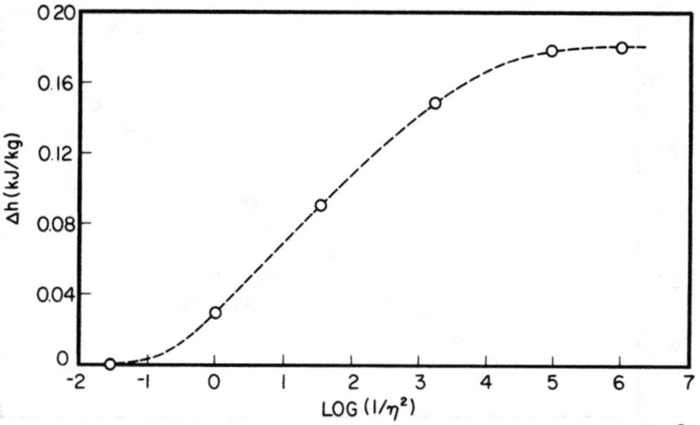

Fig. 3. Enthalpy variation as a function of $1/\eta^2$ for hot water reservoir.

The calculated fluid compressibility value is in reasonable agreement with the isenthalpic fluid compressibility ($\sim$2.01 GPa^{-1}) but is considerably different from the isoenergetic ($\sim$1.24 GPa^{-1}) and isothermal ($\sim$2.49 GPa^{-1}) compressibilities. This verifies the speculation that the isenthalpic fluid compressibility should be used in the definition for total formation compressibility. Finally, for the sake of completeness, Figure 3 shows the enthalpy variation with $1/\eta^2$. Although max $|\Delta h/h|$ (0.20/1344 $\sim 1.5 \times 10^{-4}$) is very small, the fluid does undergo enthalpy changes. Furthermore, Δh (and hence h) approaches a constant value for $1/\eta^2 > 10^4$ (or $\eta < 10^{-2}$).

Cold Water Injection Into a Hot Water Well

In this section, we consider a case wherein cold water is injected into a hot water well. The numerical grid and the formation properties for this case are identical with those employed in the preceding section. The initial formation pressure and temperature are 8.7917 MPa and 300°C, respectively. The cold fluid (temperature $\sim$ 151°C) is injected at a constant rate of 35 kg/s for $t = 5.868 \times 10^5$ s; the well is then shut-in for $\Delta t = 1.3932 \times 10^6$ s. The pressure buildup (injection) data are seen to fit a straight line (Figure 4); the slope of this straight line yields $\nu \sim 1.985$

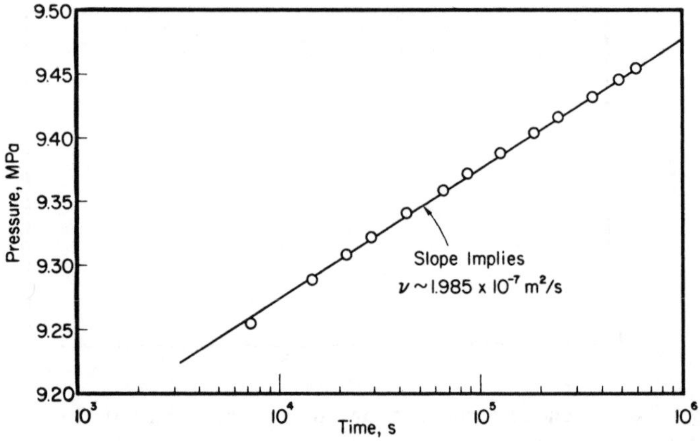

Fig. 4. Pressure buildup (injection) data for cold water injection into a hot water well (p_i = 8.7917 MPa, T_i = 300°C, $T_{injection}$ ~ 151°C).

x 10^{-7} m^2/s, which is in good agreement with the kinematic viscosity of the injected fluid (~1.955 x 10^{-7} m^2/s). Figure 5 shows the radial distribution of ν and temperature T at the end of the injection period (~5.868 x 10^5 s); the thermal front is seen to have propagated approximately 6 m into the formation. The falloff

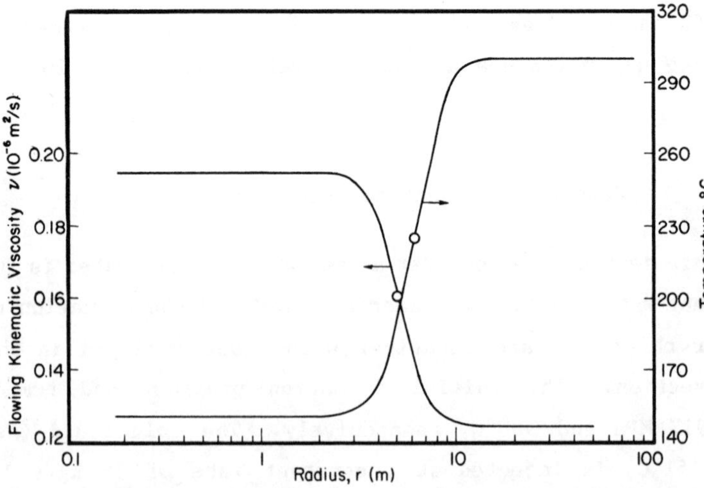

Fig. 5. Radial distribution of flowing kinematic viscosity and temperature at t ~ 5.868 x 10^5 s (end of injection period) "O" denotes the location of the front (defined as the midpoint).

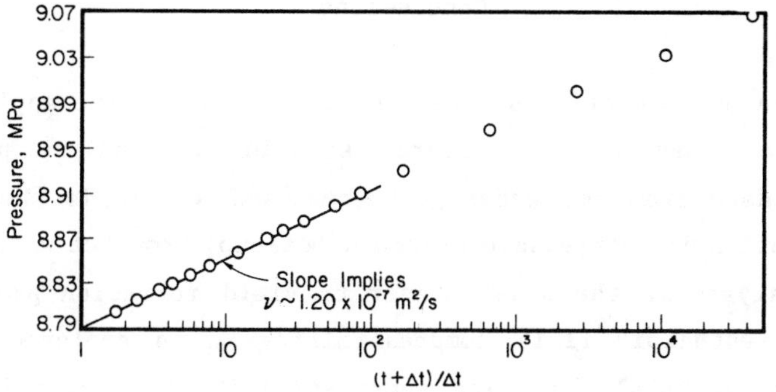

Fig. 6. Pressure falloff data for cold water injection into a hot water well.

(shut-in) data are plotted in Figure 6; the falloff data asymptote to a straight line whose slope yields a value of $v \sim 1.20 \times 10^{-7}$ m^2/s, which is in good agreement with the kinematic viscosity of hot reservoir water ($\sim 1.25 \times 10^{-7}$ m^2/s). No straight line corresponding to cold water pressure falloff can, however, be identified on Figure 6; the reason for this is tied to the relatively small radius (~6 m) affected by cold water injection. The time to investigate a particular radius r during falloff is approximately given by [see, e.g., Matthews and Russell, 1967]

$$\Delta t \simeq \frac{r_{inv}^2}{4} \frac{\phi \mu C_T}{k}$$

where

Δt falloff time, s;

r_{inv} radius of investigation, m;

μ dynamic fluid viscosity, Pa s;

C_T total formation compressibility, Pa^{-1};

k formation permeability, m^2.

With $r_{inv} = 6$ m, $\phi = 0.1$, $\mu \sim 1.8 \times 10^{-4}$ Pa s, $C_T \sim 0.075 \times 10^{-8}$ Pa^{-1} (μ and $C_T = C_h$ are evaluated at 151°C), we have $\Delta t \sim 2.4$ s. The first point on the falloff curve corresponds to a Δt of 14 s.

Conclusions

The main purpose of this paper is to investigate the applicability of the classical line source solution to analyze pressure transient data from hot water geothermal wells. It is shown that the production data (pressure drawdown/buildup) from hot water wells may be analyzed in the usual manner to yield formation properties provided isenthalpic fluid compressibility C_h is employed in the definition for total formation compressibility C_T. Numerical results presented herein also demonstrate that cold water injection data can be interpreted on the basis of the line source solution; in this case, it is, however, necessary to use different values of kinematic viscosity (i.e., cold water kinematic viscosity for injection data and hot water kinematic viscosity for falloff data) for pressure buildup (i.e., injection) and for pressure falloff data.

Acknowledgments. Work performed under subcontract to WESTEC Services, Inc. with funding provided by the U.S. Department of Energy under cooperative agreement DE-FC03-78ET27163.

References

Brownell, D. H., Jr., S. K. Garg, and J. W. Pritchett, Governing equations for geothermal reservoirs, Water Resour. Res., 13, 929-934, 1977.

Ferris, J. G., D. B. Knowles, R. H. Brown, and R. W. Stallman, Theory of aquifer tests, U.S. Geol. Surv. Water Supply Pap., 1536-E, 69-174, 1962.

Garg, S. K., and J. W. Pritchett, On pressure work, viscous dissipation and the energy balance relation for geothermal reservoirs, Adv. Water Resour., 1, 41-47, 1977.

Grant, M. A., Interpretation of downhole measurements in geothermal wells, Rep. 88, Appl. Math. Div., Dep. of Sci. and Ind. Res., Wellington, N.Z., 1979.

Matthews, C. S., and D. G. Russell, Pressure Buildup and Flow Tests in Wells, Monogr. Ser., vol. 1, Society of Petroleum Engineers, Dallas, Tex., 1967.

O'Sullivan, M., A similarity method for geothermal well test analysis, Water Resour. Res., 17, 390-398, 1981.

Pritchett, J. W., Geothermal reservoir engineering computer code comparison and validation calculations using MUSHRM and CHARGR geothermal reservoir simulators, Rep. SSS-R-81-4749, Syst., Sci. and Software, La Jolla, Calif., Nov. 1980.

Aquifer Testing for Thermal Energy Storage

A. David Parr
University of Kansas, Lawrence, Kansas 66045

Fred J. Molz and Joel G. Melville
Auburn University, Auburn, Alabama 36849

Introduction

The possibility of using confined aquifers for the temporary storage of heated water has received in-depth study for the past 5 years [Werner and Kley, 1977; Mathey, 1977; Molz et al., 1978, 1979, 1981; Papadopulos and Larson, 1978; Yokoyama et al., 1980; Tsang et al., 1981]. Related problems involving the regional gradient-induced drift of fluids from the storage zone have also been studied [Molz and Bell, 1977; Whitehead and Langhetee, 1978]. In addition, there are experiments recently completed or presently underway in Denmark, France, Germany, Sweden, and Switzerland. Up-to-date information can be obtained by consulting the various issues of the Seasonal Thermal Energy Storage (STES) Newsletter (C. F. Tsang, Editor, Earth Sciences Division, Lawrence Berkeley Laboratory, Berkeley, California 94720). The STES concept is of major interest in Europe because, should it prove workable, it may be readily integrated with existing district heating systems.

In the United States, experimental study of aquifer thermal energy storage was started by Auburn University, near Mobile, Alabama, in the summer of 1976 [Molz et al., 1978]. This and subsequent experiments funded by the U.S. Geological Survey and the U.S. Department of Energy through Oak Ridge National Laboratory and the Battelle Pacific Northwest Laboratories [Molz et al., 1979, 1981] provided data which were analyzed in part by Papadopulos and Larson [1978] and by Tsang et al. [1981].

At the beginning of an aquifer thermal energy storage project, one usually has a rough idea of a possible storage aquifer based on regional hydrology. Determination of the suitability of the specific confined aquifer requires the performance of a variety of hydraulic thermodynamic and chemical tests. Important parameters include the regional hydraulic gradient, vertical and horizontal permeability of the storage aquifer, horizontal dispersivity, vertical permeability of the upper and lower aquitards, thermal conductivities, heat capacities and chemical characteristics of the aquifer matrix, and native groundwater.

Most chemical and thermodynamic tests can be performed in the laboratory using core samples and groundwater samples. Permeability and dispersivity measurements, however, are best performed in the field by using a variety of pumping tests and data reduction procedures that are available. The data that result can serve as a basis for developing a conceptual design of a proposed aquifer storage system and estimating its thermal efficiency. Also, one can attempt to to anticipate any geochemical problems (corrosion, precipitation, solution, clay swelling, etc.) that may occur.

The purpose of this paper is to describe the hydraulic, thermodynamic, and chemical tests that were performed at the Mobile site. The procedures constitute a fairly complete program for obtaining the data necessary for determining the potential of a confined aquifer for thermal energy storage.

Aquifer Hydraulic Testing

The project site is located in a soil borrow area at the Barry Steam Plant of the Alabama Power Company, about 32 km north of Mobile, Alabama. The surface area consists of a low-terrace deposit of Quaternary age, consisting of interbedded sand and clay deposits that have, in geologic time, been recently deposited along the western edge of the Mobile River. These sand and clay deposits extend to a depth of approximately 200 ft (60 m) where the contact between the Tertiary and Quaternary geologic eras is located. Below the

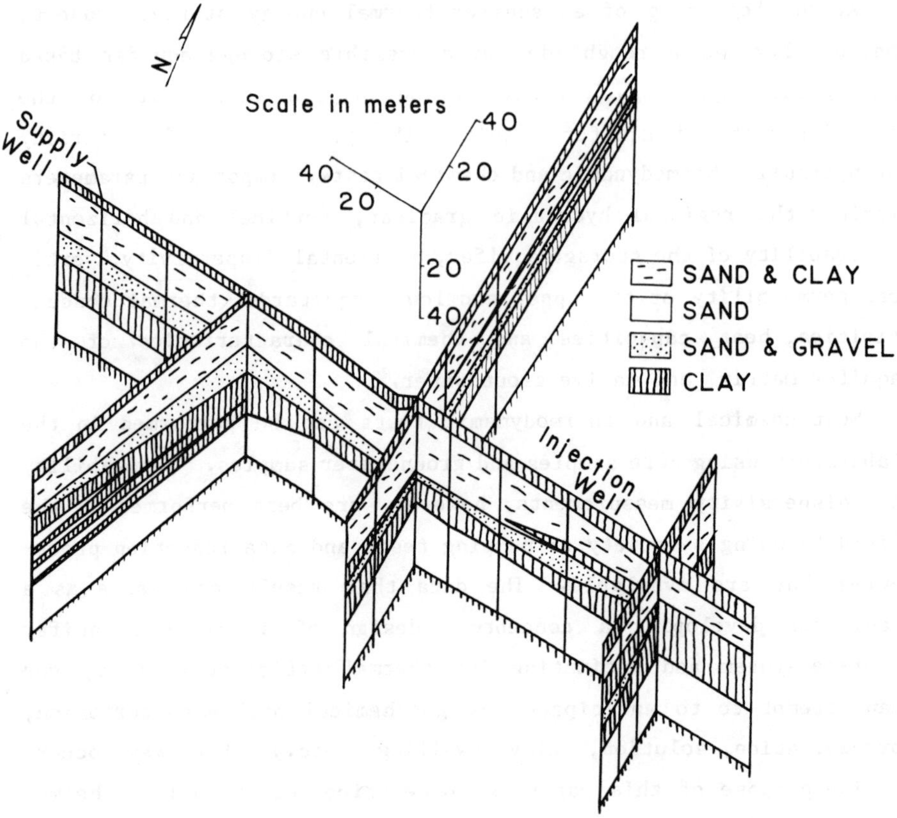

Fig. 1. Fence diagram constructed from well logs at Mobile site.

contact, deposits of the Miocene series are found that consist of undifferentiated sands, silty clays, and thin-bedded limestones extending to an approximate depth of 1000 ft (300 m).

The well field was established in the Quaternary deposits, and based on drilling logs, the fence diagram shown in Figure 1 was constructed. Each vertical line on the diagram represents a well of some type. These wells were screened in the sand formation, which extends from approximately 39 to 61 m below the land surface. This formation constitutes the confined aquifer used for thermal energy storage.

The initial hydraulic tests performed at a potential site should include a short-duration standard pumping test using a single observation well followed by a precise set of measurements of piezometric head for at least three observation wells. Such tests will provide the data needed to determine hydraulic conductivity and the hydraulic gradient by the well triangulation method [Todd, 1980]. These parameters, together with the porosity, can be used to calculate the natural pore velocity and storage zone drift. Acceptable spacings for the observation wells used in defining the piezometric surface are dependent on both the precision of the leveling instruments and on the magnitude of the hydraulic gradient at the site. The following procedure can minimize the possibility of having to construct extra observation wells because the inital wells were not spaced properly.

Storage zone drift during a time interval ot is equal to pore velocity times ot times the quantity C_{vw}/C_{va}, where C_{vw} is the volumetric heat capacity of the saturated portion of the aquifer and C_{va} is the volumetric heat capacity of the entire aquifer which includes solid and liquid. For a given injection-storage-recovery time sequence, one can decide on a maximum acceptable storage zone drift. Knowing the hydraulic conductivity and porosity, one can calculate the maximum tolerable gradient. Then two additional observation wells can be located so that a gradient equal to or greater than the maximum tolerable can be measured with available instrumentation. A more careful procedure would assure the ability to measure some fraction of the maximum tolerable gradient. The main consideration is to avoid placing the observation wells so close together that the maximum tolerable gradient cannot be measured due to exceedingly small differences in water levels.

At the Mobile site, the latest measurement indicated a regional gradient of 3.3×10^{-4} m/m. This value along with a porosity of 0.33, a hydraulic conductivity of 53.6 m/d, and a volumetric aquifer heat capacity of 661 Kcal/m^3/°C, with the water-filled portion contributing 329 Kcal/m^3/°C, yields a storage zone drift of approximately 0.8 m/month. This is to be compared with a planned storage zone radius in excess of 50 m and a 6-month injection-storage-recovery cycle.

TABLE 1. Lists of Parameters Obtainable From the Various Types of Pumping Tests Performed at the Mobile Site

Pumping Test	Parameters
Anisotropy	horizontal permeability of aquifer storage coefficient of aquifer vertical permeability of aquifer
Standard	horizontal permeability of aquifer storage coefficient of aquifer location of lateral boundaries
Leaky Aquifer	horizontal permeability of aquifer storage coefficient of aquifer vertical hydraulic diffusivity of aquitards.

After the regional gradient was determined, several types of pumping and dispersivity tests were performed and analyzed using a variety of classical and modern methods. Classical pumping test procedures [Ferris et al., 1962] are still very applicable, and several were applied. However, more recent procedures were required to determine parameters such as vertical to horizontal permeability ratio and vertical aquitard permeability. The type of tests performed and their objectives are outlined in Table 1.

It should be noted at the outset that the analysis of all tests assumed a homogeneous, anisotropic aquifer with principal axes in the coordinate directions. To a significant but unknown degree, the assumption of homogeneity is violated at the Mobile site. The analyses of most pumping tests are subject to such violations.

Anisotropy Test

The ratio of horizontal to vertical permeability is a parameter that strongly affects the degree of tilting of the thermal front for a mass of hot water injected into a confined aquifer, since substantial tilting of the thermocline induces rapid rates of energy loss to the upper confining layer and encourages mixing of hot and cold water during recovery pumping. This results in poor energy recovery. It is therefore important to determine accurately the permeability ratio at a potential thermal energy storage site. First, a method for analyzing anisotropy pumping tests will be discussed; then the method will be applied to the Mobile field data.

Weeks [1969] presented three methods whereby drawdown data in partially penetrating observation wells or piezometers near a partially penetrating well pumped at a constant rate can be analyzed to determine the permeability ratio. This paper will consider Weeks' method 2 for piezometers or observation wells screened over no more than about 20% of the aquifer thickness. The method is based on Hantush's [1961, p. 90] drawdown equation.

$$s = \frac{Q}{4\pi t}\left[W(u) + f\right]$$

$$= \frac{Q}{4\pi T}\left\{W(u) + \frac{4b}{\pi(z_w - d)} \sum_{n=1}^{\infty} \frac{1}{n} K_o\left[\frac{n\pi r}{b}\sqrt{\frac{K_z}{K_r}}\right] \left(\sin\frac{n\pi z_w}{b} - \sin\frac{n\pi d}{b}\right)\cos\frac{n\pi z}{b}\right\} \qquad (1)$$

where

Q pumping rate, m^3/d;

T transmissibility in m^2/d;

$W(u)$ well function;

$u = r^2S/4Tt$;

r distance from pumped well to piezometer, m;

S storage coefficient;

t time, days;

K_o modified Bessel function of the second kind and zero order;

K_z vertical permeability, m/d;

K_r horizontal permeability, m/d.

The rest of the terms are defined in Figure 2. The dimension z is measured from the middle of screen for observation wells. Equation (1) applies for $t > bS/2K_z$.

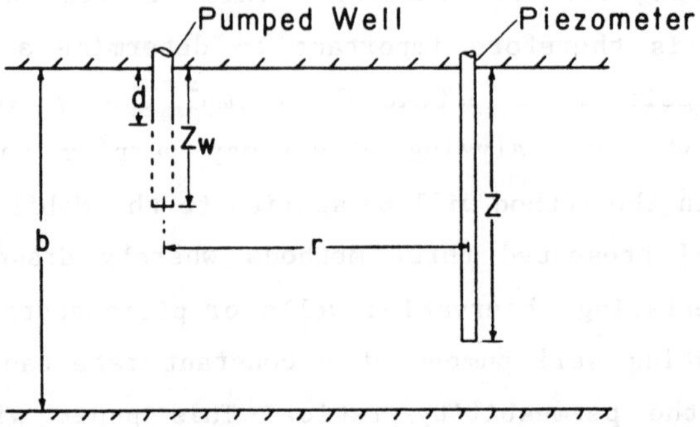

Fig. 2. Definition sketch for equation (1).

The term f in (1) accounts for the deviation in drawdown observed in a partially penetrating piezometer from that predicted for a fully penetrating observation well at the same location. The deviation is therefore given by

$$\delta s = \frac{Q}{4\pi T} \; f \tag{2}$$

where δs is in meters.

Two or more partially screened piezometers are required to perform method 2. The procedure, as given by Weeks, is paraphrased as follows:

Step 1. Determine values of T for each piezometer from the time drawdown plots using the modified nonequilibrium method.

Step 2. For a selected time, plot drawdown versus r for each of the wells on semilog paper with r on the logarithimic scale. Also draw a line of slope $\Delta s = 2.3Q/2\pi T$ beneath the data points if δs is negative (or above if δs is positive).

Step 3. Determine trial values of δs for each well by subtracting observed drawdown from the corresponding straight-line drawdown.

Step 4. Determine f for each well from equation (2) using the trial δs values obtained in step 3 and make a semilog plot of f versus r/b with f on the arithmetic scale.

Step 5. Prepare a type curve on semilog paper of f from equation (1) versus $(r/b)\sqrt{Kz/Kr} = r_c/b$ with f on the arithmetic scale.

Step 6. Match the data plot with the type curve and select a match point.

Step 7. Determine the r/b and r_c/b coordinates for the match point then calculate the permeability ratio from

$$\frac{K_r}{K_z} = \left[\frac{(r/b)}{(r_c/b)}\right]^2 \tag{3}$$

Step 8. Correct the trial f values computed in step 4 by adding algebraically the value obtained by subtracting the data curve value of f from the type curve value of f for the match point. (Note that this step seems to be misworded in Weeks' [1969] paper.)

Step 9. Determine a calculated storage coefficient S_c for each well from the time drawdown plots, assuming the wells are fully penetrating.

Step 10. Determine the true storage coefficient for each well by using the corrected f values from step 8 and the calculated storage coefficients from step 9 in the equation

$$S = S_c \exp [f \quad] \tag{4}$$

Figure 3 shows the well configuration used for the anisotropy pumping test at the Mobile site. Observation wells screened over 3.05 m were located 7.62, 15.2, and 22.9 m north of the partially screened pumped well. Throughout the pumping test, water was pumped from the confined aquifer at a constant rate of 818 m^3/d. Drawdowns, measured by pulley-float systems, are shown in Figure 4 for each of the observation wells. The effect of a boundary is noted about 20 min after startup. Regression analysis was used on the data for which $u < 0.01$ and $t < 20$ min to determine the following relationships:

$$s_1 = 8.55 + 12.9 \log t \tag{5}$$

$$s_2 = 6.83 + 13.4 \log t \tag{6}$$

$$s_3 = 6.10 + 13.1 \log t \tag{7}$$

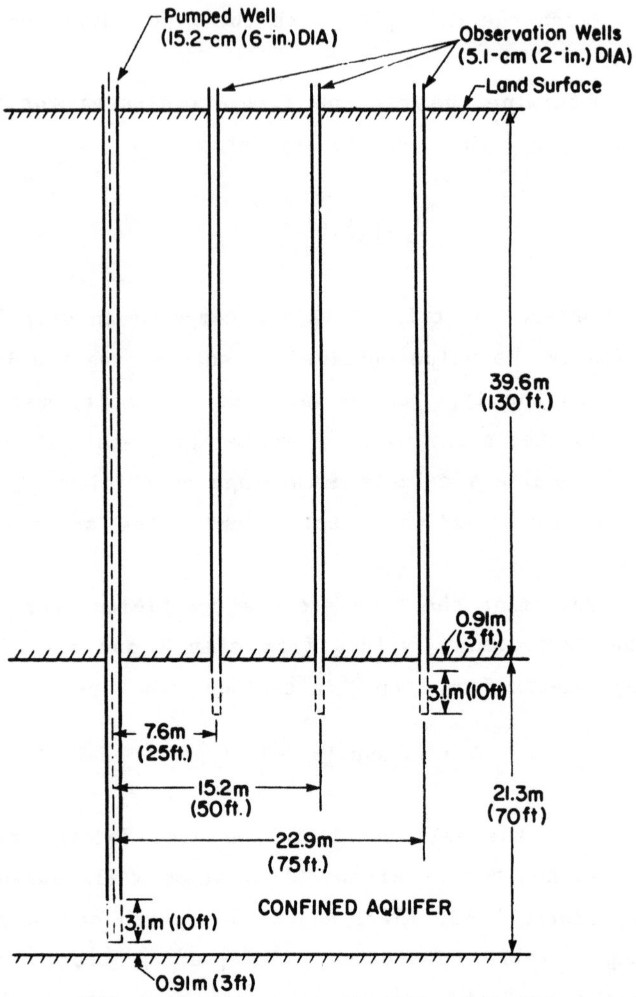

Fig. 3. Well configuration for anisotropy pumping test.

where s_1, s_2, and s_3 are drawdowns (in cm) of observation wells located at r = 7.62, 15.2, and 22.9 m, respectively, and t is in min. These equations are shown as straight lines passing through the appropriate early data in Figure 4. The data analysis by the procedure given above was performed as follows:

Step 1. The transmissibility T was determined for each well according to the modified nonequilibrium method [Jacob, 1950] by the equation

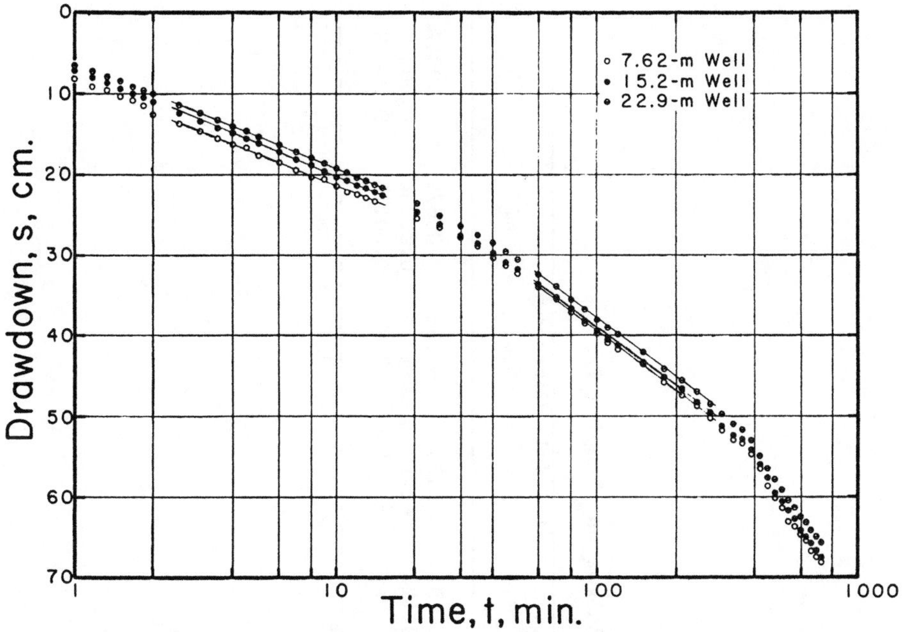

Fig. 4. Semilog plot of drawdown versus time for anisotropy pumping test.

$$T = \frac{2.3Q}{4\pi(\Delta s/\Delta(\log t))} \tag{8}$$

where $\Delta s/\Delta(\log t)$ is the slope of the s versus log t curve for $u < 0.01$ and must be expressed in units consistent with Q and T. The values of T for each well determined by this method using (5)-(7) are shown in Table 2.

Step 2. The drawdowns at t = 10 min as determined from (5)-(7) are plotted in Figure 5, together with a straight line with a slope of $2.3Q/2\pi T_{ave}$ = 0.263 m, where T_{ave} = 1140 m^2/d = average transmissibility for the three observation wells.

Step 3. The trial drawdown deviations, δs, shown on Figure 5, are given in Table 2.

Step 4. Values of f determined from (2) are shown in Table 2.

Step 5. The data and type curves are shown, overlain, in Figure 6. Note that the coordinate axes of the graphs must be parallel.

Step 6. The match point is shown on Figure 6.

TABLE 2. Parameters for Analysis of Anisotropy Pumping Test

Observation Well Number	Distance From Pumped Well r m	r/b	Transmissibility T, m^2/d	δs, m	Initial f,s	Corrected f,s	S_c	S
1	7.62	0.357	1160	0.187	-3.28	-2.68	0.00679	0.00047
2	15.2	0.714	1120	0.120	-2.10	-1.50	0.00234	0.00052
3	22.9	1.07	1140	0.084	-1.47	-0.87	0.00116	0.00049

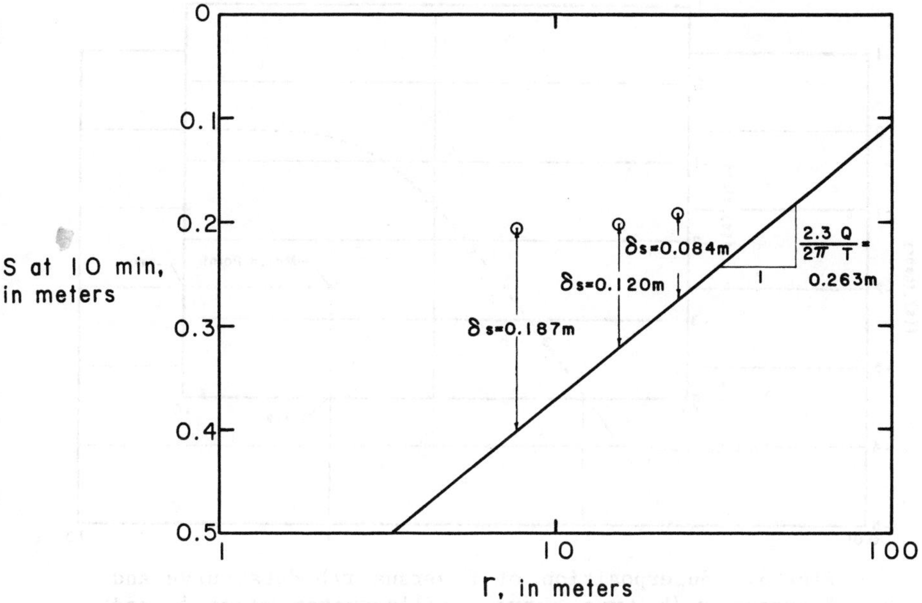

Fig. 5. Distance-drawdown plot at t = 10 min for observation wells. (Illustrates steps 2 and 3 of method 2.)

Step 7. The match-point coordinates for the abcissas are r/b = 2.59 and r_c/b = 1. Equation (3) yields

$$\frac{K_r}{K_z} = \left[\frac{(r/b)}{(r_c/b)}\right]^2 = \left[\frac{2.59}{1}\right]^2 = 6.71$$

Step 8. The correction factor is

$$\Delta f - f_{tc} - f_{dc} = -2 \text{ m} - (-2.60 \text{ m}) = +0.60 \text{ m}$$

where f_{tc} = f match-point value for type curve and f_{dc} = f match point for data curve. The corrected values of f, obtained by adding 0.60 m to the initial f values are shown in Table 2.

Step 9. The calculated storage coefficients S_c can be obtained for each of the wells from [Jacob, 1950]

$$S_c = \frac{2.25T\ t_o}{r^2} \qquad (9)$$

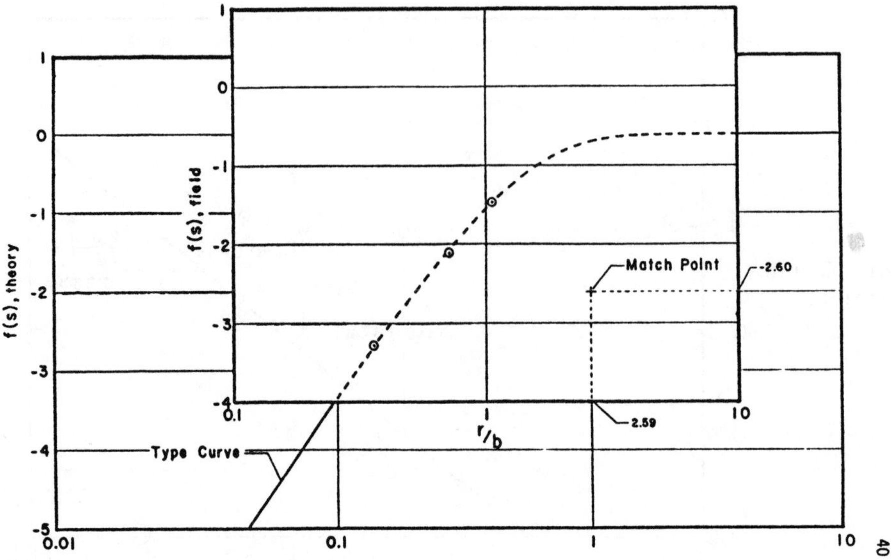

Fig. 6. Superposition of f versus r/b data curve and f versus r_c/b type curve. (Illustrates steps 5 and 6 of method 2.)

where t_0 is determined by setting s = 0 and solving for t in (5) - (7). The values of S_c are given in Table 2.

Step 10. The true value of the storage coefficient as determined by (4) is shown in Table 2.

Where the storage coefficient is known from previous pumping tests, the permeability ratio can be determined with only one partially penetrating observation well and one partially penetrating pumping well. This method entails calculating the transmissibility for the observation well by the modified nonequilibrium method, as discussed above, then determining K_r/K_z by trial and error from equation (1) for a measured drawdown at a specified time (where $u < 0.01$). Values of K_r/K_z obtained in this manner for the data shown in Figure 4 and for S = 0.0005 are 5.98, 7.20, and 6.74 for observation wells 1, 2, and 3, respectively. Weeks' method 3 can also be used to determine the permeability ratio with only one observation well. It does involve plotting a type curve, however.

It is interesting to note that for certain ranges of the geometric

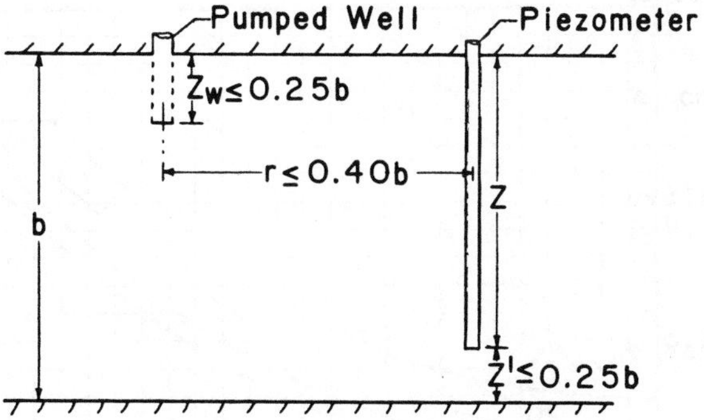

Fig. 7. Definition sketch showing conditions for which equation (10) is applicable.

parameters for the partially penetrating observation and pumped wells, equation (1) can be approximated by a much simpler expression. Specifically, when $r/b \leq 0.4$, $(z-b)/b = z'/b \leq 0.25$, $z_w/b \leq 0.25$, and $d = 0$, as shown in Figure 7, equation (1) can be approximated by

$$s = \frac{Q}{4\pi T}\left\{W(u) + \zeta(z'/b,\ z_w/b) + \ln\left[(r/b)^2(K_z/K_r)\right]\right\} \qquad (10)$$

where the function ζ is given graphically in Figure 8.

Assuming that $u < 0.01$ and the modified nonequilibrium method applies, (10) can be expressed as

$$s = \frac{Q}{4\pi T}\left\{\ln\left[\frac{2.25\ Tt}{r^2 S}\right] + \zeta + \ln\left[(r/b)^2(K_z/K_r)\right]\right\} \qquad (11)$$

which can be rearranged to give

$$K_r/K_z = \frac{2.25Tt}{b^2 S}\ \exp\left[\zeta - \left(\frac{4\pi Ts}{Q}\right)\right] \qquad (12)$$

where s is the drawdown at time t for $u < 0.01$. After determining T from the modified nonequilibrium method, (12) can be used together with Figure 8 to solve for the permeability ratio directly for a single piezometer or a short-screened observation well when the storage coefficient is known a priori. Consistent units must be used in applying (12).

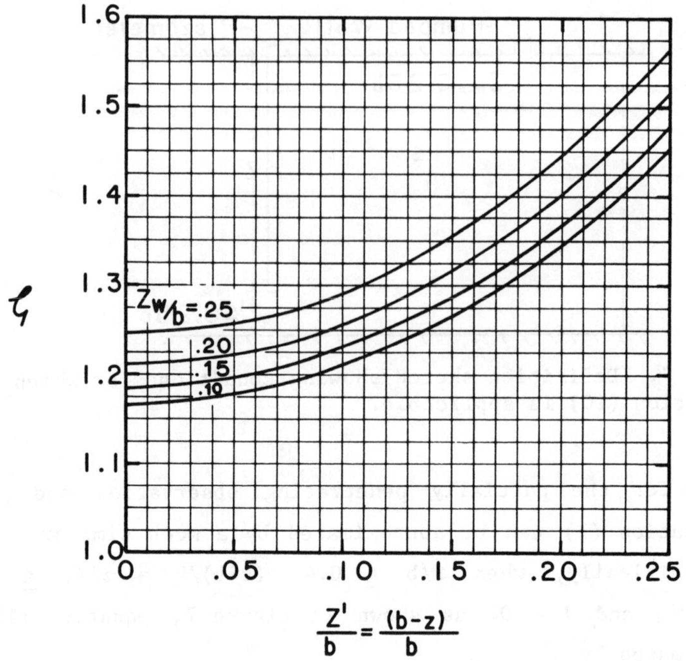

Fig. 8. Dimensionless graph of ζ versus z_w/b and z^1/b.

Although (12) is applicable over a rather limited range of r, z_w, d, and z, it is often desirable to separate vertically the screened portions of the pumped and the observation wells by as much as possible in order to maximize the drawdown deviations δs. The arrangement depicted in Figure 7 would also minimize the costs associated with installing a temporary partially screened pumped well.

Standard Pumping Test and Boundary Location

After completing the anisotropy test the temporary partially screened observation wells were removed and a permanent fully penetrating screen was installed in the injection well. A standard well test was then performed in order to determine the transmissibility and the storage coefficient. These parameters are important in determining the capacity of a potential aquifer for accepting and

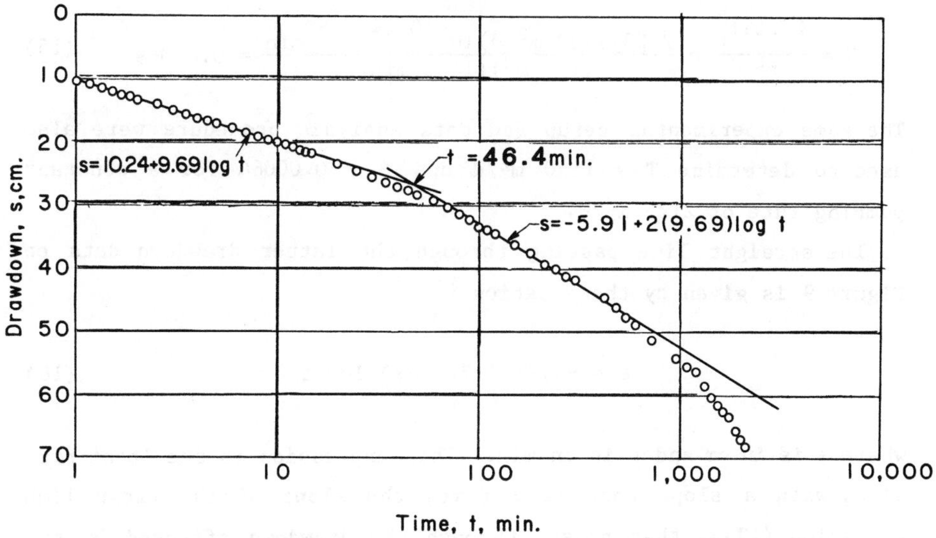

Fig. 9. Semilog plot of drawdown versus time for standard pumping test.

releasing hot water at economically acceptable pumping heads and flow rates. Water was pumped at a constant rate of 600 m^3/d from the confined aquifer throughout the test. The drawdown in a fully screened observation well located 15 m north of the pumped well is shown in Figure 9. A boundary effect was again noticed about 20 min after startup. Regression analysis was used to determine the following relationship for $u < 0.01$ and $t < 20$ min

$$s = 10.24 + 9.69 \log t \qquad (13)$$

where s is in cm and t is in min. Equation (13) is shown as a solid line passing through the early drawdown data in Figure 9. The slope of (13) can be used to determine T from (8) of the modified nonequilibrium method as follows:

$$T = \frac{2.3Q}{4\pi[\Delta s/\Delta(\log t)]} = \frac{2.3(600\ m^3/d)}{4\pi(9.69\ cm)(0.01\ cm/m)} = 1130\ m^2/d \qquad (14)$$

The storage coefficient can then be determined by the equation

$$S = \frac{2.25Tt_o}{r^2} = \frac{2.25(1130\ m^2/d)10^{-10.24/9.69}min}{(15^2\ m^2)(1440\ min/d)} = 0.00069 \quad (15)$$

The same experimental setup and data analysis procedure were also used to determine T = 1140 m^2/d and S = 0.00066 for a constant pumping rate of 2125 m^3/d.

The straight line passing through the latter drawdown data on Figure 9 is given by the equation

$$s = -5.91 + 2(9.69)\log t \quad (16)$$

where s is in cm and t is in min. This expression is the 'best fit' line, with a slope that is 2 times the slope of the first limb (equation (13)), that passes through the drawdown affected by the first boundary. A method for locating the boundary is given by Bear [1979, pp. 479-481]. The effect on drawdown of a boundary can be simulated by an image well with the same pumping rate located beyond the boundary, with the boundary face perpendicularly bisecting the line between the real and imaginary wells. The distance from the observation well to the image well is given by the formula

$$r_1 = \sqrt{\frac{r_o^2}{t_o} t_1} \quad (17)$$

where r_1 is the distance from observation well to image well; r_o is the distance from observation well to pumped well; t_o is the time corresponding to s = 0 on the first straight-line plot, or limb; and t_1 is the time at the intersection of the first and second limb. Therefore for Figure 9 the distance from the observation well to the image well is given by

$$r_i = \sqrt{(\frac{15^2\ m^2}{0.0877\ min})\ 46.4\ min} = 345\ m$$

and since the observation well is very near the pumped well, the distance from the pumped well to the image well and the boundary

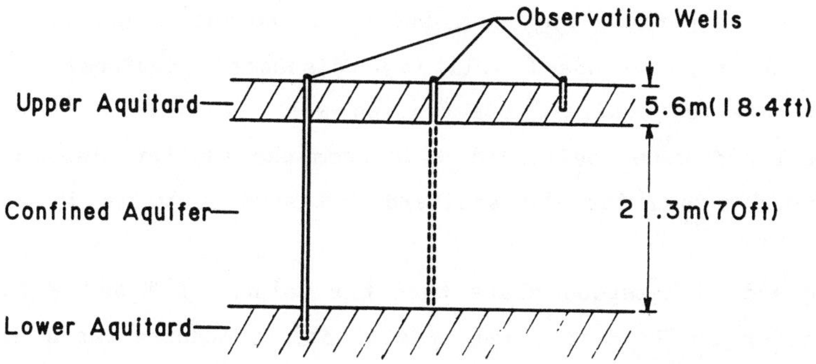

Fig. 10. Well configuration for leaky aquifer pumping test.

are roughly 350 and 175 m, respectively. Similar analyses for at least two or more wells are required, however, to determine the location of the boundary [see Todd, 1980, pp. 147-149].

The effect of another boundary is apparent from Figure 9 as the data points fall below the second limb. If the aquifer were homogeneous, a line with 3 times the slope of the first limb could be passed through the data, and the distance from the observation well to a second image well could be determined. This was not done for the drawdown data of Figure 9, however, since the steep slope of the data affected by the second boundary indicated nonhomogeneity.

Leaky Aquifer Pumping Test

One of the principal sources of energy loss in aquifer thermal energy storage systems is transport of heat by conduction and convection to the confining layers and, ultimately, to overlying or underlying aquifers. In order to estimate the extent of this process and to characterize fully the hydraulic characteristics of a proposed storage site, a leaky aquifer pumping test should be performed to determine the vertical permeability of the confining layers.

The ratio method proposed by Neuman and Witherspoon [1972] provided the basis for the design and analysis of the leaky aquifer test performed at the Mobile site. Figure 10 shows the well configuration used to perform the test, which was conducted concurrently with the 600 m^3/d standard well test discussed previously. The partially screened aquitard observation wells were 15 m from the pumped well and were positioned $\pm 45°$ from the aquifer observation well. The drawdown for the aquitard and aquifer wells is shown in Figure 11.

Neuman and Witherspoon state that the values of T and S for a leaky aquifer can be determined by standard procedures for a close well at early times. Since previous tests of the Mobile site have shown that the confining layers are classified as slightly leaky, the values of T = 1130 m^2/d and S = 0.00069 obtained by the modified nonequilibrium test discussed previously are appropriate.

The ratio method is straightforward and does not require curve matching. When

$$t \leq 0.1 S_s' b'^2 / K' \tag{18}$$

where t is the time in days, S_s' is aquitard specific storage in m^{-1}, b' is the aquitard thickness in m, and K' is the vertical aquitard permeability in m/d, the following procedure can be used to determine K' given S_s'

Step 1. Calculate s'/s at a given radial distance r at a specific time t, where s and s' are the observation well drawdowns in the aquifer and aquitard, respectively.

Step 2. Calculate $t_D = Tt/Sr^2$ for the time used in determining s'/s in step 1.

Step 3. Read a value of $t_D' = K't/S_s'z^2$ corresponding to the s'/s and t_D values from Figure 12, where z is the distance from the middle of the aquitard observation well screen to the aquifer-aquitard interface.

Step 4. Calculate the aquitard hydraulic diffusivity (or coefficient of consolidation) from

$$\alpha' = \left(\frac{z^2}{t}\right) \quad t_D' \tag{19}$$

Step 5. Determine the vertical permeability of the aquitard from

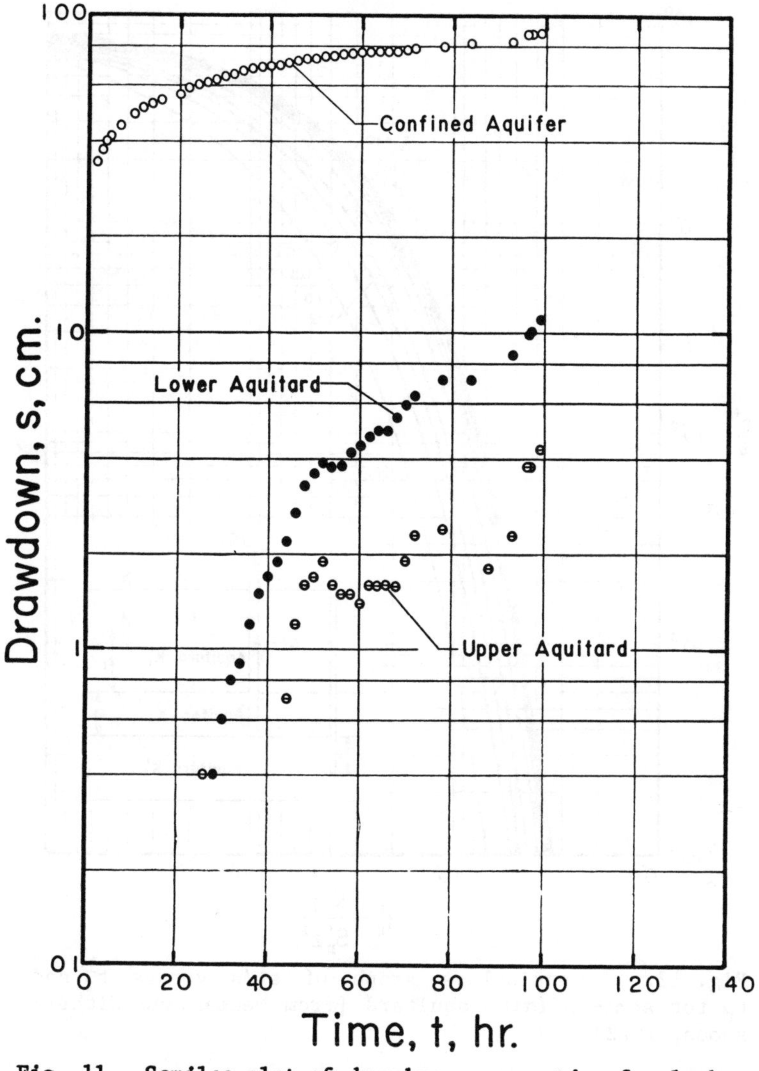

Fig. 11. Semilog plot of drawdown versus time for leaky aquifer pumping test.

$$K' = \alpha' S'_s \tag{20}$$

This procedure is quite simple to apply. It is, however, suggested that the detailed discussion of Neuman and Witherspoon [1972] be referred to in order to appreciate fully the applicability of the ratio method.

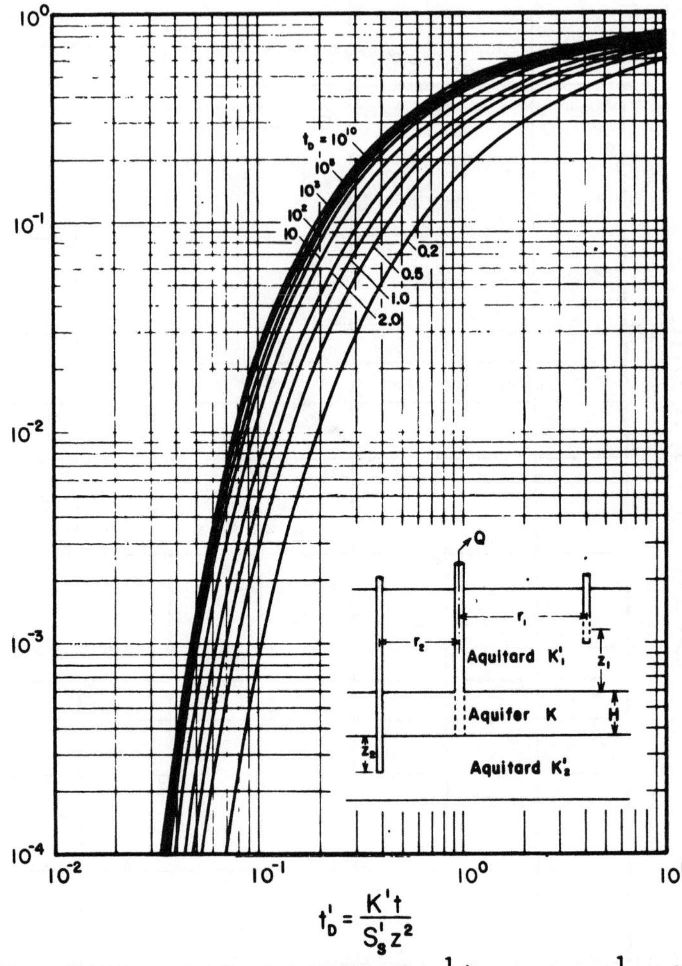

Fig. 12. Dimensionless graph of s^1/s versus t_d^1 and t_d for semi-infinite aquitard [from Neuman and Witherspoon, 1972].

The Mobile drawdown data at t = 50 hours will be considered to provide an example of the application of the ratio method. The results of the procedure given above are presented in Table 3.

Step 5 of the method has not been carried out, since the results of the consolidation test to determine S_s' for the upper and lower aquitards have not yet been received.

The drawdown data for the lower aquitard followed the shape of Neuman and Witherspoon's theoretical curves. This was because the

TABLE 3. Results of Ratio Method Analysis for Drawdowns at t = 50 Hours

	z, m	s' or s, m	s'/s	t_D	t_D'	K'/S_s', m^2/d
Upper aquitard	3.58	0.017	0.0231	15,200	0.110	0.68
Lower aquitard	4.11	0.036	0.0489	15,200	0.150	1.22
Aquifer		0.736				

lower aquitard is quite thick and it is relatively easy to place an isolated observation well (well logging to depths about 50 m below the storage formation have not identified an underlying aquifer). The average value of the lower aquitard hydraulic diffusivity $\alpha' = K'/S_s'$ for all of the data points is 1.10 m^2/d.

The upper aquitard is only about 5.6 m thick, and communication with the overlying aquifer near the top of the observation well screen may have affected the drawdown about 54 hours after pumping began. The average value of the upper aquitard hydraulic diffusivity for the first six drawdown values is 0.71 m^2/d.

Dispersivity Testing

The hydrodynamic dispersion coefficient is an important parameter which can affect the efficiency of a thermal energy storage system. In general, the smaller the dispersivity, the sharper the interface between hot and cold water. Minimal mixing of injected and native waters maximizes the recovery temperature.

In an attempt to provide a useful measure of the dispersion coefficient at the Mobile site, a conservative tracer test was performed during first cycle injection. Sodium bromide was combined with the hot injection water at a concentration of approximately 11 mg/l [Davis et al., 1980]. The resulting concentration in the storage aquifer was recorded in a tracer observation well (well 15) located 15.2 m from the injection well. This well was screened over a length of 1.52 m with the screened section located in the middle of the confined aquifer.

The sampling apparatus was a section of 2.54 cm I.D. fiberglass pipe. Holes were drilled in the pipe to coincide with the screened section of the well. Flexible tubing (3/8 in. I.D.), was used to transport the samples to the ground surface. It is essential to use vacuum tubing to eliminate collapse in the event of some clogging of the line. Plugs of silicon were injected into the fiberglass pipe to isolate the sampling section and to secure the flexible tubing. The entire fiberglass pipe and flexible tubing apparatus was lowered, by hand, into the wells with a nylon rope. A coarse sand was backfilled into the space between the fiberglass pipe and the well casing. Continuous or intermittent sampling was accomplished with variable speed peristaltic pumps. For intermittent sampling, the pumps were run at 1 l/min for 1 hour before taking a sample.

A 22.7-m^3 tank containing NaBr at 40,000 mg/l was prepared, and a diaphragm pump was used to control flow of the tracer into the injection line. Because of possible clogging in the aquifer, constant tracer flow against a variable head was a desired capability of the system. However, no significant pressure increase occurred, and the diaphragm pump did not operate consistently against the low head. Changes in field temperatures also contributed to inconsistent pump behavior. A variable speed peristaltic pump will be used in the future.

Variation in the injected water tracer concentration over the duration of the experiment (756 hours) was between 19.5 and 11.0 mg/l. This variation was due to inconsistent diaphragm pump behavior and also to several down periods necessary for boiler repairs.

Experimental results are shown in the breakthrough curve in Figure 13. During the first 100 hours of the experiment, the injection concentration c_0 was relatively constant and averaged 11.0 mg/l. As an initial estimate of longitudinal dispersivity α, the method described by Gupta et al. [1980] was applied to the four data points shown on the breakthrough curve.

This procedure is based on an approximate solution to the steady

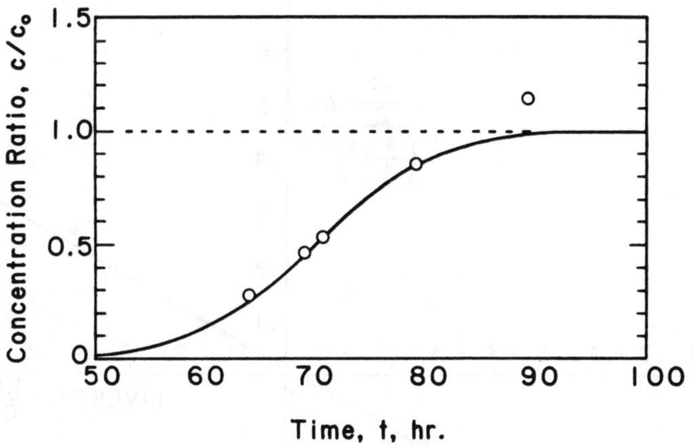

Fig. 13. Concentration ratio at tracer well 15 located at r = 15 m (C_0 - 11.0 ppm for first 100 hours of experiment).

state radial flow dispersion equation given by Hoopes and Harleman [1967]

$$c/c_0 = 0.5 \; \mathrm{erf_c} \; (u) \qquad (21)$$

where $\mathrm{erf_c}$ is the complementary error function; $u = (r^2/2 - At) / (4\alpha r^3/3)^{1/2}$; r is the radius from injection well; α is the dispersivity; t is the time; and $A = Q/2\pi bn$, where Q is the injection rate, b is the aquifer thickness, and n is the porosity. Through manipulation of (21),

$$\mathrm{erf}(u) = 1\text{-}2c/c_0 \quad \text{or inverf } (1\text{-}2c/c_0) = u \qquad (22)$$

Hence

$$\sqrt{\alpha} \; \mathrm{inverf}(1\text{-}2c/c_0) = (r^2/2\text{-}At)/(4r^3/3)^{1/2} \qquad (23)$$

Thus a plot of $\mathrm{inverf}(1\text{-}2c/c_0)$ versus $(r^2/2\text{-}At)/(4r^3/3)^{1/2}$ is a straight line with a slope equal to $\sqrt{\alpha}$. Such a plot for the Mobile tracer data which is shown in Figure 14 yielded a local, apparent dispersivity of 9.1 cm. The continuous curve in Figure 13 is based on this value for α.

If a homogeneous aquifer is assumed at the Mobile site, the arrival time of a nondispersed front of injected fluid is given by t

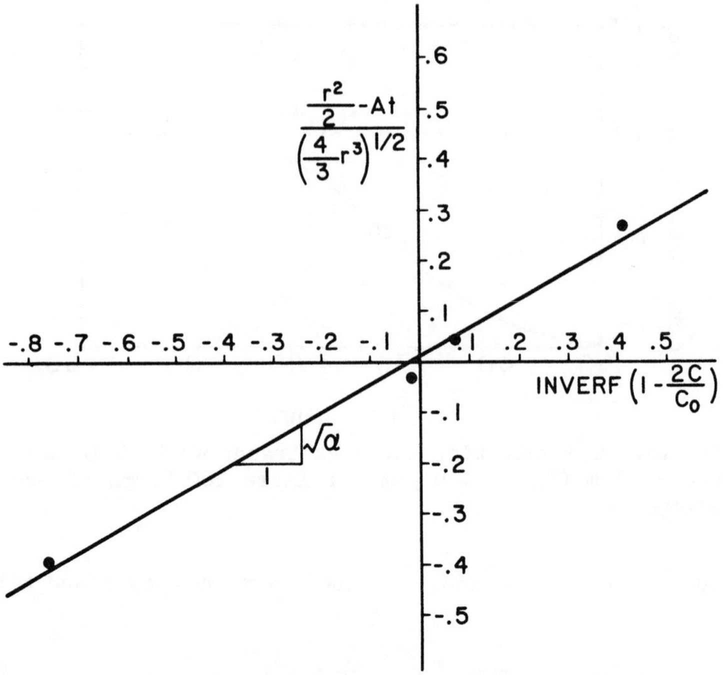

Fig. 14. Slope plot for tracer test analysis.

$= \pi r^2 nb/Q$. For r = 15.2 m, n = 0.33, b = 21.3 m, and Q = 45.4 m^3/h, this yields t = 112 hours. The corresponding time on the dispersed front of the approximate solution is the time when c/c_0 = 0.5. From the experimental data on Figure 13, it is seen that when c/c_0 = 0.5, t = 70 hours. Nonhomogeneous aquifer properties contribute to the difference between theoretical and experimental arrival times. Pumping tests have suggested an increase in aquifer transmissivity in the direction of tracer well 15. More importantly, temperature data and electric logs indicate that hydraulic conductivity is largest near the center of the confined aquifer and decreases in magnitude near the upper and lower confining layers. Thus the 9.1-cm value for α, which is among the lowest values ever obtained in the field, is almost certainly not representative of the overall dispersivity of the aquifer. Later data which might be more representative are complicated by the occurrence of free thermal convection.

Geochemical Testing

Three separate aquifer storage experiments which were performed in the United States have been plagued by geochemical problems of one type or another. In all cases, the problem led to clogging of a well during some type of injection. Such a situation may be very difficult or impossible to correct once it has occurred. Therefore the goal of geochemical testing should be to anticipate geochemical problems and, if at all possible, prevent their occurrence.

In experiments performed by Texas A & M University, water was cooled by a spray pond prior to injection [Reddell et al., 1979]. Oxygen which entered the water reacted with iron to produce iron-oxide precipitates capable of plugging the injection well. Clogging was prevented through the use of a rapid sand filter prior to injection. It was necessary to backwash the filter after each injection volume of 950 m^3.

During previous tests at the Mobile site, more serious clogging resulted due to swelling of formation clays [Molz et al., 1979, 1981]. This was caused by a moderate water quality difference between groundwater native to the storage aquifer and the injected water which was obtained from a shallow supply aquifer. Listed in Table 4 are selected chemical properties of the supply and storage aquifer waters. The storage formation is composed of a medium sand containing about 15% silt and clay by weight. Since this fraction contains smectite clays, there is clearly a potential for osmotic swelling and subsequent clay particle dispersion if water from the supply aquifer is injected into the storage aquifer [van Olphen, 1963]. This phenomenon is also called freshwater sensitivity and occurs when a clay particle containing interlayer water with a relatively high ion concentration comes in contact with water having a relatively low ion concentration. There is then a tendency, similar to osmosis, for the surrounding water to diffuse into the clay particle, causing it to swell. Such swelling has been observed many times in both the laboratory and the field [Brown and Silvey, 1977].

TABLE 4. Selected Physical and Chemical Properties of Groundwater From the Supply and Storage Aquifers

	Supply Aquifer	Storage Aquifer
Temperature	19.5°C	19.5°C
pH	7.19	7.38
Fe	0.07 mg/1	0.055 mg/1
Na	2 mg/1	9.35 mg/1
Si	7.7 mg/1	10.1 mg/1
Ca	0.27 mg/1	0.22 mg/1

As mentioned previously, clay swelling and dispersion caused serious clogging problems at the Mobile site during previous experiments. Regular backwashing of the injection well was required to maintain even minimally acceptable injection rates [Molz et al., 1979, 1981]. The problem was solved during the present experiments by obtaining supply water from the storage aquifer itself (doublet supply-injection system) and by increasing the Na ion content of the injected water by approximately 5 mg/1. This and a fully penetrating injection well increased the specific capacity by approximately a factor of 7 compared to previously observed values.

Presently, there is a cold storage experiment underway on the Stony Brook Campus of the State University of New York [Stern, 1980]. Water is being pumped from a supply well, chilled by an air conditioning system and injected into the same aquifer through a well about 85 m from the supply well.

The first injection went smoothly with no apparent problems. However, when water was recovered from the injection well for reinjection through the supply well, serious clogging of the supply well developed. The problem is being studied, but no explanation has been developed to date.

The previously discussed case histories support the contention that careful geochemical testing must be performed as part of the design of an aquifer thermal energy storage system. Even if potential problems involving changes in oxygen content, biological activity, and water quality differences are eliminated, problems can develop related solely to heating the injection water. Specifically, a temperature increase would affect (1) the chemical equilibrium between the minerals of the aquifer matrix and their concentrations in the groundwater solution, (2) the ion exchange capacity and selectivity of clays, (3) the distribution of hydrated water, and (4) the rates of chemical and physical reactions. Each of these phenomena will be discussed briefly.

Equilibrium Changes

All minerals which compose an aquifer matrix are involved in some type of chemical reaction with the surrounding groundwater. Normally, the reactions of greatest interest are those involving calcium, magnesium, silica, and the carbonate system (alkalinity). However, many of the other chemical species may indirectly affect solubility calculations when speciation and ionic strength adjustments are considered [Kramer, 1967]. Because few simple chemical phase equilibria exist which involve chloride and sulfate, these substances are not considered further. Phosphates usually occur in negligible concentrations when compared with the carbonate system, and therefore their effect on equilibrium conditions is usually small. Sodium and potassium are normally very soluable and do not influence equilibrium calculations other than through electroneutrality. Thus the reactions of greatest interest are those involving calcium, magnesium, silica, and alkalinity (carbonate system).

The major reactions for establishing equilibrium of calcium involve calcium carbonate. Changes in relevant chemical equilibrium constants with temperature are described by the Van't Hoff equation, which can be written as

$$2.303 \log \frac{K_{T2}}{K_{T1}} = - \frac{\Delta H}{R} \left(\frac{1}{T_2} - \frac{1}{T_1}\right) \qquad (24)$$

where K_{T2} and K_{T1} are equilibrium constants at absolute temperatures

T_1 and T_2, respectively, R is the universal gas constant (R = 1.986 cal/°C/mol), and ΔH is the standard enthalpy of formation for the reaction in question. This equation has been used to estimate the effect of temperature over a limited range for a number of chemical reactions. However, inherent in the use of this equation is the assumption that ΔH remains constant. In many reactions of interest, this is not the case, and in such situations care should be taken when adjusting the equilibrium constant for temperature variations. Empirical relationships are used to adjust equilibrium constants for temperature variations for those reactions which do not follow the Van't Hoff relationship.

The dissolution reaction for calcite has the form

$$CaCO_{3(s)} \rightleftharpoons Ca^{2+}_{(aq)} + CO^{2-}_{3(aq)} \tag{25}$$

The solubility product equation for this reaction is

$$K_s = [Ca^{2+}_{(aq)}]\,[CO^{2-}_{3(aq)}] \tag{26}$$

An empirical relationship often used to adjust the solubility product constant in equation (26) for temperature is

$$pK_s = 0.01183T + 8.03 \tag{27}$$

where T represents reaction temperature °C. Equation (27) is valid within the temperature range 0°C to 80°C.

Calcite is less soluble as the temperature increases. For example, increasing the temperature from 20°C to 80°C changes the solubility product constant from $10^{-8.27}$ to $10^{-8.98}$. Thus there is a five-fold decrease in the solubility of $CaCO_3$ over this temperature range.

Similarly, the solubilities of dolomite ($CaMg(CO_3)_2$) and aragonite ($CaCO_3$) decrease with increasing temperature. If the solution becomes saturated, these materials probably precipitate as very fine solid particles.

Furthermore, because CO_2 is less soluble at higher temperatures, pH tends to increase with increased temperature. This change in pH would shift the carbonate equilibria toward $CO_3^{=}$, and thus all carbonate compounds, including those containing $FeCO_3$ and $MnCO_3$, would tend to form and precipitate as temperature is raised [Krauskopt, 1967].

Thermochemical data relating the silicate system to temperature changes are not as well understood as the carbonate system [Kramer, 1967]. However, many researchers [Bostrom, 1967] have indicated that silica increases in solubility at elevated temperatures if the following reactions are considered:

$$SiO_2 + H_2O + OH^- \rightleftharpoons H_3SiO_4^- \qquad (28)$$

or

$$SiO_2 + 2H_2O \rightleftharpoons Si(OH)_4 \qquad (29)$$

Considering Van't Hoff's equation for the latter reaction, the solubility of SiO_2 can change from about 10 ppm SiO_2 at 20°C to about 40 ppm SiO_2 at 80°C. The solubility of SiO_2 is further enhanced by a rise in pH brought about by decreasing CO_2 concentration. Thus an increase in temperature would be expected to increase the concentration of $Si(OH)_4$ dissolved in the water and also increase the alkalinity.

The equilibrium concentration of $Si(OH)_4$ may affect the distribution of aquifer clay type. A typical reaction relates K-feldspar (aluminum silicates) and the clay kaolinite:

$$\underset{\text{(K-feldspar)}}{4K\ AlSi_3O_8} + 22H_2 \rightleftharpoons 4K^+ + 4OH^- + \underset{\text{(Kaolinite)}}{Al_4Si_4O_{10}(OH)_8} + 8Si(OH)_4 \qquad (30)$$

Thus an increase in dissolved silica may decrease the amount of kaolinite found. This, in turn, could affect the distribution of muscovite, montomorillonite, and illite. Other clay materials may be affected in a similar manner.

Distribution of organic compounds could also be significantly altered at elevated temperatures. These compounds would tend to decompose, reducing the solid fraction and forming dissolved gasses and inorganic compounds.

Ion Exchange Reactions of Clays

Like chemical equilibrium constants, temperature dependence of ion exchange equilibria can be described by the Van't Hoff equation. However, ion exchange does not involve primary chemical bonds and, as a rule, does not evolve or absorb significant heat [Helfferich, 1962]. For a reaction

$$A + \bar{B} \rightleftharpoons B + \bar{A} \tag{31}$$

where the bar over the species represents association with clay sites, standard enthalpy changes are usually smaller than 2 kcal/mol. Consequently, the temperature dependence of ion exchange equilibria is usually small.

On the other hand, selectivity resulting from processes such as complex formation may have considerable enthalpy changes. These types of processes are usually discouraged by an increase in temperature. Thus selectivity decreases with increasing temperature. As the selectivity is changed, the character of clays could also change to reflect the chemical composition of the contact solution. This could alter the swelling characteristics of the clays.

Hydrated Water

Occluded and bound water (i.e., water associated with solids as water of crystallization or as water occluded in the interstices of crystals) would not be completely removed at temperatures less than 100°C. However, any increase of temperature would tend to drive out the bound water and thus reduce the solid volume of crystals. This effect is not likely to be important, especially in a negative sense.

Rates of Reaction

The rates of most chemical reactions increase with increasing temperature. The rate constant k, of the reaction, normally changes according to the Arrhenius equation

$$\frac{dLnk}{dT} = \frac{E_a}{RT^2} \tag{32}$$

where E_a is the Arrhenius activation energy, T is absolute temperature, and R is the universal gas constant. Thus as the solution temperature increases, reactions tend to reach equilibrium faster.

Summary

Obviously, the previous discussion of heat-induced chemical changes constitutes an introduction to a complex problem that is very site specific. Anything can happen, from nothing to clay swelling to calcium carbonate precipitation to solution of quartz grains or cementing agents. The latter could be of particular importance in some consolidated aquifers. Presently, the Battelle Pacific Northwest Laboratories at Richland, Washington, is attempting to develop rigorous field and laboratory test procedures for determining the suitability of a confined aquifer for thermal energy storage based on geochemical considerations [Stottlemyre et al., 1980].

Limited water quality studies at the Mobile site during the previous set of experiments (Table 4) indicated that the supply aquifer water was of high quality and undersaturated with most natural occurring minerals such as $CaCO_3$ and SiO_2. It had a very low alkalinity and, consequently, was poorly buffered. The primary effect of heating this water up to 100°C would only be to hasten kinetically the equilibrium reactions.

The storage aquifer water had a significantly greater alkalinity, silica concentration, and hence overall ion content. The calcium concentration was low, suggesting that this water also was undersaturated with respect to $CaCO_3$. An increase in temperature would probably cause some minor dissolution of silica materials and perhaps affect the clay type distribution. The major potential problem due to mixing of supply and storage water was for osmotic swelling of storage formation clays which, as discussed previously, did occur.

Aquifer Thermodynamic Testing

The major thermodynamic quantities which must be measured or at least estimated are the thermal conductivity and heat capacity of the aquifer and confining layers. These quantities are subject to much less natural variation than the hydraulic properties which were discussed previously. Therefore they can normally be estimated or measured in the laboratory by using core samples obtained during construction of the various exploratory and/or test wells.

The specific heats of many common dry rock materials are in the relatively narrow range of 0.19 to 0.22 kcal/kg/°C [Bear, 1972]. Using values for pure materials obtainable from standard tables, one can estimate the effective heat capacity of a water-saturated porous medium on a volumetric basis using the equation

$$C_{va} = n\rho_w C_w + (1-n)\rho_s C_s \tag{33}$$

where C_{va} is the aquifer volumetric heat capacity; ρ_w, ρ_s are the densities of water and solid, respectively; C_w, C_s are the specific heat of water and solid, respectively; and n is the porosity. A porosity in the range of 20% to 60% would yield an effective heat capacity between about 500 and 800 kcal/m^3/°C. Typical porosity ranges for natural materials may be found in the work by Todd [1959]. At the Mobile site with an estimated porosity of 0.33, a volumetric heat capacity of 661 kcal/m^3/°C was calculated.

The thermal conductivity of most saturated, porous, sedimentary materials will fall in the range of 0.75 to 3 kcal/(m hr °C) depending mainly on composition and porosity [Mitchell and Tsung, 1978].

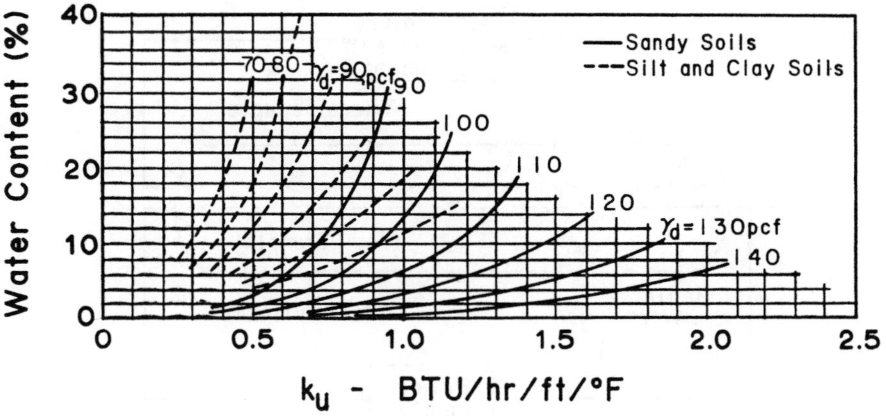

Fig. 15. Thermal conductivity of sandy and silty clay soils as a function of water content and dry unit weight. (For conversion purposes, one BTU/h/ft/°F is equivalent to 1.49 kcal/h/m°C. Also, one lb/ft^3 is equivalent to 16 kg/m^3.) After Mitchell and Tsung [1978].

If either or both of these properties are known, the graph presented by Mitchell and Tsung [1978] and reproduced as Figure 15, can be used to obtain an estimate that may be adequate for many applications. If a particular value cannot be chosen with an acceptable degree of certainty, an alternative is to base calculations on an upper and lower bound.

Several laboratory procedures are available for direct measurement of the thermal conductivity of unconsolidated porous media. Two prominent methods are the thermal needle technique which was studied in some detail by Mitchell and Tsung [1978] and the line-source method which as developed by van der Held and van Drunen [1949] and studied further by Nix et al. [1969].

The line-source method was used to measure the thermal conductivity of the storage aquifer and upper aquitard at the Mobile site. (J. Goodling of the Mechanical Engineering Department at Auburn University supervised the measurements.) Specimens were placed in glass cylinders 20.3 cm long and 5.1 cm in diameter (Figure 16). The heater wire, which runs down the center line of the specimen, was composed of constantan and placed across the

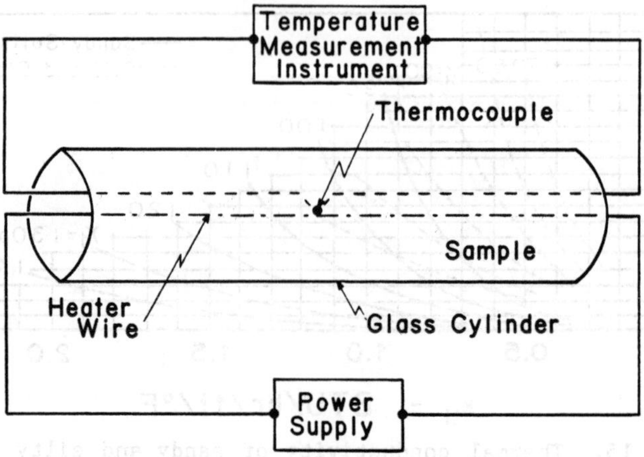

Fig. 16. Schematic diagram of the line-source method for measuring thermal conductivity.

terminals of a direct current power supply. Heater wire temperature as a function of time was measured with an iron-constantan thermocouple placed as shown in Figure 16. This device was calibrated and several runs were made. (Details are available from the authors upon request.) The results indicated an aquifer thermal conductivity of 1.97 ± 0.16 kcal/(m h °C) and an aquitard conductivity of 2.20 ± 0.13 kcal/(m h °C). Using a porosity of 33% and a solids density of 2.6 g/cm^3, the graph in Figure 15 [Mitchell and Tsung, 1978] yields a thermal conductivity of about 1.93 kcal/(m h °C), which is an excellent estimate of our measured aquifer value.

Summary and Conclusions

Fairly extensive testing is required in order to evaluate the potential of an aquifer for thermal energy storage. Important parameters include the regional gradient, vertical and horizontal permeability of the storage aquifer, horizontal dispersivity, vertical permeability of the upper and lower aquitards, thermal conductivities, heat capacities, and chemical characteristics of the aquifer matrix and native groundwater.

At the Mobile site, chemical and thermodynamic tests were performed in the laboratory using core samples and groundwater samples. The chemical analyses indicated that there was a potential for clay particle swelling and loss of permeability in the storage aquifer if relatively high-quality (pure) water was heated and injected. This phenomenon was observed in previous studies when water from a shallow supply aquifer was heated and pumped into the storage aquifer. The problem was eliminated in the present study by obtaining supply water from the storage aquifer itself.

For many purposes, it appears that thermodynamic parameters such as heat capacities and thermal conductivities can be estimated without actually performing measurements. The specific heats of many common dry rock materials are in the relatively narrow range of 0.19 to 0.22 kcal/kg/°C. Therefore effective volumetric heat capacity, which depends on porosity, will usually fall in the range of 500 to 800 kcal/m^3/°C. At the Mobile site a volumetric heat capacity of 661 kcal/m^3/°C was calculated for the storage aquifer for an estimated porosity of 0.33.

The thermal conductivity of most saturated, porous, sedimentary materials will fall in the range of 0.75 to 3 kcal/(m h °C) depending mainly on composition and porosity. If either or both of these properties are known, the graph reproduced as Figure 15, can be used to obtain an estimate that may be adequate for many applications. Measurements made using the line source method indicated an aquifer thermal conductivity of 1.97 $\pm$ 0.16 kcal/(m h °C). Using a porosity of 33% and a solids density of 2.6 g/cm^3, the graph in Figure 15 yields a thermal conductivity of about 1.93 kcal/(m h °C), which is an excellent estimate of the measured value.

Unlike thermodynamic and chemical properties, the determination of hydraulic parameters requires the performance of extensive field testing. A series of new and existing observation wells were used at the Mobile site to conduct pumping tests in which the storage coefficients, the vertical and horizontal permeabilities of the storage aquifer, and the upper and lower confining layer hydraulic diffusivities were determined.

Temporary partially screened observation and pumping wells were installed in the aquifer for the performance of the anisotropy test. The pumped well was screened over a 3.1-m section near the bottom of the 70-ft-thick aquifer and the observation wells, located 7.6, 15.2, and 22.9 m, respectively, from the pumped well, were screened over 3.1-m sections near the top of the aquifer. Water was withdrawn at a constant rate of 818 m^3/d for the test. Drawdown in the observation wells was affected by a boundary about 20 min after pumping began. Consequently, the data analysis was based on early data. The average transmissibility and storage coefficient for the test were 1140 m^2/d and 0.00049, respectively, and the ratio of horizontal to vertical permeability as determined by a method described by Weeks [1969] was 6.71.

Standard pumping tests were performed for pumping rates of 600 and 2125 m^3/d using fully penetrating pumping and observation wells. Analysis of the early drawdown data by the modified nonequilibrium method resulted in values for the transmissibility and the storage coefficient of 1130 m^3/d and 0.00069, respectively, for the low pumping and 1140 m^2/d and 0.00066, respectively, for the high pumping rate. The drawdown deviation from the Theis curve was analyzed to locate a boundary about 150 m from the pumped well.

Partially screened observation wells were located 15 m from the fully penetrating pumped well in the upper and lower aquitards for the leaky aquifer test. This pumping test was performed concurrently with the standard pumping test at the withdrawal rate of 600 m^3/d. The drawdown in the aquitard wells and in a fully penetrating aquifer observation well located 15 m from the pumped well was analyzed by the ratio method of Neuman and Witherspoon [1972]. Values of the ratio of vertical permeability to specific storage were 0.67 and 1.21 for the upper and lower aquitards, respectively.

This series of pumping tests at the Mobile site emphasized the importance of obtaining good early drawdown data for each of the well tests. Leakage or boundary effects can cause drawdown data to deviate from the Theis curve very soon after pumping begins for confined aquifers. The principal data for evaluating the basic hydraulic parameters at the Mobile site were taken from 2 to 15 min after pumping began.

A dispersivity field test was performed during the first injection cycle at the Mobile test site. Sodium bromide was injected at an average rate of 11 mg/l into the supply line to the injection well. The average hot water injection rate was 45.4 m^3/h. Water samples withdrawn from a well located 15 m from the pumped well were analyzed throughout the injection. A method outlined by Gupta et al. [1980] was applied to determine a local, apparent hydrodynamic dispersion coefficient of 9.1 cm at the Mobile site. This value is thought to be unrepresentative of the overall aquifer.

Acknowledgments. This work was made possible by financial support from the U.S. Department of Energy. Support was provided to Auburn University through the Battelle Pacific Northwest Laboratories (contract B-67770-A-0) and to a lesser extent through Oak Ridge National Laboratory (contract 7338). The help of David King, Sam Jones, Ernest Stokes, and James Warman in the performance of several of the tests described herein is gratefully acknowledged.

References

Bear, J., Dynamics of Fluids in Porous Media, Elsevier, New York, 1972.

Bear, J., Hydraulics of Groundwater, McGraw-Hill, New York, 1979.

Bostrom, K., Some pH-controlling redox reactions in natural waters, in Equilibrium Concepts in Natural Water Systems, p. 286, American Chemical Society, Columbus, Ohio, 1967.

Brown, D. L., and W. D. Silvey, Artificial recharge to a freshwater-sensitive brackish-water sand aquifer, Norfolk, Virginia, U.S. Geol. Surv. Prof. Pap., 939, 1977.

Davis, S. N., G. N. Thompson, H. W. Bentley, and S. Gary, Groundwater tracers--A short review, Ground Water, 18, 18-23, 1980.

Ferris, J. G., D. B. Knowles, R. H. Brown, and R. W. Stallman, Theory of aquifer tests, U.S. Geol. Surv. Water Supply Pap., 1536-E, 174 pp., 1962.

Gupta, S. K., R. E. Batta, and R. N. Pandey, Evaluating hydrodynamic dispersion coefficients, J. Hydrol., 47, 369-372, 1980.

Helfferich, F., Ion Exchange, McGraw-Hill, New York, 1962.

Hoopes, J. A., and D. R. F. Harleman, Dispersion in radial flow from a recharge well, J. Geophys. Res., 72, 3595-3607, 1967.

Jacob, C. F., Flow of groundwater, in Engineering Hydraulics, edited by H. Rouse, pp. 321-386, John Wiley, New York, 1950.

Kramer, J. R., Equilibrium models and composition, in Equilibrium Concepts in Natural Water Systems, p. 255, American Chemical Society, Columbus, Ohio, 1967.

Krauskopt, K. B., Introduction to Geochemistry, McGraw-Hill, New York, 1967.

Mathey, B., Development and resorption of a thermal disturbance in a phreatic aquifer with natural convection, J. Hydrol., 34, 315-333, 1977.

Mitchell, J. K., and C. K. Tsung, Measurement of soil thermal resistivity, J. Geotech. Eng. Div. Am. Soc. Civ. Eng., 104(GE10), 1307-1320, 1978.

Molz, F. J., and L. C. Bell, Head gradient control in aquifers used for fluid storage, Water Resour. Res., 13, 795-798, 1977.

Molz, F. J., J. C. Warman, and T. E. Jones, Aquifer storage of heated water, I, A field experiment, Ground Water, 16, 234-241, 1978.

Molz, F. J., A. D. Parr, P. F. Andersen, V. D. Lucido, and J. C. Warman, Thermal energy storage in a confined aquifer: Experimental results, Water Resour. Res., 15, 1509-1514, 1979.

Molz, F. J., A. D. Parr, and P. F. Andersen, Thermal energy storage in a confined aquifer: Second cycle, Water Resour. Res., 17, 641-645, 1981.

Neuman, S. P., and P. A. Witherspoon, Field determination of the hydraulic properties of leaky multiple aquifer systems, Water Resour. Res., 8, 1284-1298, 1972.

Nix, G. H., R. I. Vachon, G. W. Lowery, and T. A. McCurry, The line-source method: Procedure and iteration scheme for combined determination of conductivity and diffusivity, in Proceedings of 8th Conference on Thermal Conductivity, Plenum, New York, 1969.

Papadopulos, S. S., and S. P. Larson, Aquifer storage of heated water, II, Numerical simulation of field results, Ground Water, 16, 242-248, 1978.

Reddell, D. L., R. R. Davison, and W. B. Harris, Cold water aquifer storage, Proceedings of Fourth Annual Thermal Energy Storage Review Meeting, DOE Publ. CONF-791232, U.S. Dep. of Energy, Washington, D.C., 1979.

Stern, L. E., Conceptual design of aquifer thermal energy storage system demonstration, in Proceedings of Mechanical, Magnetic, and Underground Energy Storage 1980 Annual Contractor's Review, pp. 28-33, National Technical Information Service, Springfield, Va., 1980.

Stottlemyre, J. A., Equilibrium geochemical modeling of a seasonal thermal energy storage aquifer field test, Proceedings of Fourth Annual Thermal Energy Storage Review Meeting, DOE Publ. CONF-791232, U.S. Dep. of Energy, Washington, D.C., 1979.

Stottlemyre, J. A., C. H. Cooley, and Gary J. Banik, Physiochemical properties analyses in support of the seasonal thermal energy storage program, in Proceedings of the Mechanical, Magnetic, and Underground Energy Storage 1980 Annual Contractor's Review, pp. 90-95, National Technical Information Service, Springfield, Va., 1980.

Todd, D. K., Ground Water Hydrology, John Wiley, New York, 1959.

Todd, D. K., Ground Water Hydrology, 2nd ed., John Wiley, New York, 1980.

Tsang, C. F., T. Buscheck, and C. Doughty, Aquifer thermal energy storage--A numerical simulation of Auburn University field experiments, Water Resour. Res., 17, 647-658, 1981.

van der Held, E. F. M., and F. G. van Drunen, A method for measuring the thermal conductivity of liquid, Physics, 15, 00-00, 1949.

van Olphen, H., An Introduction to Clay Colloid Chemistry, John Wiley, New York, 1963.

Weeks, E. P., Determining the ratio of horizontal to vertical permeability by aquifer test analysis, Water Resour. Res., 5, 196-214, 1969.

Werner, D., and W. Kley, Problems of heat storage in aquifers, J. Hydrol., 34, 35-43, 1977.

Whitehead, W. R., and E. J. Langhetee, Use of bounding wells to counteract the effects of pre-existing groundwater movement, Water Resour. Res., 14, 273-280, 1978.

Yokoyama, T., H. Umemiya, T. Teraoka, H. Watanabe, K. Katsuragi, and K. Kasahara, Seasonal thermal sttorage in aquifer for utilization, Bull. Jpn. Soc. Mech. Eng., 23, 1646-1654, 1980.

4 MODELING

Marked progress has been made in the last two decades in the application of groundwater hydraulics to the problems of predictively simulating responses of aquifer systems to stress. This progress has been related to the advances in computer technology and the ready availability of this technology to the groundwater hydrologist. The six papers in this chapter cover a variety of subjects related to modeling and reflect the advance capability of the groundwater scientist to apply effectively theoretical aspects in the modeling of complex groundwater problems. These papers cover such diverse subjects as use of programmable calculators and desk top computers to solve the analytical model equations, the difficult problem of groundwater modeling of complex fractured rocks systems, finite element transport modeling of groundwater restoration for in situ solution mining of uranium, etc. The papers in this chapter clearly demonstrate the advanced stage of the state of the art and appropriately reflect the marked progress that the groundwater hydrologist has made in the last two decades in this phase of groundwater science. The papers also give us insight into some of the areas where future progress will have to be made.

Analytical Groundwater Modeling With Programmable Calculators and Hand-Held Computers

William C. Walton
Geraghty & Miller, Inc., Savoy, Illinois 61874

Introduction

Significant progress in modeling using analytical solutions continues, and the inventory of models is impressive. Analytical models simulating the flow of groundwater to and from wells and streams, mass and heat transport in aquifer systems, and land subsidence due to artesian pressure decline have been developed for many aquifer system, well, and stream conditions. Applications of these models to various aquifer conditions is advancing through equivalent section, incremental, and successive approximation techniques.

Programmable calculators and pocket computers are available for rapid, accurate, and inexpensive solution of analytical model equations. Polynomial and other approximations of well functions simplify programming of calculators and computers. Problems involving analytical models with boundaries and multiple-well systems may be solved utilizing the image well theory and the x, y coordinate system capability of calculators and computers. Although not as versatile as numerical digital computer models, analytical models continue to play an important role in groundwater resource evaluation.

Available Models

Available analytical groundwater models are most useful in the analysis of aquifer test data, simplified aquifer system evaluation, and design and verification of numerical digital computer models. Aquifer system parameter data bases largely depend upon the analysis of aquifer test data with analytical models. This situation is not likely to change appreciably in the near future even though automated parameter estimation techniques are advancing at a rapid rate.

Numerical models are more realistic and adaptable than analytical models. However, models should be in tune with the data base and the scale of decisions to be affected by model results. In some cases, basic data and the scale of decisions are not sufficient to warrant a rigorous description of aquifer systems, and analytical models may be more appropriate than numerical models.

Significant progress in analytical modeling continues and the inventory of analytical models is impressive [Walton, 1979]. A large number of analytical models are structured to solve partial differential equations governing groundwater flow, solute transport, heat transport, and land subsidence due to artesian pressure decline.

Analytical models simulating flow to and from wells describe unsteady state time drawdown and distance drawdown in nonleaky, leaky, and water table aquifer systems. Both uniformly porous and fractured rock aquifer system models are available. In some cases, isotropic conditions are assumed to prevail, and in other cases, anisotrophy is taken into consideration. Either negligible aquitard storage or aquitard storage release is assumed with leaky artesian conditions. Isothermal and constant groundwater density and viscosity conditions are assumed to prevail.

Production and injection wells can be modeled as having infinitesimal diameters and no storage capacity or finite diameters and storage capacity. Both fully and partially penetrating aquifer wells and partially penetrating aquitard wells are considered. A few analytical models involving single-boundary or multiboundary aquifer systems have been developed. Flow models based in part on the assumption of infinite areal extent of aquifer systems are modified to cover finite areal extent situations involving hydrogeologic boundaries with the image well theory.

Analytical models describing flow to and from streams in nonleaky artesian and water table systems simulate changes in groundwater levels caused by stream stage changes, groundwater level declines due to stream discharge, groundwater level changes and stream discharge changes caused by uniform or sudden increments of recharge from precipitation on stream drainage basins, groundwater contribution to streamflow, and bank storage. Analytical models are available for estimating the rate and volume of stream depletion by nearby production wells and the cone of depression created by a production well near a stream.

Solute transport analytical models simulate advection and dispersion with various injection and production well conditions. Steady state groundwater flow with or without regional flow components and isothermal conditions are assumed to prevail. Dispersion in one direction is considered dominant. The density and viscosity of the injected solute are assumed to be the same as those of the native groundwater. Solutes of given concentration are introduced into an aquifer at a constant rate or as a slug. With the addition of a retardation factor, conservative solute models may be used to simulate migration of nonconservative solutes. Radioactive contaminant decay can also be simulated. Solute transport models have been developed to simulate advection and dispersion from a stream, advection from a stream to a production well, upconing of salt water below a partially penetrating production well, and saltwater intrusion.

Analytical models simulating heat transport from heated-water injection wells are available. These models describe convection and conduction without dispersion. Steady state groundwater flow without regional flow components is assumed to prevail. Heat conduction in deposits overlying the aquifer and heat convection in the aquifer are assumed to dominate heat flow conditions. The density and viscosity of the injected heated water are assumed to be the same as those of the native groundwater. Heat transport models have been developed to simulate conduction and convection from a stream and conduction and convection through an aquitard.

Analytical models simulating land subsidence due to artesian pressure decline are available. These models are based on the theory of one-dimensional consolidation of a linearly elastic soil, two-dimensional depth effective pressure head increases, and effective pressure head increases which are time dependent in aquitards and clay interbeds.

Simulation Techniques

Recognized departures from ideal conditions do not necessarily dictate that these analytical models be rarely used. Such departures emphasize the need for sound professional judgment in the application of analytical models to existing hydrogeologic conditions and in properly qualifying results according to the extent of departures. With appropriate recognition of hydrogeologic controls, there are many practical ways of circumventing analytical difficulties posed by complicated field conditions. Application of analytical models is advancing through equivalent parameter or section, incremental, and successive approximation techniques.

For example, consider an aquifer consisting of several horizontal layers, each with different thicknesses and permeabilities. The multilayer aquifer can be imitated with a single-layer aquifer model. An equivalent horizontal permeability of the single-layer aquifer model is computed as the sum of the products of individual layer permeabilities and thicknesses divided by the total aquifer thickness.

Analytical models often require straight-line boundary demarcations and uniform width, length, and thickness. The variability of the areal extent of an aquifer can be converted to an equivalent uniform area to meet this requirement. In addition, hydrogeologic boundaries of aquifers can be idealized to fit comparatively elementary geometric forms such as wedges and infinite or semi-infinite rectilinear strips.

Nonhomogeneous conditions can be simulated by varying analytical model parameters incrementally with time. Suppose a cone of depression encounters changing transmissivities as it expands. Keeping tract of the effective radius of the cone of depression with time, one transmissivity can be used to determine drawdown at the end of the first time step and another transmissivity can be used to determine the change in drawdown between the first and second time steps, etc.

Analytical models for aquifer test analysis generally assume that wells completely penetrate the aquifer. In cases where wells only partially penetrate the aquifer, observed drawdowns must be corrected for partial penetration effects before they are used to estimate aquifer parameters. However, the partial penetration correction depends upon the vertical permeability-horizontal permeability ratio and, in addition, upon transmissivity. Both the ratio and transmissivity are unknown prior to the analysis of drawdown data. Thus successive approximations must be employed in correcting observed drawdowns for the effects of partial penetration. The procedure is to estimate initially the ratio and transmissivity based on an analysis of observed drawdown data. Then values of partial penetration corrections are computed. Observed drawdowns are then corrected for partial penetration effects. Corrected values of drawdown are analyzed to determine the ratio and transmissivity. These values are compared with initially assumed values, and if the two values are the same, the solution is declared valid. Otherwise, the process is repeated until the values of the ratio and transmissivity used in adjusting observed drawdowns are the same as the values computed with corrected drawdowns.

The equivalent parameter or section, incremental, and successive approximation techniques described above often can be applied to complicated field conditions with minor sacrifice in accuracy of analysis of some problems.

Programmable Calculators and Pocket Computers

Programmable calculators and pocket computers are available for rapid, accurate, and inexpensive solution of analytical groundwater equations [see Warner and Yow, 1979, 1980]. They replace the need for volumes of table and graphs. Polynomial fits and other approximations of functions available from the National Bureau of Standards [1964] handbook or generated with integral-estimating techniques such as Gauss-Legendre three-point quadrature simplify programming of calculators and computers. Problems involving analytical models with boundaries and multiple wells may be solved utilizing x, y coordinate systems. An optional printing unit is available. Instruction manuals are provided by manufacturers of calculators or computers which make programming and program execution relatively simple. Magnetic cards or tapes are available to retain programs for repeated use.

Polynomial fits or other approximations for several frequently used functions in analytical groundwater equations are as follows:

<u>W(u)</u>

When $0 < u \leq 1$,

$$W(u) = -\ln u + a_0 + a_1 u + a_2 u^2 + a_3 u^3 + a_4 u^4 + a_5 u^5 + \varepsilon(u)$$

$$|\varepsilon(u)| < 2 \times 10^{-7}$$

$a_0 = -0.57721566$ $\qquad$ $a_3 = 0.05519968$

$a_1 = 0.99999193$ $\qquad$ $a_4 = -0.00976004$

$a_2 = -0.24991055$ $\qquad$ $a_5 = 0.00107857$

When $1 \leq u < \infty$

$$W(u) = \left[\frac{u^4 + a_1 u^3 + a_2 u^2 + a_3 u + a_4}{u^4 + b_1 u^3 + b_2 u^2 + b_3 u + b_4}\right] \Big/ u \exp(u) + \varepsilon(u)$$

$$|\varepsilon(u)| < 2 \times 10^{-8}$$

$a_1 = 8.5733287401$ $b_1 = 9.5733223454$

$a_2 = 18.0590169730$ $b_2 = 25.6329561486$

$a_3 = 8.6347608925$ $b_3 = 21.0996530827$

$a_4 = 0.2677737343$ $b_4 = 3.9584969228$

W[u(r/B)] [Hantush and Jacob, 1955]

When $u > 1$ and $(r/B) \leq 2$

$$W(u,\tfrac{r}{B}) = I_o(\tfrac{r}{B})\, E_1(u) - \exp(-u) \left[0.5772 + \ln(\frac{r^2}{4B^2u}) + E_1(\frac{r^2}{4B^2u}) - (\frac{r^2}{4B^2u}) + \left(\frac{I_o(r/B) - 1}{u}\right)\right] + \exp(-u) \sum_{n=1}^{6} \sum_{m=1}^{n} \frac{(-1)^{n+m}(n-m+1)!}{(n+2)!^2 u^{n-m}} (\frac{r^2}{4B^2})^n$$

When $u \leq 1$ and $\frac{r}{B} \leq 2$

$$W(u, \tfrac{r}{B}) = 2K_o(\tfrac{r}{B}) - \left\{I_o(\tfrac{r}{B})\, E_1(\frac{r^2}{4B^2u}) - \exp(\frac{r^2}{4B^2u}) \left[0.5772 + \ln(u) + E_1(u) - u + \frac{I_o(r/B) - 1}{r^2/4B^2u}\right] + u \sum_{n=1}^{6} \sum_{m=1}^{n} \frac{(-1)^{n+m}(n-m+1)!}{(n+2)!^2 u^{m-n}} (\frac{r^2}{4B^2})^m\right\}$$

$$[W(\tfrac{r}{m}\sqrt{P_v/P_h}, \tfrac{1}{m}, \tfrac{d}{m}, \tfrac{y}{m})] \quad [W(\tfrac{r}{m}\sqrt{P_v/P_h}, \tfrac{1}{m}, \tfrac{d}{m}, \tfrac{y}{m})]^{-1} = [4m/\pi(1-d)] \sum_{n=1}^{\infty} (1/n)\,[\sin(n\pi 1/m) - \sin(n\pi d/m)]\, \cos(n\pi y/m)\, K_o\,[(n\pi r/m)\,(P_v/P_h)^{\frac{1}{2}}]$$

$K_o(x)$

When $0 < x \leq 2$

$$K_o(x) = -\ln(x/2) I_o(x) - 0.57721566 + 0.42278420\,(x/2)^2 + 0.23069756\,(x/2)^4 + 0.03488590\,(x/2)^6 + 0.00262698\,(x/2)^8 + 0.00010750\,(x/2)^{10} + 0.00000740\,(x/2)^{12} + \varepsilon(x) \quad |\varepsilon(x)| < 1 \times 10^{-8}$$

when $2 \leq x < \infty$

$$K_0(x) = [1.25331414 - 0.07832358\ (2/x) + 0.02189568\ (2/x)^2 - 0.01062446\ (2/x)^3 + 0.00587872\ (2/x)^4 - 0.00251540\ (2/x)^5 + 0.00053208\ (2/x)^6]/\ x^{\frac{1}{2}}\ \exp(x) + \varepsilon(x)\ |\varepsilon(x)| < 1.9 \times 10^{-7}$$

$\underline{I_0(x)}$

when $-3.75 \leq X \leq 3.75$

$$I_0(x) = 1 + 3.5156229\ (x/3.75)^2 + 3.0899424\ (x/3.75)^4 + 1.2067492\ (x/3.75)^6 + 0.2659732\ (x/3.75)^8 + 0.0360768\ (x/3.75)^{10} + 0.0045813\ (x/3.75)^{12} + \varepsilon(x)\ |\varepsilon(x)| < 1.6\mathrm{x}10^{-7}$$

$\underline{\mathrm{erf}(x)}$

$$\mathrm{erf}(x) = 1 - 1/(1 + a_1\ x + a_2\ x^2 + \ldots a_6\ x^6)^{16} + \varepsilon(x)\ |\varepsilon(x)| \leq 3\mathrm{x}10^{-7}$$

$a_1 = 0.0705230784$ $a_4 = 0.0001520143$

$a_2 = 0.0422820123$ $a_5 = 0.0002765672$

$a_3 = 0.0092705272$ $a_6 = 0.0000430638$

Codes

The Texas Instruments programmable TI-59 calculator has up to 100 registers available for data storage and up to 960 registers for program storage. Codes can be developed for the TI-59 calculator and the following analytical groundwater models: (1) infinite nonleaky artesian aquifer with single well or multiple wells, (2) infinite leaky artesian aquifer with single well or multiple wells, (3) two mutually leaky artesian infinite aquifers with single well, (4) infinite nonleaky artesian aquifer with multiple wells and specified drawdown, (5) infinite nonleaky artesian aquifer

TABLE 1. User Instructions: TI-59

Step	Procedure	Enter	Press	Display
1	Read side 1 of card	1		1
2	Read side 2 of card	2		2
3	Enter distance from production well, ft	r	STO 01	r
4	Enter storage coefficient	S	STO 02	S
5	Enter transmissivity, gpd/ft	T	STO 03	T
6	Enter time after discharge started, min	t	STO 04	t
7	Enter production well discharge, gpm	Q	STO 05	Q
8	Execute program		RST R/S	s

with multiple finite line sinks, (6) nonleaky artesian aquifer with parallel boundaries, (7) infinite nonleaky artesian aquifer with partially penetrating well, (8) infinite nonleaky artesian aquifer with advective mass transport, (9) nonleaky artesian aquifer with advective mass transport from stream to nearby production well, (10) infinite nonleaky artesian aquifer with upconing of salt water beneath a production well, (11) nonleaky artesian coastal aquifer with a production well near a saltwater front, and (12) infinite nonleaky artesian aquifer with conductive and convective heat transport.

The infinite nonleaky artesian aquifer with multiple-wells model when used in context with the image well theory simulates finite areal extent situations. Boundaries are replaced by imaginary wells which produce the same disturbing effects as the boundaries.

As an example, the program description, user instructions, and code for the infinite nonleaky artesian aquifer with single-well model and the gallon-day-foot system of units are presented in Tables 1 and 2.

Program Description

Wells fully penetrate an artesian aquifer overlain and underlain by aquicludes. The nonleaky aquifer is homogeneous, isotropic,

TABLE 2. Code: TI-59

Loc	Code	Key	Loc	Code	Key	Loc	Code	Key
000	02	2	047	85	+	094	07	7
001	06	6	048	93	.	095	06	6
002	09	9	049	09	9	096	00	0
003	03	3	050	09	9	097	00	0
004	65	X	051	09	9	098	04	4
005	43	RCL	052	09	9	099	65	X
006	01	01	053	09	9	100	43	RCL
007	33	X^2	054	01	1	101	06	06
008	65	X	055	09	9	102	45	Y^X
009	43	RCL	056	03	3	103	04	4
010	02	02	057	65	X	104	85	+
011	55	÷	058	43	RCL	105	93	.
012	53	(	059	06	06	106	00	0
013	43	RCL	060	75	−	107	00	0
014	03	03	061	93	.	108	01	1
015	65	X	062	02	2	109	00	0
016	43	RCL	063	04	4	110	07	7
017	04	04	064	09	9	111	08	8
018	54	)	065	09	9	112	05	5
019	95	=	066	01	1	113	07	7
020	42	STO	067	00	0	114	65	X
021	06	06	068	05	5	115	43	RCL
022	01	1	069	05	5	116	06	06
023	32	X⇄T	070	65	X	117	45	Y^X
024	43	RCL	071	43	RCL	118	05	5
025	06	06	072	06	06	119	95	=
026	22	INV	073	33	X^2	120	42	STO
027	77	GE	074	85	+	121	07	07
028	33	X^2	075	93	.	122	61	GTO
029	61	GTO	076	00	0	123	24	CE
030	34	$\sqrt{X}$	077	05	5	124	76	LBL
031	76	LBL	078	05	5	125	34	$\sqrt{X}$
032	33	X^2	079	01	1	126	53	(
033	43	RCL	080	09	9	127	43	RCL
034	06	06	081	09	9	128	06	06
035	23	LNX	082	06	6	129	45	Y^X
036	94	+/-	083	08	8	130	04	4
037	75	−	084	65	X	131	85	+
038	93	.	085	43	RCL	132	08	8
039	05	5	086	06	06	133	93	.
040	07	7	087	45	Y^X	134	05	5
041	07	7	088	03	3	135	07	7
042	02	2	089	75	−	136	03	3
043	01	1	090	93	.	137	03	3
044	05	5	091	00	0	138	02	2
045	06	6	092	00	0	139	08	8
046	06	6	093	09	9	140	07	7

TABLE 2. (continued)

Loc	Code	Key	Loc	Code	Key	Loc	Code	Key
141	04	4	187	07	7	233	06	6
142	00	0	188	07	7	234	65	X
143	01	1	189	07	7	235	43	RCL
144	65	X	190	03	3	236	06	06
145	43	RCL	191	07	7	237	33	X^2
146	06	06	192	03	3	238	85	+
147	45	Y^X	193	04	4	239	02	2
148	03	3	194	03	3	240	01	1
149	85	+	195	95	=	241	93	.
150	01	1	196	55	÷	242	00	0
151	08	8	197	53	(	243	09	9
152	93	.	198	43	RCL	244	09	9
153	00	0	199	06	06	245	06	6
154	05	5	200	45	Y^X	246	05	5
155	09	9	201	04	4	247	03	3
156	00	0	202	85	+	248	00	0
157	01	1	203	09	9	249	08	8
158	06	6	204	93	.	250	02	2
159	09	9	205	05	5	251	07	7
160	07	7	206	07	7	252	65	X
161	03	3	207	03	3	253	43	RCL
162	00	0	208	03	3	254	06	06
163	65	X	209	02	2	255	85	+
164	43	RCL	210	02	2	256	03	3
165	06	06	211	03	3	257	93	.
166	33	X^2	212	04	4	258	09	9
167	85	+	213	05	5	259	05	5
168	08	8	214	04	4	260	08	8
169	93	.	215	65	X	261	04	4
170	06	6	216	43	RCL	262	09	9
171	03	3	217	06	06	263	06	6
172	04	4	218	45	Y^X	264	09	9
173	07	7	219	03	3	265	02	2
174	06	6	220	85	+	266	02	2
175	00	0	221	02	2	267	08	8
176	08	8	222	05	5	268	54	)
177	09	9	223	93	.	269	54	)
178	02	2	224	06	6	270	55	÷
179	05	5	225	03	3	271	53	(
180	65	X	226	02	2	272	43	RCL
181	43	RCL	227	09	9	273	06	06
182	06	06	228	05	5	274	65	X
183	85	+	229	06	6	275	43	RCL
184	93	.	230	01	1	276	06	06
185	02	2	231	04	4	277	22	INV
186	06	6	232	08	8	278	23	LNX

TABLE 2. (continued)

Loc	Code	Key	Loc	Code	Key	Loc	Code	Key
279	54	)	287	04	4	294	43	RCL
280	95	=	288	93	.	295	07	07
281	42	STO	289	06	6	296	55	÷
282	07	07	290	65	X	297	43	RCL
283	76	LBL	291	43	RCL	298	03	03
284	24	CE	292	05	05	299	95	=
285	01	1	293	65	X	300	91	R/S
286	01	1						

infinite in areal extent, and constant in thickness throughout. Water is released instantaneously from storage by the compaction of the aquifer and its associated beds and by the expansion of the water itself.

The equation being solved is [see Theis, 1935]

$$s = Q\ W(u)/4\pi T$$

where

$$u = r^2 S/4Tt$$

See Tables 1 and 2.

Codes can also be developed with BASIC language for the Radio Shack TRS-80 pocket computer. This computer has a program capacity of 1424 steps and 26 memories with memory safeguard. As an example,

TABLE 3. User Instructions: TRS-80

Step	Procedure	Enter	Press	Printer Display
1	Load program file			
2	Enter run mode	Run	<Enter>	Statement 10
3	Prompt response	r	<Enter>	Radius (ft) =
4	Prompt response	T	<Enter>	T (gpd/ft) =
5	Prompt response	S	<Enter>	Storage Coef. =
6	Prompt response	Q	<Enter>	Q (gpm) =
7	Prompt response	t	<Enter>	Time (min) =
				u =
				W (u) =
				Drawdown (ft) =

TABLE 4. Code: TRS-80

Statement Number	Statement	Statement Number	Statement
10	BEEP 3: PRINT "NON-LEAKY ARTESIAN, 1 WELL":BEEP 1	80	PRINT"U=";USING"##.####^";U
		90	A=U^2:B=U^3:C=U^4: D=U^5
20	INPUT "RADIUS(FT)="; R:BEEP 1	100	IF U>IGOTO 130
25	PRINT "RADIUS(FT)="; USING"##.####^";R	110	W=-LN U- .57721566 +.99999193*U- .24991055*A +.05519968*B- .00976004*C +.00107857*D
30	INPUT "T(GPD/FT)=";T: BEEP 1	120	GOTO 160
35	PRINT "T(GPD/FT)="; USING"##.####^";T	130	M=C+8.5733287401*B +18.0590169730*A +8.6347608925*U +.2677737343
40	INPUT "STORAGE COEF. = ";S:BEEP 1		
		140	N=M/(C+9.5733223454*B +25.6329561486*A +21.096530827*U +3.9584969228)
45	PRINT"STORAGE COEF.="; USING"##.####^";S		
50	INPUT"PUMPING RATE(GPM)="; Q:BEEP 1	150	W=N(U*EXP U)
55	PRINT"PUMPING RATE(GPM)="; USING"##.####^";Q	160	BEEP 2:PRINT"W(U)="; USING"##.#####^";W
60	INPUT"TIME(MIN)=";Z:BEEP 1	170	H=114.6*Q*W/T:BEEP 2
65	PRINT"TIME(MIN)=";USING" ##.####^";Z	180	PRINT "DRAWDOWN(FT)="; USING"##.####^";H
70	U=2693*S*R^2/(T*Z):BEEP 2		

the user instructions and code for the nonleaky artesian infinite aquifer with single-well model and the gallon-day-foot system of units are presented in Tables 3 and 4.

Conclusions

In conclusion, analytical modeling is no longer a laborious exercise thanks to programmable calculators and pocket computers.

These devices are important groundwater resource evaluation tools that can effectively bridge the gap between hand computations and sophisticated computer programs.

Notation

d	depth from top of aquifer to unscreened or open portion of production well, L.
$E_1(x)$	exponential integral, dimensionless.
$erf(x)$	error function.
$I_0(x)$	modified bessel function of the first kind and order zero, dimensionless.
$K_0(x)$	modified bessel function of the second kind and order zero, dimensionless.
l	depth from top of aquifer to bottom of production well, L.
P_h	aquifer horizontal permeability, L/T.
P_v	aquifer vertical permeability, L/T.
Q	production well discharge, L^3/T.
r	distance from production well, L.
s	drawdown, L.
S	aquifer storage coefficient, dimensionless.
t	time after pumping started, T.
T	aquifer transmissivity, L^2/T.
$W(u)$	well function for infinite nonleaky artesian aquifer, dimensionless.
$W[u,(r/B)]$	well function for infinite leaky artesian aquifer, dimensionless.
y	depth from top of aquifer to bottom of observation well, L.

References

Hantush, M. S., and C. E. Jacob, Nonsteady radial flow in an infinite leaky aquifer, Eos Trans. AGU, 36(1), 95-100, 1955.

National Bureau of Standards, Handbook of Mathematical Functions With Formulas, Graphs, and Mathematical Tables, Appl. Math. Ser., vol. 55, U.S. Government Printing Office, Washington, D.C., 1964.

Theis, C. V., The relation between the lowering of piezometric surface and the rate and duration of discharge of a well using ground-water storage, Eos Trans. AGU, 16, 519-524, 1935.

Walton, W. C., Progress in analytical groundwater modeling, J. Hydrol., 43, 149-159, 1979.

Warner, D. L., and M. G. Yow, Programmable hand calculator programs for pumping and injection wells, I, Constant or variable pumping rate, single or multiple fully penetrating wells, Ground Water, 17(6), 532-537, 1979.

Warner, D. L., and M. G. Yow, Programmable hand calculator programs for pumping and injection wells, II, Constant pumping (injection) rate, single fully penetrating well, semiconfined aquifer, Ground Water, 18(2), 126-133, 1980.

Numerical Treatment of Leaky Aquifers in the Short-Time Range

Benito Chen and Ismael Herrera
Instituto de Investigaciones en Matematicas Aplicadas y en Sistemas
Universidad Nacional Autonoma de Mexico A.P.20-726,
Mexico City, Mexico

Introduction

Mathematically, leaky aquifers are characterized by the assumption of vertical flow in the aquitards, which is well established for most cases of practical interest [Neuman and Witherspoon, 1969a, b]. There are two main approaches for the numerical modeling of systems of leaky aquifers: one treats the basic equations in a direct manner without any further development [Chorley and Frind, 1978], and one applies a transformation to obtain an equivalent system of integrodifferential equations [Herrera, 1976; Herrera and Rodarte, 1973; Herrera and Yates, 1977; Hennart et al., 1981].

The latter procedure offers considerable advantages [Herrera et al., 1980], both from the point of view of computing time and capacity required and from the point of view analysis. In the first approach, the aquitard must be discretized, while in the integro-differential one the evolution of the aquitards is obtained by means of a series expansion [Herrera and Yates, 1977]. The accuracy in the first procedure depends on the number and distribution of the nodes used in the aquitard, while in the second one it depends on the number of terms used in the series expansion. Due to this fact, it is easier and more economic to control the errors and achieve a desired accuracy.

Specially delicate from this point of view is the treatment of leaky aquifers in the short-time range. This corresponds approximately to that defined by Hantush [1960], although it is more precise to say that the short-time range is that on which the aquitard can be approximated by a layer of infinite thickness [Herrera and Rodarte, 1973]. When the latter point of view is adopted, the definition of the short-time range depends on the error that one is willing to accept. For example, if the admissible error for the approximation of the aquitard behavior is 10%, then the short-time range is $t'<0.27$, where t' is the dimensionless time (see notation section).

We must warn that the use of the term 'short-time range' may be misleading because it refers to dimensionless time and not to physical time. For a layer of clay, for example, in the Mexico City area (Texcoco Lake) [Herrera et al., 1974], having a thickness of 32 m, specific storage of 5.2×10^{-2} m^{-1} and permeability of 0.47×10^{-3} m/d, short-time range means any time before the first 83.8 years of operation, i.e., the whole life span of the well field.

The numerical modeling of the short-time range is difficult. One can understand the nature of such difficulties by looking more carefully into the actual physical situation. When pumping starts in the main aquifer (Figure 1), the effects are first manifested there and then slowly propagate into the aquitard. The short-time range corresponds to that period of time at which the part of the aquitard that is responding to pumping is much smaller than the total thickness of the aquitard. If a uniform mesh is used on the whole aquitard, it would be required to have a very refined one; thus too many nodes would be introduced.

When the integrodifferential equation approach is used, these facts are reflected in the number of terms that are required in the series expansion of the memory functions. This article is devoted to presenting a procedure that avoids, to a large extent, these difficulties.

The method is based on the observation that in the short-time range the response of the aquitard is approximately the same as if the thickness of the aquitard is infinity, i.e., it does not depend on the actual thickness of the aquitard. In the integrodifferential approach, this gives rise to a one-parameter family of possible representations for the memory function, and in this paper a procedure

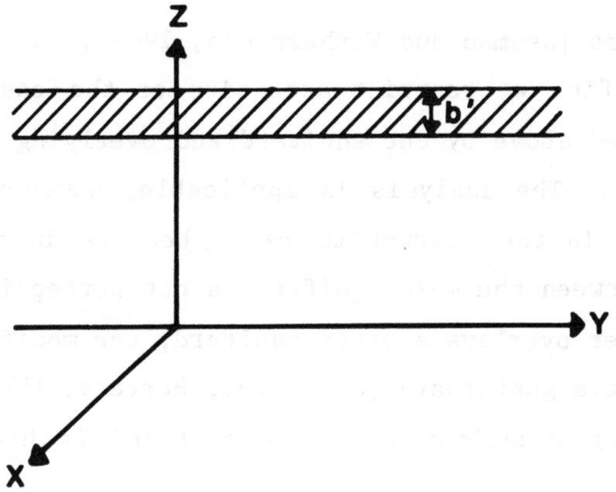

Fig. 1. A single leaky aquifer overlaid by one aquitard.

is given to adjust that parameter in a manner that the number of terms, in the series expansion, is decreased to a minimum. In addition, a criterium that permits defining in advance the number of terms required to obtain a given level of accuracy is supplied. Examples of application to actual field situations are also given.

It seems worthwhile to recall that the same basic idea can also be used to simplify the numerical treatment of leaky aquifers in the short-time range, when the aquitard is discretized. The procedure for this case, however, is not discussed here in detail.

Examples of application to actual field situations are also given. The results obtained for them exhibit substantial reductions in the number of terms required for the series expansions. At the worst, this number is 12, but it would have been 128 if the transformation had not been used.

Preliminary Notions

Leaky aquifers are characterized by the fact that the permeabilities of the semipervious layers (aquitards) separating the main aquifers (Figure 1) are very small. When the contrast of permeabilities between the main aquifers and the aquitards is one order of magnitude or more, the flow can be taken as vertical in the aquitards [Neuman and Witherspoon, 1969a,b].

To be specific, we restrict attention to the case of one main aquifer, limited above by one aquitard and overlying an impermeable bed (Figure 1). The analysis is applicable, however, to a multi-aquifer system in the short-time range because in this range the interaction between the main aquifers is not perceptible. Also, if the main aquifer overlays another aquitard, the modification of the arguments is straightforward [see, e.g., Herrera, 1970].

The governing equations when the aquitard is homogeneous, are

$$\frac{\partial}{\partial x}(T\frac{\partial s}{\partial x}) + \frac{\partial}{\partial y}(T\frac{\partial s}{\partial y}) + K'\frac{\partial s}{\partial z}(x,y,0,t) + Q = S\frac{\partial s}{\partial t} \tag{1}$$

in the aquifer and

$$\frac{\partial^2 s'}{\partial z^2} = \frac{1}{\alpha'}\frac{\partial s'}{\partial t} \tag{2}$$

in the semipervious bed. If the system is in hydrostatic equilibrium initially,

$$s(x,y,0) = s'(x,y,z,0) = 0 \tag{3}$$

The boundary conditions for s' are

$$s'(x,y,b',t) = 0 \tag{4a}$$

$$s'(x,y,0,t) = s(x,y,t) \tag{4b}$$

The system of (1) to (4) is equivalent [Herrera and Rodarte, 1973] to the integrodifferential equation

$$\frac{\partial}{\partial x}(T\frac{\partial s}{\partial x}) + \frac{\partial}{\partial y}(T\frac{\partial s}{\partial y}) - \frac{K'}{b'}\int_0^t \frac{\partial s}{\partial t}(x,y,t-\tau)f(\alpha'\tau/b'^2)\partial\tau \tag{5}$$

$$+ Q = S\frac{\partial s}{\partial t}$$

subject to

$$s(x,y,0) = 0 \tag{6}$$

When these equations are supplemented by the usual boundary conditions in the horizontal boundaries, one can obtain the drawdown s in the main aquifer without computing the drawdown s' in the aquitard. When this is required, however, it can be easily derived from the drawdown s [Herrera and Rodarte, 1973; Herrera and Yates, 1977].

The memory function f(t') has two alternative representations [Herrera and Rodarte, 1973]:

$$f(t') = 1 + 2 \sum_{n=1}^{\infty} e^{-n^2\pi^2 t'} = \frac{1}{(\pi t')^{1/2}} \left(1 + 2 \sum_{n=1}^{\infty} e^{-n^2/t'}\right) \tag{7}$$

It can be used to define an apparent storage coefficient $S_a(t)$, which for a leaky aquifer changes with time. Consider (5) without horizontal flow; it is

$$S\frac{\partial s}{\partial t} + \frac{K'}{b'}\int_0^t \frac{\partial s}{\partial t}(t-\tau)\ f\ (\alpha'\tau/b'^2)\ d\tau = Q \tag{8}$$

Let the drawdown s(t) be a unit step function, i.e., it is initially zero and suddenly rises to be 1, keeping it fixed at this value. Then,

$$\frac{\partial s}{\partial t}(t) = \delta(t) \tag{9}$$

where $\delta(t)$ is Dirac's delta function. By substitution of (9) into (8), one gets

$$Q(t) = S\ \delta(t) + \frac{K'}{b'}\ f\ (\alpha' t/b'^2) \tag{10}$$

and the total yield given by the system aquifer aquitard is

$$\int_0^t Q(\tau)d\tau = S + \frac{K'}{b'}\int_0^t f(\alpha'\tau/b'^2)d\tau = S + S'F(\alpha' t/b'^2) \tag{11}$$

where F is defined by

$$F(t') = \int_0^{t'} f(\tau)d\tau \tag{12}$$

It is natural to define the apparent storage coefficient S_a of the system aquifer-aquitard as the total volume of water per unit area yielded by the system under a unit decline of piezometric head. Hence

$$S_a(t) = \int_0^t Q(\tau)d\tau = S + S'F(\alpha' t/b'^2) \tag{13}$$

which is an increasing function of the time elapsed since the application of the unit decline of piezometric head. If there is no aquitard, $S' = 0$ and $S_a(t) = S$, which is independent of time and exhibits the consistency of our definition.

Since $f(t')$ is defined by an infinite series, it is desirable to obtain a good approximation to $F(t')$ using only a small number of terms. We consider two obvious approximations.

First, truncate

$$F(t') - t' + 2 \sum_{n=1}^{\infty} \int_0^{t'} e^{-n^2\pi^2\tau} d\tau \tag{14}$$

to N terms

$$F_N(t') = t' + 2 \sum_{n=1}^{N} \int_0^{t'} e^{-n^2\pi^2\tau} d\tau = F(t') - 2 \sum_{n=N+1}^{\infty} \int_0^{t'} e^{-n^2\pi\tau} d\tau \tag{15}$$

The second one [Herrera and Yates, 1977] is to integrate (14) first, sum exactly one of the resulting series, and truncate the other one:

$$\begin{aligned} F(t') &= t' - \frac{2}{\pi'} \left(\sum_{n=1}^{\infty} \frac{e^{-n^2\pi^2 t'}}{n^2} - \sum_{n=1}^{\infty} \frac{1}{n^2} \right) \\ F(t') &= t' + \frac{1}{3} - \frac{2}{\pi^2} \sum_{n=1}^{\infty} \frac{e^{-n^2\pi^2 t'}}{n^2} \end{aligned} \tag{16}$$

Define then

$$F_N(t') = t' + \frac{1}{3} - \frac{2}{\pi^2} \sum_{n=1}^{N} \frac{e^{-n^2\pi^2t'}}{n^2} = t' + A_N + \sum_{n=1}^{N} \frac{e^{-n^2\pi^2t'}}{n^2} \quad (17)$$

with

$$A_N = F_N(0) = \frac{1}{3} - \frac{2}{\pi^2} \sum_{n=1}^{\infty} \frac{1}{n^2} = \frac{2}{\pi^2} \sum_{n=N+1}^{\infty} \frac{1}{n^2} \quad (18)$$

Equation (17) can be written alternatively as

$$F_N(t') = F(t') - \frac{2}{\pi^2} \sum_{n=N+1}^{\infty} \frac{e^{-n^2\pi^2t'}}{n^2} = F(t') + 2 \sum_{n=N+1}^{\infty} \int_{t'}^{\infty} \quad (19)$$

$$e^{-n^2\pi^2\tau} \, d\tau$$

Of the two approximations (15) and (17), the second one yields better results. First, for any N, it preserves total yield [see Herrera and Yates, 1977]. Second, we are only truncating $\sum_{n=1}^{\infty} (e^{-n^2\pi^2t'}/n^2)$ which converges very fast for positive t'. In (15) the series $2/\pi^2 \sum_{n=N+1}^{\infty} 1/n^2$ dominates the error, in contrast to $2/\pi^2 \sum_{n=N+1}^{\infty} (e^{-n^2\pi^2t'})/n^2$, which is the error in (17). Third, when we integrate the problem numerically, the largest error is in the first time step $\Delta t'$, $2 \sum_{n=N+1}^{\infty} \int_{\Delta t'}^{\infty} e^{-n^2\pi^2\tau} d\tau$, and it diminishes rapidly for further time steps. So it is very easy to control the error for any time t'. The same is not true for (15), where the error increases with time.

Approximation of the Apparent Storage Coefficient S_a

From (7), it follows that for every $\lambda > 0$,

$$f(\frac{t'}{\lambda^2}) = \frac{\lambda}{(\pi t')^{1/2}} (1 + 2 \sum_{n=1}^{\infty} e^{\frac{-(n\lambda)^2}{t'}}) = 1 + 2 \sum_{n=1}^{\infty} e^{-n^2\pi^2t'/\lambda^2} \quad (20)$$

Furthermore, define

$$d(t') = 1 + 2 \sum_{n=1}^{\infty} e^{-n^2/t'} \quad (21)$$

Then

$$\frac{(f(t')}{f(t'/\lambda^2)} = \frac{1}{\lambda} \quad \frac{(d(t')}{d(t'/\lambda^2)} \tag{22}$$

Let t" be

$$t'' = (S')^2 t' \tag{23}$$

Then t" is nondimensional because the storage coefficient of the aquitard $S' = S_s'b'$ and t' are nondimensional. Therefore if λ is replaced by λ^2/S', it is obtained:

$$f(t') = \frac{S'}{\lambda} \frac{d(t''/S'^2)}{d(t''/\lambda^2)} f(t''/\lambda^2) \tag{24}$$

Define the parameter θ by

$$\theta = (S'/\lambda)^2 \tag{25}$$

and $\delta(\theta,t')$ by

$$\delta(\theta,t') \equiv \frac{d(t')}{d(\theta t')} - 1 \tag{26}$$

Using this notation, equation (24) becomes

$$f(t') = \sqrt{\theta} \; [1 + \delta(\theta,t')] \quad f(\theta t') \tag{27}$$

Hence

$$F(t') = \theta^{-1/2} F(\theta t') + \Delta(\theta,t') \tag{28}$$

where

$$\Delta(\theta,t) = \sqrt{\theta}\int_0^{t'} \delta(\theta,\tau) f(\theta\tau)d\tau \tag{29}$$

A bound for Δ can be given when a bound for δ is known. Indeed, if

$$|\delta(\theta,\tau)| \leq \varepsilon_1 \text{ for } 0 \leq \tau \leq t' \tag{30}$$

then

$$|(\Delta,t')| \leq \theta^{-1/2} \varepsilon_1 F(\theta t') \tag{31}$$

Suppose that a numerical model of a leaky aquifer is going to be implemented in the range $0 \leq t' \leq t'_{max}$. Notice that (21) and (26) imply, when $\theta > 1$, that

$$|\delta(\theta,t')| = 1 - \frac{d(t')}{d(\theta t')} < 1 - \frac{d(t'_{max})}{d(\theta t'_{max})} \tag{32}$$

because $d(t')/d(\theta t')$ is a decreasing function of t' (if $\theta > 0$). Recalling that

$$1 - \frac{d(t')}{d(\theta t')} = -2 \sum_{n=1}^{\infty} e^{-n^2/t'} - \sum_{m=1}^{\infty} (-2 \sum_{n=1}^{\infty} e^{-n^2/\theta t'})^m$$
$$- 2 \sum_{n=1}^{\infty} e^{-n^2/t'} \sum_{m=1}^{\infty} (-2 \sum_{n=1}^{\infty} e^{-n^2/\theta t'})^m \tag{33}$$

it can be seen that

$$|\delta(\theta,t')| \leq 2 \sum_{n=1}^{\infty} e^{-\frac{n^2}{\theta t'_{max}}} \tag{34}$$

Choosing

$$\theta = t'_c/t'_{max} > 1 \tag{35}$$

where $t'_c > t'_{max}$, and defining

$$\varepsilon_1 = 2 \sum_{n=1}^{\infty} e^{-\frac{n}{t'_c}} \tag{36}$$

then

$$|\delta(\theta,\tau)| \leq \varepsilon_1 \tag{37}$$

whenever $0 < \tau < t'_{max}$.

Define the approximation

$$F_a(t') = \theta^{-1/2} F_N(\theta t') \tag{38}$$

where F_N is given by (19):

$$F_N(t') = F(t') + \frac{2}{\pi^2} \sum_{N+1}^{\infty} \frac{e^{-n^2\pi^2 t'}}{n^2} = F(t') + 2 \sum_{N+1}^{\infty} \int_{t'}^{\infty} e^{-n^2\pi^2\tau}\, d\tau \tag{39}$$

We recall that $F_N(t')$ can be recognized as the approximation that has been used in the numerical implementation of the integrodifferential equations approach to leaky aquifers [Herrera and Yates, 1977].

From definition (12) of $F(t')$, it follows that

$$F(t') = t' + 2 \sum_{n=1}^{\infty} \int_{0}^{t'} e^{-n^2\pi^2\tau}\, d\tau \tag{40}$$

Therefore

$$F(t') - F_a(t') = -\frac{2\theta^{-\frac{1}{2}}}{\pi^2} \sum_{N+1}^{\infty} \frac{e^{-n^2\pi^2\theta t'}}{n^2} + \Delta(\theta,t') \tag{41}$$

Let ε be the relative error

$$\varepsilon = \frac{|F(t')-F_a(t')|}{|F(t')|} \tag{42}$$

Using (31) and (42), it is seen that an estimate of ε is given by

$$\varepsilon \le \max \{\pi^3\theta t')^{-1/2} \sum_{n=N+1}^{\infty} \frac{e^{-n^2\pi^2\theta t'}}{n^2}\}+\varepsilon_1 \tag{43}$$

Here the approximation

$$F(t') \approx 2\sqrt{t'/\pi} \tag{44}$$

implied by (7) was used. When carrying out numerical integration, the maximum value achieved the first term in (44) is attained when the value of t' is $\Delta t'$. This is due to the monotonically decreasing character of the function occurring there. Hence

$$\varepsilon \leq (\pi^3\theta t')^{-\frac{1}{2}} \sum_{n=N+1}^{\infty} \frac{e^{-n^2\pi^2\theta\Delta t'}}{n^2} + 2 \sum_{n=1}^{\infty} e^{-(n^2/t'_c)} \tag{45}$$

From (13) it is seen that the approximation used corresponds to have replaced the storage coefficient S' by

$$S'\sqrt{\frac{t'_{max}}{t'_c}} < S' \tag{46}$$

everywhere, after truncation of the series, in the manner implied by (39).

The main advantage is due to the fact that when $\theta > 1$,

$$\sum_{N+1}^{\infty} \frac{e^{-n^2\pi^2\theta\Delta t'}}{n^2} < \sum_{N+1}^{\infty} \frac{e^{-n^2\pi^2\Delta t'}}{n^2} \tag{47}$$

which shows that the error when (39) is directly used is larger. This permits using fewer terms in the series expansion to achieve a desired accuracy.

The procedure can be carried out as follows. Let $\varepsilon > 0$ be the admissible error. If t'_{max} and $\Delta t'$ are given, then one can choose t'_c so that

$$2 \sum_{n=1}^{\infty} e^{-n^2/t'_c} = \varepsilon/2 \tag{48}$$

This equation can be solved for t'_c, using a bisection type scheme. For example, if $\varepsilon = 0.1$, then

$$t'_c = 0.27 \tag{49}$$

Once t'_c has been defined, one needs to choose N, so that

$$(\pi^3\theta\Delta t')^{-1/2} \sum_{n=N+1}^{\infty} \frac{e^{-n^2\pi^2\theta\Delta t'}}{n^2} < \varepsilon < (\pi^3\theta\Delta t')^{-1/2} \sum_{n=N}^{\infty} \frac{e^{-n^2\pi^2\theta\Delta t'}}{n^2} \tag{50}$$

TABLE 1. Valley of Mexico A

ε	t'_c	θ	N	N(θ = 1)
0.01	0.16690	3.38822	9	17
0.05	0.22821	4.63268	6	13
0.10	0.27108	5.50312	5	11
0.15	0.30455	6.18262	4	10
0.20	0.33379	6.77617	4	9
0.25	0.36064	7.32118	3	9

$S' = 4.8$, $T' = 1.816 \times 10^{-5}$ km ^{2}/yr, $b' = 4.8 \times 10^{-2}$ km, $t_{max} = 30$ years, $\Delta t = 0.5$ year or in nondimensional form, $t'_{max} = 4.926 \times 10^{-2}$, $\Delta t' = 8.210 \times 10^{-4}$.

TABLE 2. Valley of Mexico B

ε	t'_c	θ	N	N(θ = 1)
0.01	0.16690	3.38822	9	14
0.05	0.22821	4.63268	6	11
0.10	0.27108	5.50312	5	9
0.15	0.30455	6.18262	4	8
0.20	0.33379	6.77617	4	8
0.25	0.36064	7.32118	3	7

$S' = 2.4$, $T' = 3.333 \times 10^{-6}$ km^2/yr, $b' = 2.4 \times 10^{-2}$ km, $t_{max} = 30$ years, $\Delta t = 0.5$ year or in dimensionless form, $t'_{max} = 7.226 \times 10^{-2}$, $\Delta t' = 1.200 \times 10^{-3}$.

TABLE 3. Guaymas

ε	t'_c	θ	N	N(θ = 1)
0.01	0.16690	3.38822	12	128
0.05	0.22821	4.63268	8	97
0.10	0.27108	5.50312	6	83
0.15	0.30455	6.18262	5	75
0.20	0.33379	6.77617	5	69
0.25	0.36064	7.32118	4	65

$S' = 0.75$, $T' = 1.230 \times 10^{-7}$ km^2/yr, $b' = 7.5 \times 10^{-2}$ km, $t_{max} = 50$ years, $\Delta t = 0.5$ year or in dimensionless form, $t'_{max} = 1.457 \times 10^{-3}$, $\Delta t' = 1.457 \times 10^{-5}$.

TABLE 4. Aquifer with Fictitious Properties

ε	t_c'	θ	N	N(θ = 1)
0.01	0.16690	3.38822	9	78
0.05	0.22821	4.63268	6	59
0.10	0.27108	5.50312	5	51
0.15	0.30455	6.18262	4	46
0.20	0.33379	6.77617	4	42
0.25	0.36064	7.32118	3	39

$S' = 4.$, $T' = 3.1536 \times 10^{-6}$ km^2/yr, $b' = 1. \times 10^{-1}$ km, $t_{max} = 30$ years, $\Delta t = 0.5$ year, or in dimensionless form, $t'_{max} = 2.365 \times 10^{-3}$ $\Delta t' = 3.942 \times 10^{-5}$.

Numerical Examples

Two of the leaky aquifers that have been extensively studied in Mexico are the ones under the Valley of Mexico [Herrera et al., 1974] and Guaymas. The different properties of the aquifer have a wide range of variation. Some of the values reported in the literature are only local values that do not correspond to average properties.

To exemplify the efficiency of the procedure, we have used two sets of values from the Valley of Mexico that we consider representative, one set from Guaymas, and another one from an aquifer with fictitious properties.

For each aquifer the computations were done for both the θ 'optimum' given by (35) and for $\theta = 1$ for a wide range of relative errors. A result with a relative error of 10% is usually very satisfactory.

The results are represented in Tables 1-4.

Notation

$A_N = (2/\pi^2) \sum_{n=N+1}^{\infty} (n^2)^{-1}$.

b' thickness of aquitard, L.

d(t') function defined by (21).

f(t') memory function, equal to $1 + 2 \sum_{n=1}^{\infty} e^{-n^2\pi^2 t'}$

$F(t') = \int_0^{t'} f(\tau)d\tau$.	
$F_a(t')$	approximation to $F(t')$, defined by (38).
$F_N(t')$	approximation to $F(t')$, defined by (39).
K'	permeability of aquitard, L/T.
$Q(t')$	pumping rate from aquifer, L^3/T.
s	drawdown in aquifer, L.
s'	drawdown in aquitard, L.
S	storage coefficient of aquifer.
S'	storage coefficient of aquitard.
$S_a(t)$	apparent storage coefficient of system, defined by (13).
t	time, T.
t'	dimensionless time, equal to $\alpha' t/b'^2$.
t''	dimensionless time, equal to $(\delta')^2 t'$.
t'_c	upper bound of short-time range.
t'_{max}	maximum value of t' we are interested in.
T	transmissibility of aquifer, L^2/T.
T'	transmissibility of aquitard, L^2/T.
x,y,z	coordinates, L.
$\alpha' = T'/S'$, L^2/T.	
$\delta(t')$	Dirac's delta function.
$\delta(\theta, t')$	function defined by (26).
$\Delta(\theta, t')$	function defined by (29).
ε	relative error, defined by (42).
ε_1	error defined by (36).
λ	positive parameter.
θ	parameter defined by (25).

References

Chorley, D. W., and E. O. Frind, An iterative quasi-three-dimensional finite element model for heterogeneous multiaquifer systems, Water Resour. Res., 14, 943-952, 1978.

Hantush, M. S., Modification of the theory of leaky aquifers, J. Geophys. Res., 65(11), 3713-3725, 1960.

Hennart, J. P., R. Yates, and I. Herrera, Extension of the integrodifferential approach to inhomogeneous multiaquifer systems, Water Resour. Res., 17(4), 1044-1050, 1981.

Herrera, I., Theory of multiple leaky aquifers, Water Resour. Res., 6(1), 185-193, 1970.

Herrera, I., A review of the integrodifferential equations approach to leaky aquifer mechanics, Adv. Groundwater Hydrol., 29-47, 1976.

Herrera, I., and L. Rodarte, Integrodifferential equations for systems of leaky aquifers and applications, 1, The nature of approximate theories, Water Resour. Res., 9(4), 995-1005, 1973.

Herrera, I., and R. Yates, Integrodifferential equations for systems of leaky aquifers, 3, A numerical method of unlimited applicability, Water Resour. Res., 13(4), 725-732, 1977.

Herrera, I., J. Alberro, J. L. León, and B. Chen, Análisis de asentamientos para la construcción de los lagos del plan Texcoco, Rep. 340, Inst. de Ing., Univ. Nac. Auton. de Mex., Mexico City, 1974.

Herrera, I., J. P. Hennart, and R. Yates, A critical discussion of numerical models for multiaquifer systems, Adv. Water Resour., 3(4), 159-163, 1980.

Neuman, S. P., and P. A. Witherspoon, Theory of flow in a confined two-aquifer system, Water Resour. Res., 5(4), 803-816, 1969a.

Neuman, S. P., and P. A. Witherspoon, Transient flow of ground water to wells in multiple aquifer systems, Geotech. Eng. Rep. 69-1, Univ. of Calif., Berkeley, Jan. 1969b.

Groundwater Modeling of Detailed Systems

Particularly in Fractured Rock

H. E. Skibitzke and Justin M. Turner
Hydro-data Inc., Tempe, Arizona 85282

Introduction

The computer techniques used to analyze groundwater conditions have evolved through three principal stages since the first crude computer models were built in 1954. Analog methods were the first to be applied to groundwater modeling. Active element analogs were used in a few cases, but passive element analogs were standard in the early models. As is the case with most newly developed technological applications, there were inherent problems in analog modeling.

With continued research, improved computer methods were developed. As their size was reduced and their speed increased, digital computers became useful for analyzing groundwater problems. Finite difference models, first used with analog methods, were also utilized in the digital computers. Now however, both finite difference and finite element models using digital techniques are widely used. But digital techniques have their drawbacks also. There are mathematical limitations on model construction by finite element techniques, and computer costs for detailed solutions to complex problems using finite difference or finite element techniques are excessive.

Characteristics of Analog and Digital Techniques

A comparison of analog and digital techniques shows that both have highly desirable characteristics for use in groundwater engineering as well as features that detract from their usefulness. The technology for both methods is constantly being improved.

The construction of the first generation of analog models required the facilities of an electronics shop, and the processing required electronic equipment, such as arbitary function generators and oscilloscopes. The output, simulating water level data, was presented as an electronic signal on the face of an oscilloscope and photographed. The data were then manually plotted into graphic format, a time-consuming operation.

Digital models, however, are produced in an office environment by specialists in data preparation, key punching, programming, and analysis. The output from the digital system is in ready-to-use printed or graphic format that requires no manual preparation.

The passive element analog models provided inherently stable and convergent solutions. By contrast, the solutions derived by digital models may be unstable for many problems, converging poorly or not at all. Inherently stable model techniques can be very expensive for small time steps and large arrays. The instability in a digital model may not be evident until near the end of a long and expensive computer run, and nonconvergence may not be identified until the computed computer run is analyzed. Each unproductive run can be costly.

The digital techniques that replaced analog methods in groundwater studies featured large memory cores and speedy computations. Matrix techniques facilitated the handling of factors such as evapotranspiration, recharge, and losses to rivers and streams. Matrices were printed out in a useful and easily understood format. Costs for computer analyses for the less complex problems began to decline. Recent innovations have made it possible to present the final computer output in graphic pen and ink format. The plotting techniques enable trial and error solutions to conform computer-derived information to data measured in the field. All in all, digital techniques have served well. However, more complex issues are now being raised.

Development of a Hybrid System

A new approach was needed to evaluate groundwater conditions for industries such as mining. Where the conditions are complex, large models, finely detailed, are required to describe the geologic parameters of affected aquifers, and long computer runs are required to simulate pumping over long periods of time. When faulted or fractured rock areas are involved, the model becomes even more complicated. In multilayered aquifers where layers are interconnected due to faulting and fracturing, the matrix bandwidths become very large. Using digital techniques, instability is very likely to occur at some point in the analysis, but this is not a factor in an analog analysis. However, a return to the analog techniques of two decades ago, even though they provided stable and convergent solutions, is unthinkable. By using the best of both systems, the more efficient and precise input and output functions of the digital techniques and the stable computations of the analog methods, it is possible to produce detailed solutions to groundwater problems at lower costs. The integration of analog and digital techniques amounts to replacing analog hardware with digital software, a common practice in the development of computer technology. In order to provide the information needed without exorbitant computer costs, the author found the hybrid approach not only desirable, but necessary.

An example of a complex groundwater evaluation for a mining operation was presented by a client recently. The area to be studied comprised 20 miles square (52 km^2). It was to be broken into very small elements measuring 500 ft (150 m) on a side. The mining plan called for pits approximately 500 by 200 ft (150 x 60 m) in size to be excavated below the water table. The pits were to be arranged irregularly, sometimes scooped out in a line and at other times to be scattered over different parts of the aquifer. To analyze the problem in detail, a finely meshed model was necessary. Forecasts were required of groundwater conditions to

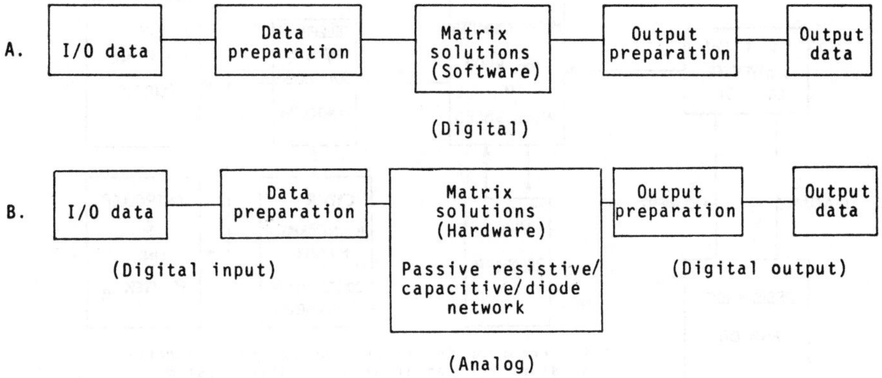

Fig. 1. Flowchart of (a) a digital aquifer model and (b) an analog model.

be expected for each 6 months over a period of 35 years. Logarithmic increments of time could not be used for the time periods covered. Thus the lengthy computations that would be required for the analysis by digital methods, which tend to instability, would cost more than $2000 for computer time alone.

This type of problem was recurring often enough that a new approach had to be considered. The incidence of instability and nonconvergence of the solutions that were derived in the course of the analyses brought to mind the advantages of the early analog models in this respect. The question then was how to incorporate the qualities of the analog method into the digital system.

The hybrid system comprises three phases: digital input of data into the analog model through digital-to-analog techniques, simultaneous analog measurements converted by analog-to-digital techniques, and data output and processing by digital techniques. The sequence is shown in Figure 1.

Hybrid computations are performed by using the same input considerations and data preparation methods that are used to prepare the data for matrix solutions with the digital system. This is illustrated in Figure 2. In the all-digital system the matrices are solved by software prepared for such computations. The results are then prepared to produce line-printed or line-plotted output.

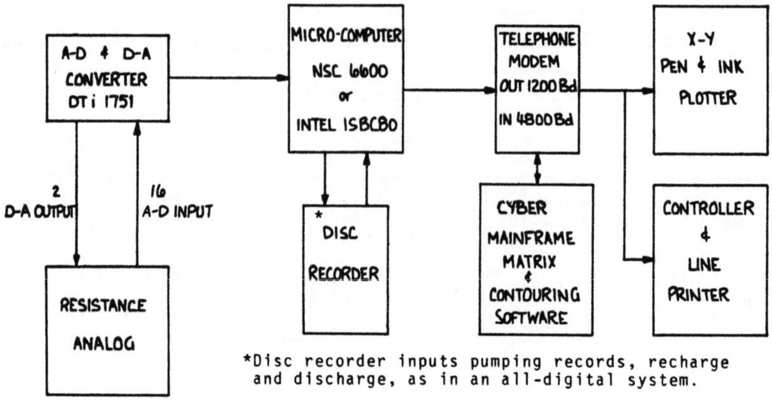

Fig. 2. Block diagram of hybrid computing system.

Essentially, the hybrid computations are accomplished the same way except that the data preparation phase includes digital-to-analog conversion and processing by analog network. The output preparation in the case of the hybrid system includes analog-to-digital conversion so that the output can be further processed by the digital computer system.

The analog solution to the matrices can be done by either (or both) active or passive element systems. Because the passive element analog system is inherently stable and converging, that technique is considered here. The passive element system is composed of continuous laminae of resistive and capacitive elements interconnected vertically with finite element resistors. It is also possible, however, to construct the model by finite elements of resistance and capacitance only. Much of the technique here depends upon high-speed digital-to-analog and analog-to-digital techniques recently developed. In addition, modern technology in component miniaturization allows a much greater versatility in analog computer components than has been available in the past.

The technology for digital-to-analog and analog-to-digital conversions is advancing rapidly and effectively so that the potential for solving problems, such as the mining company faced, is virtually without limit.

Advanced technology has also provided miniaturized components that have rendered the bulky electrical networks of the early analog models a thing of the past. They have been replaced by inexpensive microcircuits. Using the microcircuits in as many layers as needed, models can be built that would allow trial-and-error computations for large, complicated problems. The miniaturized models are integrated into the hardware of the digital system, as shown in Figure 2. Data input to the system by normal digital methods are converted by digital-to-analog techniques, and the resulting analog output is in turn fed into the miniaturized model for processing. A number of points within the model can be measured simultaneously. Given a brief lapse of time, the measured points return to zero so that another group of points can be processed. The resulting analog solutions of the problem are then transferred by analog-to-digital conversion techniques to the main core of a high-speed digital system. Once the data enter the memory of the digital computer, the analog portion of the operation is completed; it is now up to the digital computer to process the data.

The digital computer system allows sparse matrix records from the analog to be analyzed by gridding techniques. This allows the randomly distributed points of measurement to be replotted in a uniform grid format. Since this is possible, it is not necessary to measure all grid points; instead, only enough output points to compose a well-represented grid are necessary. Finally, contour maps of the grid points are computer plotted. The entire process is rapid, and the output is ready for use. Several other factors that influenced the development of the hybrid techniques are worthy of mention.

It is difficult and costly to make digital computer measurements for a few discrete points or isolated times within a problem involving a large area and a long period of time because the entire program must be run for each solution. On the other hand, the analog process allows selective measurements to be made as often as desired without increasing costs.

Another significant feature involves the plotting of the isolated point measurements. Surface gridding techniques can be applied to data measurement points derived from the analog-to-digital converter from the output of the analog model of the aquifer. Gridding techniques that are available on virtually all computers allow a few (or many) randomly located points to be compiled into a regular interval grid. The analysis is generally done by least squares techniques which determine a smooth function describing the entire region.

The gridded data then allow the plotting of excellently drafted contour maps. The resulting data can be in many forms, such as perspective view, plan view, or normal planimetric view contour maps.

The number of points that must be measured can be materially reduced through the use of gridding techniques. The output is identical in format to that derived by solving the matrix equations and is ready for any mathematical or graphics approach, just as though the solution had been reached completely by digital processes.

Contour maps can be made for each time step in the pumping regime. Some errors are bound to appear in any gridding process. Using the stepwise approach, however, the errors are confined to the particular map being gridded; they are not cumulative because the analog uses all points in the solution. The analog-derived solution for each step is complete in time and space, even if only a few points are used in the gridding process. Also, each time step utilizes the same basic data as all the other steps. The differences are in the points of measurement. The inability of the digital system to accommodate variations in local parameters being changed has limited computer analyses of large, detailed arrays where detail is desired locally over many time steps.

Microcomponents are responsible for the ease and low cost of a new generation of analog methods. The small printed circuits using microcomponents take the place of the huge soldered networks of resistors, capacitors, and diodes that once were painstakingly assembled. The printed circuits are stacked in layers, forming

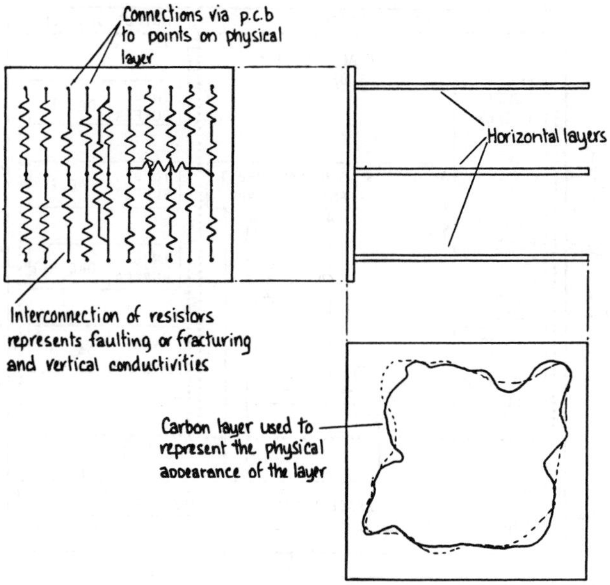

Fig. 3. Diagram of the construction of a three-layer problem.

individual units that may be analyzed separately, and are connected to make the complete model. Once the basic framework is in place, no soldering is required. A typical approach is shown in Figure 3. Changing to a new model or varying the model being used can be accomplished by unplugging and replacing microcomponents.

Illustrative Model

A simple model to illustrate this report was constructed using analog-to-digital (A-D) equipment manufactured by Data Translation, Inc. A diagram of the conversion process is shown in Figure 4. One device such as that in the diagram has two digital-to-analog (D-A) outputs, while the input circuit can measure the analog time-dependent voltage at 18 points. The particular arrangement used in Data Translation's equipment is shown in Figure 5. A large number of analog units are not required to compute a time sequence or a spatial sequence of data because one unit can be used repeatedly. The repetitive network voltages will relax to the zero value in

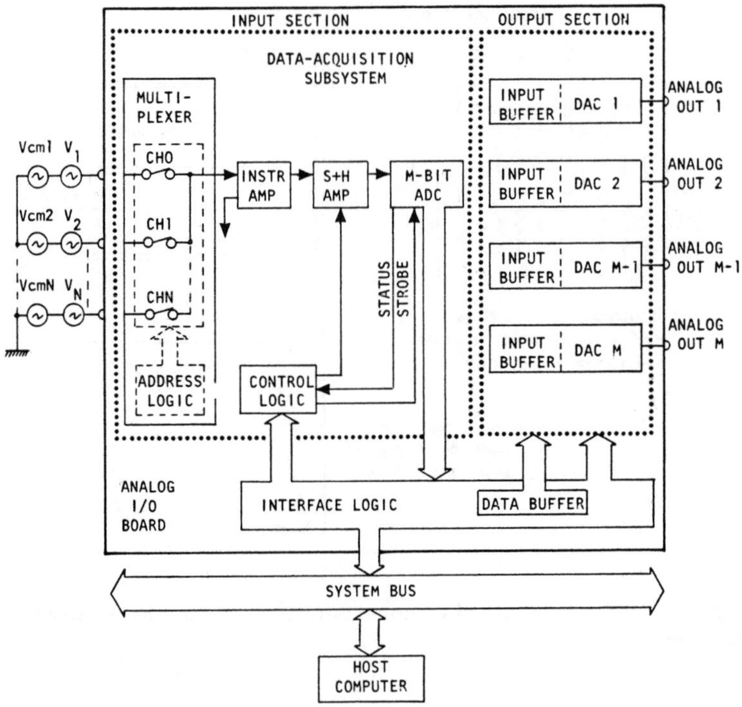

Fig. 4. Diagram showing the analog-to-digital and digital-to-analog processes.

seconds; so by the time the new points for measurements are determined, the model is stable at the zero level and ready to simulate another solution, thus allowing a number of groups of points to be measured. However, the gridding software described earlier, which uses data from only a few random points, can interpolate the mesh size into any uniform grid size that is desired. Thus only a small number of measurements are actually required. This method is less accurate than the all-digital techniques for a single time step, but the diminished accuracy of a single step is compensated for by the absence of accumulated error. Each step in the analog process is a portion of the complete solution analyzed in detail.

The measured data are returned from the analog-to-digital converter to a microcomputer and stored on a disk. The data are then transferred from the microcomputer to a large mainframe computer

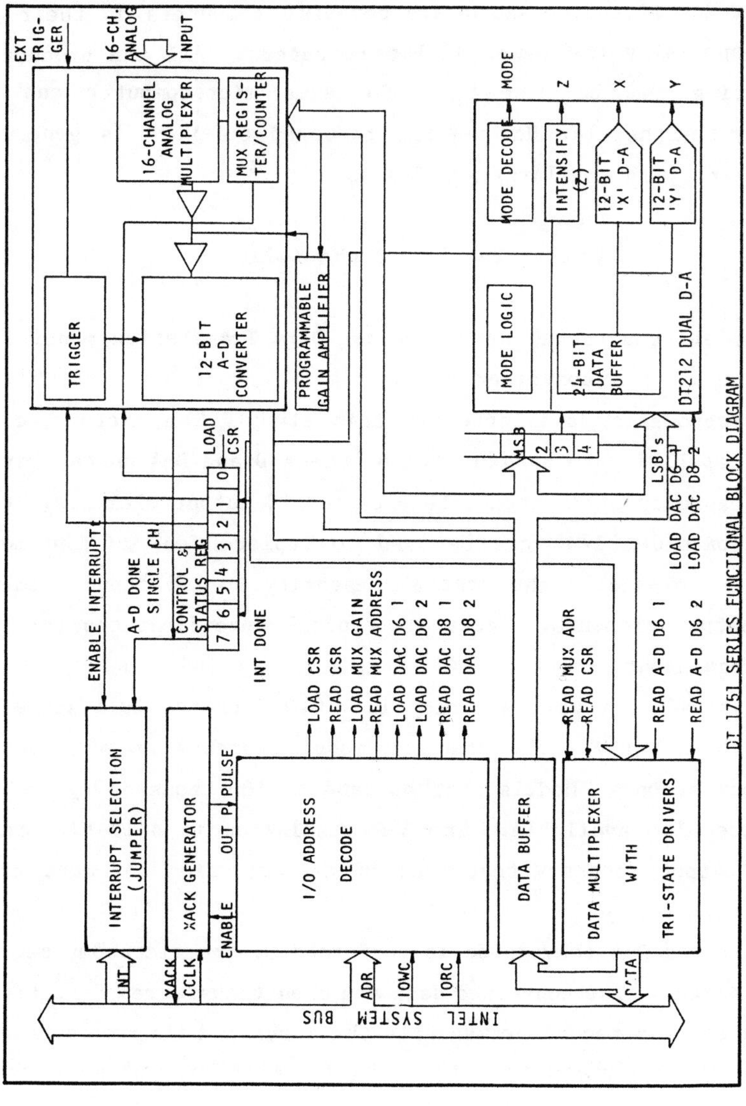

Fig. 5. A common D-A and A-D converter, manufactured by Data Translation, Inc.

where the contouring and gridding process is generated. In the model shown here, only a single Data Translation unit was utilized. Each unit can measure 18 points, and five such units can be connected in most microcomputers.

Commercial microprocessors are powerful computers in their own right, especially the newer 16-bit computers. All the processing and plotting can be done with the same microcomputer and the necessary peripherals. However,the required software is generally available only in the large mainframes.

Construction of the Model

The model constructed here used the Data Translation model 1751 as the A-D and D-A converter because it is compatible with the NSC 6600 microprocessor available for this study. The microprocessor has the capability to handle five of the A-D and D-A units, giving a total capacity of 80 input elements and 10 output elements. Data Translation model 1742 can be used to replace four of the model 1751 units, expanding the system's capacity to 256 input elements and 32 output elements. These are typical specifications for this type of equipment.

The system's hardware accuracy is $\pm$ 0.03% of the full-scale input, which is 10 V. The standard model operates at 48 thousand cycles per second. Models that operate at 100 thousand cycles per second are also available. The A-D resolution is a 12-bit binary with an output accuracy that exceeds the criterion for most model generation.

The A-D and D-A conversion is performed by the microcomputer operating system. The converted data are then transferred directly to a floppy disk to record permanently the input and output waveforms.

The complete linkage with the Data Translation system is shown in Figure 2. The NSC 6600 may be used as a direct replacement for the Intel SBC 80/1000 microcomputer shown in the diagram. The xy plotter represents a local plotter that produces the individual hydrographs connected to the microcomputer, but for these studies

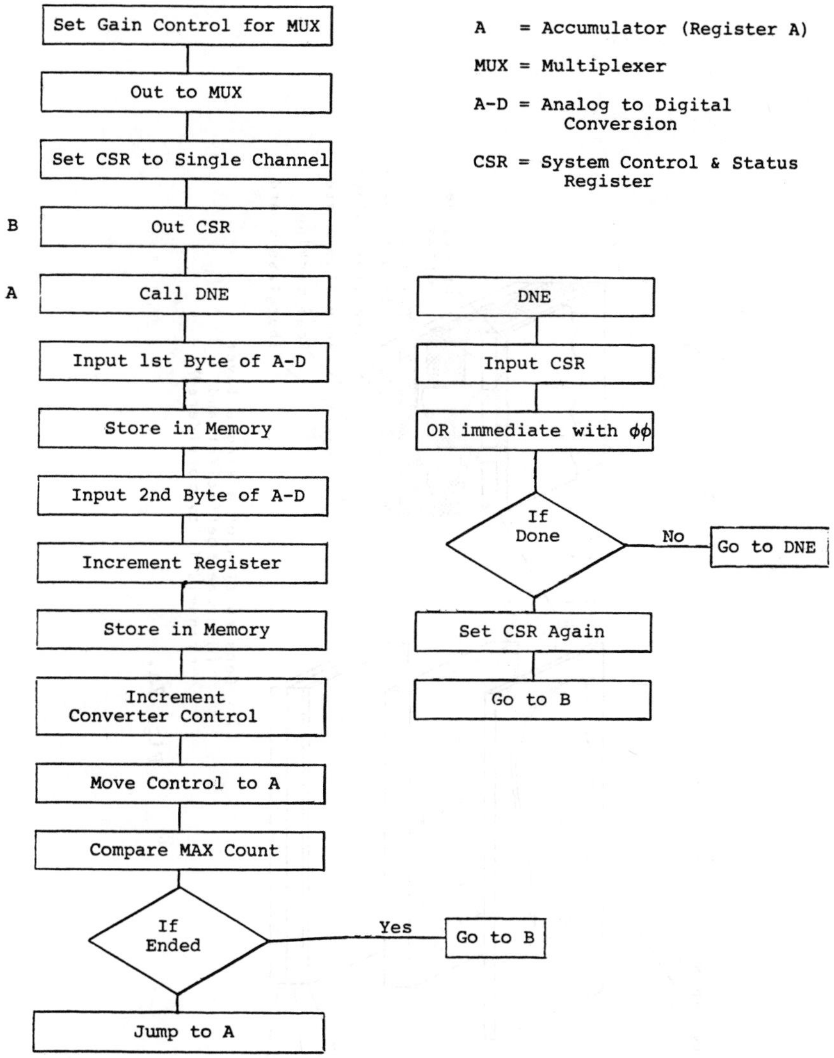

Fig. 6. Flowchart of analog-to-digital portion of hybrid computer unit.

the plotting vectors were determined in a large mainframe. The communications link used for this study was through disk storage to a Cyber 176 mainframe computer.

The analog conversion is performed as shown in Figure 4, a system typical of many that are available. The host computer system provides the input hydrographs to the output section, which furnishes

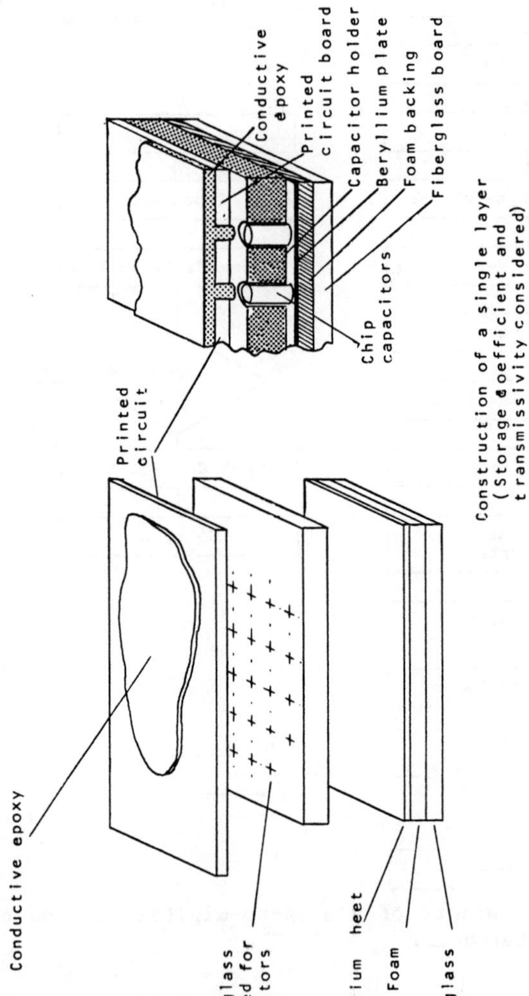
Conductive epoxy
Fiberglass drilled for capacitors
Beryllium heet
Foam
Fiberglass
Printed circuit
Conductive epoxy
Printed circuit board
Capacitor holder
Beryllium plate
Foam backing
Fiberglass board
Chip capacitors
Construction of a single layer
(Storage coefficient and
transmissivity considered)

Fig. 7a.

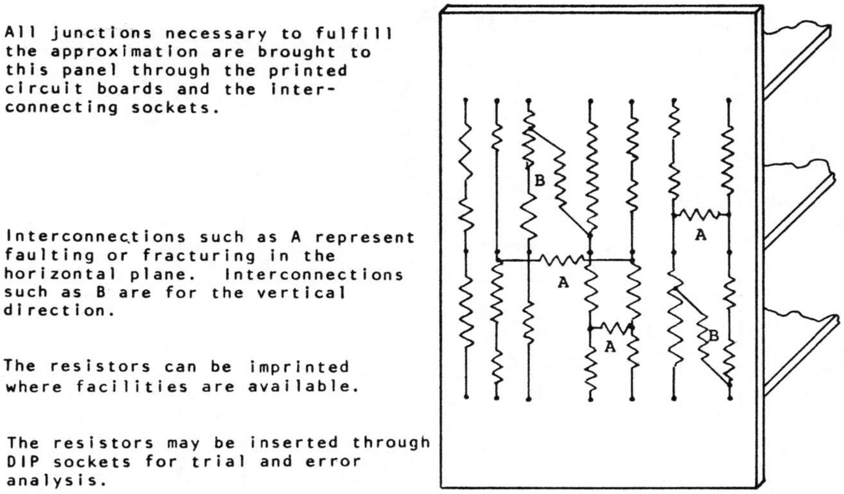

All junctions necessary to fulfill the approximation are brought to this panel through the printed circuit boards and the inter-connecting sockets.
Interconnections such as A represent faulting or fracturing in the horizontal plane. Interconnections such as B are for the vertical direction.
The resistors can be imprinted where facilities are available.
The resistors may be inserted through DIP sockets for trial and error analysis.
A
B
A
A
B

Fig. 7b.

Fig. 7. Diagram of the construction of the analog portion of a hybrid model.

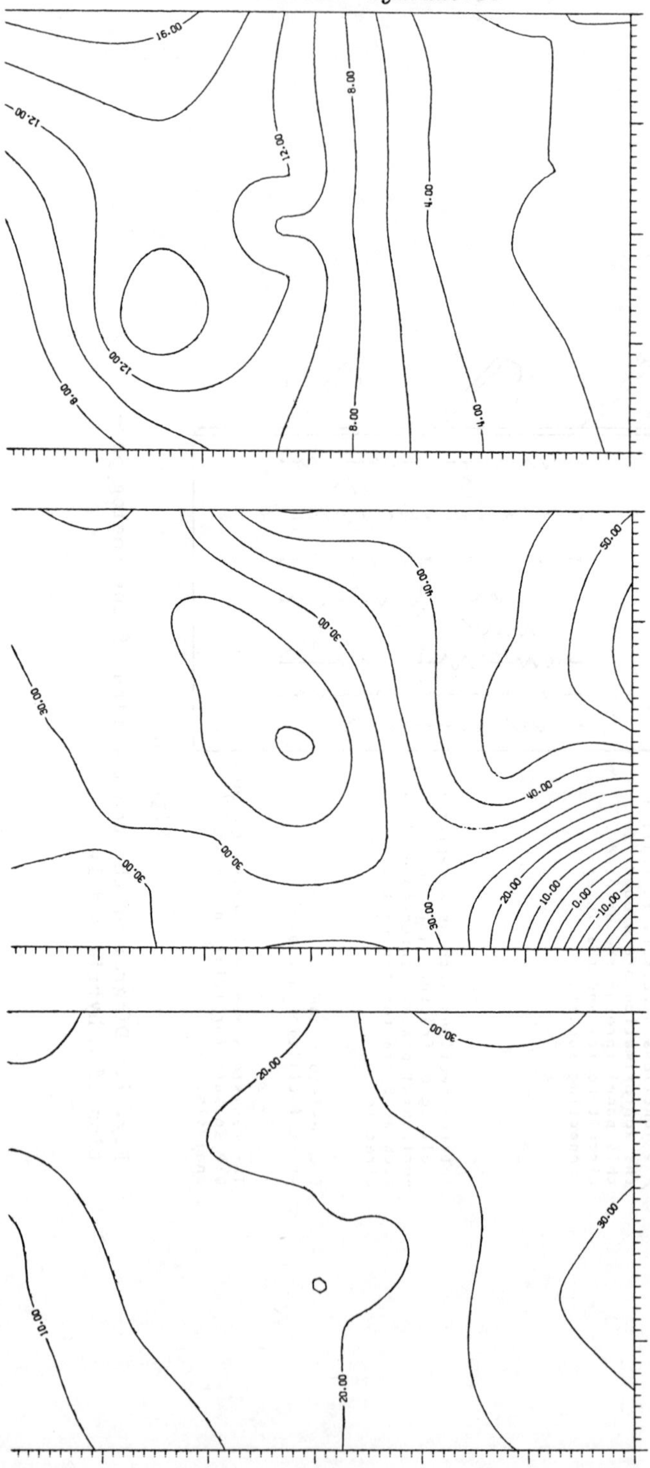

Fig. 8a.

Fig. 8b.

Fig. 8c.

Fig. 8. Contour plot representative of the final solution of a three-layer problem.

the current or voltage waveform to the model. In the meantime, the input section is stepping through a system of measurements at each of the 18 points multiplexed at 48 thousand cycles per second. The measurements are then stored for transmission to the Cyber 176 mainframe. The microcomputer functions required by the conversion unit are shown in Figure 5.

The microcomputer functions can be stepped by a simple timing loop coupled to the A-D and D-A unit. Figure 6 is a flowchart of one approach.

With the rapid advancements in microcircuitry, all of the components soon will be available on the small chips that are being developed. However, for the present, a simple way to construct the hardware for such a model is shown in Figure 7. Basically, each layer is constructed as a printed circuit board that forms the elements to connect the vertical resistor network. The assembled layers are shown in the bottom part of Figure 7. In detail, the individual boards contain the resistor elements which describe the region for that layer. The printed circuitry goes to a 200-element plug-in board for assembling the vertical conductance between layers. The storage coefficient is affected by using chip capacitors assembled through a holder for pressing the capacitors onto the resistance board. No soldering is used. Vertical conductivity is assembled using resistors, as described in Figure 3.

The effects of faulting and fracturing can be inserted by coupling resistors across the vertical conductance circuits and by applying finite values of resistance to the back of the printed circuit boards. This allows consideration of a wide range of complex parameters of the type encountered in fractured and faulted aquifers. There is no possibility of constructing an unstable or nonconverging system.

The individual plots of Figure 8, which represent the solutions to a three-layer problem, are typical of the final output product. The plots represent a model with fault lines incising a three-layer aquifer.

On the Formulation of Models Based on the Average Characteristics of a Heterogeneous Aquifer

Hillel Rubin[1] and Bent A. Christensen
University of Florida, Gainesville, Florida 32611

Introduction

This article is carried out in the framework of investigations concerning mineralization processes taking place in the Floridan aquifer. The Floridan aquifer is one of the principal sources of potable water in Florida [Hyde, 1965; Pascale, 1975; Lichtler, 1972; Bermes et al., 1963; Bentley, 1977]. The thickness of the Floridan aquifer is generally large, leading to very high transmissivities and wells yielding high discharges with small drawdowns. According to various investigators [e.g., Parker et al., 1955; Puri and Vernon, 1964] the Floridan aquifer is assumed to be consisted of several zones having variable permeabilities. In various locations, salt water seeps through semiconfining formations underlying the aquifer into the freshwater zones [Rubin and Christensen, 1982a,b]. Models capable of simulating the migration of salt in the heterogeneous aquifer are required. This paper intends to supply basic ideas for the formulation of such simplified models.

Heterogeneity of an aquifer is represented by spatially variable characteristics. Considering flow in a heterogeneous aquifer as a plane flow, Gheorghitza [1972] reviewed various approaches that can be used for the performance of the flow field simulation. In that particular discussion, variable hydraulic conductivity demonstrates

---

[1]On leave from Technion-Israel Institute of Technology, Haifa, Israel.

heterogeneity of the porous medium. In complex processes associated with transport phenomena in porous media, other characteristics like variable heat diffusivity should be considered as well. Although the flow in the aquifer is usually a plane flow, transport of certain properties like heat and solute very often occurs in the vertical direction too.

Generally, variable characteristics of the aquifer lead to non-linearities in the mathematical models used for the simulation of transport processes. In such cases, numerical calculations usually require large quantities of computer time and memory. These calculations are also subject to convergence and stability limitations.

The objective of the present study is to develop a general methodology for the formulation of the mathematical models concerning transport processes in aquifers. These models should simulate overall transport processes while reducing significantly the requirements for computer time and memory.

Basic Equations

Various combinations of the equations of continuity, motion, heat transport, solute transport, and the equation of state are the basic equations used for the analysis of transport phenomena in aquifers. In cases of flow through heterogeneous incompressible porous media these equations assume the following forms, respectively,

$$\nabla \cdot \vec{q} = 0 \tag{1}$$

$$\nabla P - \rho g \vec{k} + \frac{\mu}{K} q = 0 \tag{2}$$

$$\gamma \frac{\partial T}{\partial t} + \vec{q} \cdot \nabla T = \nabla \cdot (\kappa \nabla T) \tag{3}$$

$$\phi \frac{\partial C}{\partial y} + \vec{q} \cdot \nabla C = \nabla \cdot (\overline{\overline{e}} \cdot \nabla C) \tag{4}$$

$$\rho = \rho_o \left[1 - \alpha (T - T_o) + \alpha_s (C - C_o)\right] \tag{5}$$

where

$\vec{q}$ specific discharge;
P pressure;
g gravitational acceleration;
$\vec{k}$ unit vertical vector in the downward direction;
K permeability;
ρ fluid density;
μ viscosity;
T temperature;
κ heat diffusivity;
ϕ porosity;
γ ratio between heat capacity of the unsaturated porous matrix and that of pure water;
α coefficient of volumetric thermal expansion;
α_s coefficient relating solute concentration to density;
C solute concentration.

The coefficient γ is represented by the following expression

$$\gamma = \phi + (1 - \phi) \rho_s c_s/(\rho_w c_w) \tag{6}$$

Here ρ_s is the solid density and c_s and c_w are the solid and fluid specific heat, respectively.

The expression μ/K represents the hydraulic resistivity r, which is a scalar characteristic of the saturated porous medium. Variations in the pore size and porosity lead to variations in the hydraulic resistivity. Sometimes nonuniform distribution of temperature and solute concentration attributes to variable hydraulic resistivity.

Heat diffusivity is generally a scalar characteristic of the saturated porous formation. If we assume that the saturated porous medium consists of parallel layers of solid and fluid materials, then the heat diffusivity of the formation is given by [Lagarde, 1965]

$$\kappa = \phi\ [\lambda_s\ (1 - \phi) + \phi\ \lambda_w]/(\rho\ c_w) \tag{7}$$

where λ_s and λ_w are the heat conductivity of the solid and fluid fractions, respectively.

If we assume a model consisting of a series of fluid and solid layers, then we get

$$\kappa = \phi\ \lambda_s\ \lambda_w/\{\rho\ c_w\ [(1 - \phi)\ \ _w + \phi\lambda_s]\} \tag{8}$$

According to (7) and (8), variations of κ in the aquifer are mainly attributed to changes in λ_s, as λ_w and ϕ are almost constant for the whole saturated porous formation.

If the aquifer is cavernous and the fluid is subject to gradients of the piezometric head, then the coefficient of diffusivity should be replaced by the heat dispersion tensor. However, heat dispersion is not a common feature of an aquifer subject to heat transfer processes.

Coefficients of solute dispersion can be represented by a scalar quantity provided that the fluid is almost stationary. This scalar quantity is represented by the following expression [Saffman, 1960]:

$$e = 0.7\ \phi\kappa_s \tag{9}$$

where κ_s is the molecular solute diffusivity in the fluid fraction.

If the fluid is subject to gradients of the piezometric head, then the dispersion tensor is a second-order tensor represented by the following expression:

$$\bar{\bar{e}} = \bar{\bar{e}}_2\ \bar{\bar{I}} + (e_1 - e_2)\ \vec{\vec{qq}}/q^2 \tag{10}$$

where

e_1, e_2 coefficients of longitudinal and transverse dispersion, respectively;

$\bar{\bar{I}}$ unit matrix;

$\vec{\vec{qq}}$ correlation tensor.

The coefficients of transverse and longitudinal dispersion depend on the absolute value of the specific discharge (or local mean velocity) and the local scalar characteristics of the porous medium. According to the value of the Peclet number, the coefficients e_1 and e_2 can be linear or power functions of the absolute value of the specific discharge.

As represented by (10), the principal directions of the dispersion tensor are parallel and perpendicular to the specific discharge vector.

Referring to (1)-(5), heterogeneity of the aquifer may stem from variable values of either one of the scalar or tensorial coefficients appearing in these equations, namely, ρ, r, γ, κ, ϕ, and $\underline{\underline{e}}$.

Variability of the fluid density is represented by (5). The fluid density for quite wide ranges of temperature and salinity can be represented as a linear function of these variables. However, more general expressions can be used as well.

The aquifer's porosity is almost constant even if the aquifer consists of heterogeneous formations. Even the porosity of impermeable formations, like clay, is often similar to that of the aquifer.

The coefficient γ can also be considered as having a constant value for the whole aquifer. The difference between its value and the porosity is usually minor.

We may conclude that heterogeneity of the aquifer is generally represented by variable values of the coefficients originated by the proportion between certain specific fluxes in the flow field and certain gradients of scalar properties leading to these specific fluxes. These coefficients include the hydraulic resistivity, the thermal diffusivity, and the dispersion tensor.

If the local characteristics are variable, then any transport simulation problem associated with the utilization of either one of the equations (1)-(4) is involved with the solution of nonlinear problems. Therefore in general cases, complete numerical schemes should be used for the simulation. Sometimes it is possible to apply separation of variable techniques and use variational methods for the generation of approximate analytical solutions [e.g., Rubin, 1981]. Another approach is associated with the removal of the time derivative from the differential equation by applying the Laplace transformation. Then the resulting system is treated as a steady state problem that can be solved by a numerical or variational method in the transformed domain. The desired solution is obtained after the transform is inverted. All these techniques that may lead to analytical or numerical-analytical solutions are associated with laborious calculations. Each variable characteristic increases the complexity of the set of equations (1)-(4) leading to more and more complicated analysis. Therefore it is worthwhile to consider further approaches for the simplification of the basic equations, even though they lead to certain inaccuracies in the calculations. Such an approach is represented in the next section.

Reference to the Aquifer's Average Characteristics

The basic equations associated with transport phenomena in porous media represented by (1)-(4) are obtained by averaging variables and parameters in finite spaces of the flow field. The approach of the aquifer's average characteristic searches for an extension of that process. The aquifer is divided into several layers or assumed to consist of a single layer whose horizontal extent is much larger than their thickness. The local characteristics of the porous formation frequently change more significantly with the depth than along the horizontal coordinate. Our calculations start with a stratified aquifer whose local characteristics vary only with the depth. Such variations are usually attributed to the depositional process by which sedimentary rocks are formed. If the structure of the porous medium is completely random locally, then the formation is locally isotropic, and local characteristics depending on the features of the saturated formation alone are scalar parameters. If the local characteristics depend on the presence of a flow in addition to the formation's structure, then these characteristics are second-order tensors.

Scaler Local Characteristics

We refer to a single saturated porous layer representing a simplified model of the aquifer or a segment of the aquifer. This model is shown in Figure 1. Through the saturated porous layer a certain property is transported. As the medium is locally isotropic, the relationship between the local specific flux vector $\vec{n}_\ell$ and the local gradient vector $\vec{J}_\ell$, leading to this flux, is given by

$$\vec{n}_\ell = k\ \vec{J}_\ell \tag{11}$$

where k is the local mobility of the porous medium.

The gradient $\vec{J}_\ell$ represents the gradient of the piezometric head, provided that $\vec{n}_\ell$ represents the local specific discharge. It may represent the temperature gradient or solute concentration gradient if $\vec{n}_\ell$ represents specific diffusive heat flux or solute flux, respectively.

Due to the way of deposition of sediments and formation of sedimentary rocks we assume that $k = k(z)$, where z is a vertical coordinate, as shown in Figure 1.

If $\vec{J}_\ell$ is a horizontal vector, then $\vec{n}_\ell$ is also horizontal. In this case the proportion represented by (11) is applicable even for large horizontal distances, provided that the cross-sectional area perpendicular to $\vec{n}_\ell$ is infinitely small. If $\vec{J}_\ell$ is a vertical vector, then $\vec{n}_\ell$ is also vertical. However, in this case the proportion represented by (11) is applicable only for an infinitesimal vertical distance, no matter how large is the cross-sectional area perpendicular to $\vec{n}_\ell$. Our objective is to develop a general method of averaging of fluxes and gradients that considers these differences. We may define an average horizontal specific flux as being the total flux in that direction per unit width of the aquifer divided by the thickness of the aquifer. The average vertical gradient may be defined as the difference between the value of the parameter at the top of the aquifer and its value at the bottom of the aquifer divided by the thickness of the aquifer. As the

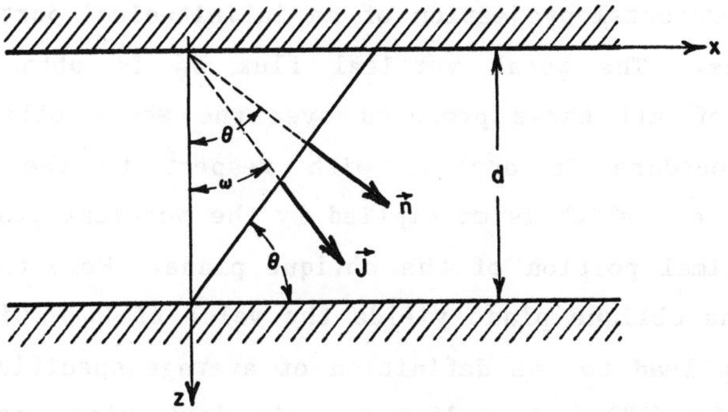

Fig. 1. Schematic of an aquifer with the average diffusive specific flux vector and the average gradient vector.

porous medium is stratified, the average specific discharge in the horizontal direction is proportional to the gradient in that direction. For the same reason the average gradient in the vertical direction is proportional to the specific discharge in that direction. These relationships are represented as follows:

$$n_h = J_h \int_0^1 kdz' \qquad n_v = J_v / \int_0^1 dz'/k \tag{12}$$

where

$$z' = z/d \qquad n_h = n_x \qquad n_v = n_z \qquad J_h = J_x \qquad J_v = J_z \tag{13}$$

For the purpose of briefing the text whenever a reference is made to the aquifer it also considers a finite segment of the aquifer.

We may construct in the aquifer an oblique plane perpendicular to the vector $\vec{n}$ representing the average specific flux vector, as shown in Figure 1. The total flux per unit width passing through that plane is represented by

$$N = N_v + N_h = n\ d/(\cos\theta) \tag{14}$$

The local specific flux vector $\vec{n}_\ell$ is not necessarily perpendicular to the oblique plane shown in Figure 1. At each point of this plane we may decompose the local specific flux into horizontal and vertical components. We multiply the vertical component of $\vec{n}_\ell$ by the horizontal projection of an infinitesimal portion of the oblique plane. The total vertical flux N_v is obtained by an integration of all these products over the whole oblique plane. The same procedure is applied with respect to the horizontal component of $\vec{n}_\ell$, which is multiplied by the vertical projection of an infinitesimal portion of the oblique plane. Here the integration over the oblique plane yields the value of N_h. Given values of N_v and N_h lead to the definition of average specific discharge represented by (12). According to this definition, as shown in Figure 1, the average specific discharge forms an angle θ with the vertical z axis. The average gradient $\vec{J}$ leading to the specific discharge of the property forms an angle ω with this axis. The angles θ and ω are not necessarily identical, as the proportion between the horizontal components of $\vec{n}$ and $\vec{J}$ is not necessarily identical to the proportion between the vertical components of these vectors. Therefore the proportion between $\vec{n}$ and $\vec{J}$ is represented by the tensor of the average mobility of the aquifer. The proportions between n_h and J_h and n_v and J_v lead to the conclusion that the horizontal and vertical axes are the principal directions of the average mobility tensor. The principal components of this tensor are represented by

$$k_h = \int_0^1 k \, dz' \qquad k_v = 1/\int_0^1 dz'/k \tag{15}$$

The ratio between the horizontal and vertical mobility is given by

$$k_h/k_v = \left(\int_0^1 k \, dz'\right)\left(\int_0^1 dz'/k\right) \geq 1 \tag{16}$$

This expression is obtained by applying the Schwartz-Cauchy's inequality [e.g., Rektory, 1969].

The horizontal and vertical components of the vectors n and J are given as follows:

$$n_h = n \sin \theta \qquad n_v = n \cos \theta$$
$$J_h = J \sin \omega \qquad J_v = J \cos \omega \tag{17}$$

By applying (12), (15), and (17) we obtain

$$\cos \omega = (\cos \theta)/[(\cos^2 \theta) + (\sin^2 \theta)(k_v/k_h)^2]^{\frac{1}{2}} \tag{18}$$

Introducing (16) into (18), we get

$$\cos \omega \geq \cos \theta \qquad \omega \leq \theta \tag{19}$$

This result can be interpreted through the following example: if $J_x = J_z$, then $\omega = 45°$; however, as the average horizontal mobility is larger than or equal to the vertical one, we get $n_x \geq n_z$, namely, $\theta \geq 45°$.

According to (18) the angles θ and ω are identical in two cases as follows:

$$\theta = 0° \qquad \text{or} \qquad \theta = 90° \tag{20}$$

The first case refers to purely vertical flux; the latter one refers to purely horizontal flux. This result is, of course, consistent with our previous calculations considering that the principal directions of the average mobility tensor are the vertical and horizontal directions. Using (18), performing certain trigonometric substitutions, and differentiating, we obtain that the maximum difference between θ and ω occurs when

$$\theta = \text{arc tan } [(k_h/k_v)^{0.5}] \tag{21}$$

In the extreme case of $k_v = 0$ we obtain from (12), (17), and (18) that $\theta = 90°$ for any value of ω.

The average hydraulic resistivity is defined as the inverse matrix of the hydraulic mobility. Therefore the principal components of the average hydraulic resistivity are represented as follows:

$$r_h = 1/\int_0^1 dz'/r \qquad r_v = \int_0^1 r dz' \tag{22}$$

Here the Schwartz-Cauchy inequality yields

$$r_h/r_v \leq 1 \tag{23}$$

Provided that mechanical dispersion has negligible effect on heat transfer through the aquifer, the aquifer's average heat diffusivity is a second-order tensor having principal component in the horizontal and vertical directions as follows:

$$\kappa_h = \int_0^1 \kappa dz' \qquad \kappa_v = 1/\int_0^1 dz'/\kappa \tag{24}$$

Summarizing our calculations, every local scalar characteristic of the porous medium which varies with depth generates an average tensorial characteristic of the aquifer whose principal directions are parallel and perpendicular to the gradient vector of the local variable characteristic.

Tensorial Local Characteristics

If the aquifer is locally isotropic, then the only local tensorial characteristic associated with transport phenomena is the mechanical dispersion tensor, leading to solute dispersion and possibly heat dispersion. The mechanical dispersion is induced by the flow of the fluid. Therefore the dispersion tensor has principal directions parallel and perpendicular to the velocity vector as represented by (10). Considering that the aquifer has an infinite horizontal extent, we may assume that in common cases the fluid flows almost horizontally in each point of the aquifer.

Heterogeneity of the aquifer stems mainly from variations of the local characteristics pore size with the depth. These variations lead to variable local hydraulic resistivity and dispersivities.

Variations of the local hydraulic resistivity with the depth cause variations in the absolute value of the flow velocity with the depth. Variability of the local dispersivities leads to variations in the proportion between the absolute value of the local velocity and the local dispersion coefficients. Due to the impermeable top and bottom of the aquifer the major flow in the aquifer takes place in the horizontal direction. In this case, at all points of the aquifer the local dispersion tensors have the same principal directions. Considering the mineralization of a leaky aquifer, the upward movement of the salt water migrating into the aquifer may sometimes be considered as a vertical flow [Rubin and Christensen, 1982a]. Also in this case, at all points of the aquifer the local dispersion tensors have the same principal directions.

The relationship between the local specific dispersed flux vector $\vec{n}_\ell$ and the local gradient vector $\vec{J}_\ell$ is given by the following expression:

$$\vec{n}_\ell = \overline{\overline{e}} \cdot \vec{J}_\ell \tag{25}$$

where $\overline{\overline{e}}$ is the local dispersion tensor as represented by (10).

Considering an aquifer in which the absolute value of the local velocity and the dispersivities are functions of the depth, we obtain $\overline{\overline{e}} = \overline{\overline{e}}(z)$. Note that $\overline{\overline{e}}$ depends on the value of the local velocities and dispersivities. The general expression for its calculation is represented by (10).

Figure 2 shows a simplified model of the aquifer, consisting of a stratified layer of porous medium whose thickness is d. We define a coordinate system whose axis is parallel to the local velocity vectors. Therefore this coordinate system represents the principal directions of the local dispersion tensors in all points of the particular portion of the aquifer.

We may define average specific dispersive fluxes in the x and y directions as being the total dispersive fluxes in these directions divided by the vertical plane through which these fluxes pass. The average vertical gradient may be defined as the differ-

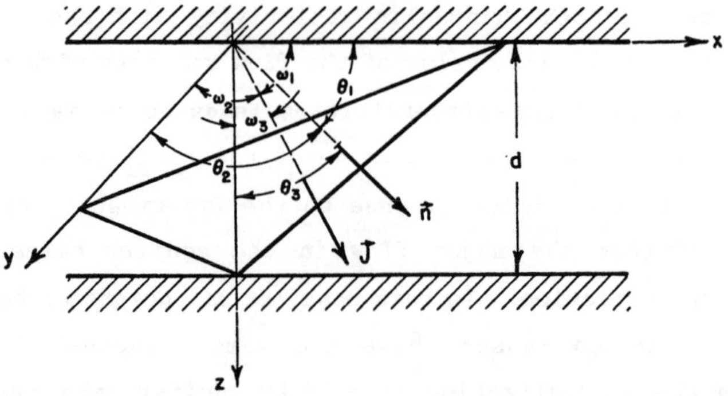

Fig. 2. Schematic of an aquifer with the average dispersive specific flux vector and the average gradient vector.

ence between the value of the parameter at the top of the aquifer and its value at the bottom of the aquifer. Applying calculations similar to those presented in the previous subsections, we obtain

$$n_x = J_x \int_0^1 e_{xx}\, dz' \qquad n_y = J_y \int_0^1 e_{yy}\, dz' \qquad n_z = J_z / \int_0^1 dz' e_{zz} \tag{26}$$

We construct in the aquifer an oblique plane perpendicular to $\vec{n}$, as shown in Figure 2. The total flux passing through this plane is represented by

$$N = N_x + N_y + N_z = n\, d^2 \cos \theta_3 / (\cos \theta_1 \cos \theta_2) \tag{27}$$

where θ_i ($i = 1, 2, 3$) are the angles which the vector $\vec{n}$ forms with the coordinates x, y, and z, respectively. The average gradient vector $\vec{J}$ forms angles ω_i ($i = 1, 2, 3$) with the coordinate axes.

According to (26) the average dispersion tensor stemming from local dispersion effects and expressing the proportion between $\vec{n}$ and $\vec{J}$ has the following principal components:

$$E_{xx} = \int_0^1 e_{xx}\, dz' \qquad E_{yy} = \int_0^1 e_{yy}\, dz' \qquad E_{zz} = 1 / \int_0^1 dz'/e_{zz} \tag{28}$$

Various experimental investigations performed with isotropic homogeneous porous media showed that the coefficient of longitudinal dispersion is much larger than the coefficient of transverse dispersion [e.g., Pfannkuch, 1963]. These studies imply that $e_{xx} \geq e_{yy} = e_{zz}$. Therefore by applying the Schwartz-Cauchy inequality we obtain

$$E_{xx} \geq E_{yy} \geq E_{zz} \tag{29}$$

It should be mentioned that the average dispersion tensor represented by (28) stems from averaging dispersion effects in each infinitesimal layer of the stratified aquifer. If the solute transport equation refers to the average specific discharge for the calculation of solute convection, then we obtain additional dispersion effects due to the velocity stratification in the aquifer. This component of the mechanical dispersion can be calculated according to the methods developed by Taylor [1953, 1954] and Aris [1956]. Therefore E_{xx} may attain values which are much larger than specified by (28).

We decompose the average specific flux and average gradient into three components as follows:

$$\begin{aligned} n_x &= n \cos \theta_1 \qquad n_y = n \cos \theta_2 \qquad n_z = n \cos \theta_3 \\ J_x &= J \cos \omega_1 \qquad J_y = J \cos \omega_2 \qquad J_z = J \cos \omega_3 \end{aligned} \tag{30}$$

Applying (26), (28), and (30), we obtain

$$\begin{aligned} \cos \omega_1 &= (\cos \theta_1)/[(\cos^2 \theta_1) + (\cos^2 \theta_2)\,(E_{xx}/E_{yy})^2 \\ &\quad + (\cos^2 \theta_3)\,(E_{xx}/E_{zz})^2]^{1/2} \\ \cos \omega_2 &= (\cos \theta_2)/[(\cos^2 \theta_1)\,(E_{yy}/E_{xx})^2 + (\cos^2 \theta_3)\,(E_{yy}/E_{zz})^2 \\ &\quad + (\cos^2 \theta_2)]^{1/2} \\ \cos \omega_3 &= (\cos \theta_3)/[(\cos^2 \theta_1)/(E_{zz}/E_{xx})^2 + (\cos^2 \theta_2)\,(E_{zz}/E_{yy})^2 \\ &\quad + (\cos^2 \theta_3)]^{1/2} \end{aligned} \tag{31}$$

Introducing the inequalities represented by (29) into (31), we obtain

$$\omega_1 \geq \theta_1 \qquad \omega_3 \leq \theta_3 \tag{32}$$

The angles θ_2 and ω_2 can be identical provided that the following condition is satisfied:

$$\frac{\cos^2 \theta_1}{\cos^2 \theta_3} = \frac{(1/E_{zz})^2 - (1/E_{yy})^2}{(1/E_{yy})^2 - (1/E_{xx})^2} \tag{33}$$

At moderate Peclet numbers, $e_{xx}/e_{yy} = 10 \div 30$ [Pfannkuch, 1963]. In such cases, (33) can be approximated by the following expression:

$$(\cos^2 \theta_1)/(\cos^2 \theta_3) = (E_{yy}/E_{zz})^2 - 1 \tag{34}$$

If (33) or (34) is satisfied, then the specific dispersive flux vector and the gradient vector are located on the surface of a circular cone whose centerline is the y axis.

According to (31) the direction cosines of the average dispersive specific flux vector $\vec{n}$ and those of the average gradient $\vec{J}$ are identical in either one of the following cases:

$$\theta_1 = 0° \quad \text{or} \quad \theta_2 = 0° \quad \text{or} \quad \theta_3 = 0° \tag{35}$$

namely, if the vector $\vec{n}$ is parallel to either one of the principal directions of the average dispersion tensor which are identical to those of the local ones.

Considering the transport of solute materials in the aquifer, for a wide range of Peclet numbers the local coefficients of dispersion are proportional to the velocity vector. Therefore if the flow field is subject to a local horizontal gradient J of the piezometric head, then the local components of the specific discharge and the dispersion tensor are given as follows:

$$q = J/r$$

$$e_{xx} = a_1 q \qquad e_{yy} = a_2 q \qquad e_{zz} = a_2 q \tag{36}$$

where a_1 and a_2 are the local dispersivities in the longitudinal and transverse directions, respectively.

Integrating these expressions over the aquifer, we get

$$E_{xx} = J \int_0^1 (a_1/r)\, dz' \qquad E_{yy} = J \int_0^1 (a_2/r)\, dz'$$

$$E_{zz} = J/[\int_0^1 (r/a_2)\, dz'] \tag{37}$$

Therefore the average dispersion coefficients are proportional to the average gradient of the piezometric head.

Again, the expression of E_{xx} in (37) does not include the effect of mechanical dispersion caused by the reference to the average flow in the aquifer. Calculations of the migration of solute materials should, of course, incorporate this component of the mechanical dispersion as well as those represented by (37).

According to (37) the coefficients determining the proportion between the gradient J and the average solute dispersion tensor form a second-order tensor depending on the distribution of the local hydraulic resistivity and dispersivities in the aquifer.

Summarizing our calculations, if the local tensorial characteristic of the porous medium varies with the depth and the vertical direction is one of its principal directions, then the average characteristic approach yields an average tensorial characteristic of the aquifer whose principal directions are identical to those of the local characteristics of the porous medium.

Discussion

The average characteristic approach may supply information about overall transport phenomena in the aquifer. It indicates total fluxes through the saturated formation but does not simulate values of local variables and parameters.

The integrals represented by (12) and (28) are performed in the

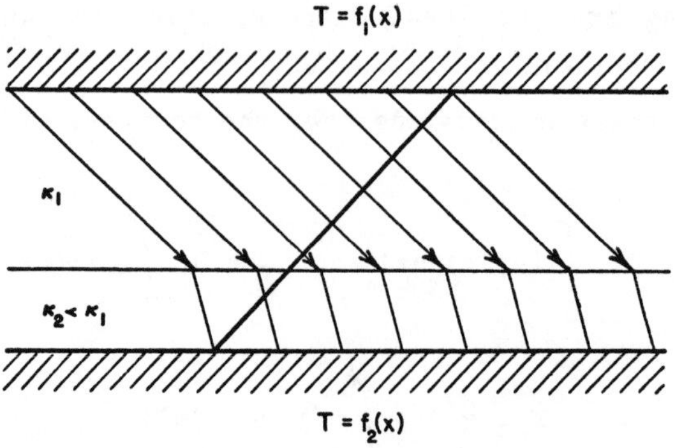

Fig. 3. Relationships between boundary conditions, streamlines, and the plane defining the average specific flux vector in a heterogeneous aquifer.

inclined planes identifying the average specific flux vector. The principal components of the average flux and those of the average gradient vector can be time dependent. They can also depend on the horizontal coordinates. Such a possibility is represented by Figure 3. This figure refers to an aquifer consisting of two parallel layers, each consisting of a saturated porous medium, having different heat diffusivities. The top and bottom of this aquifer are kept in temperatures depending on the horizontal coordinate x. We construct in this field a system of streamlines, namely, continuous lines drawn through the field so that they are tangent to the local specific diffusive heat flux and the local temperature gradient. If the temperature at the top of the aquifer varies with x differently from the variation of the bottom temperature, then the streamlines are not straight or parallel lines. At the interface between the two layers the streamlines undergo changes in the direction of the local specific heat flux. The streamlines cross the inclined plane used for the definition of the average specific flux. Generally, they are not perpendicular to this plane.

With minor modifications the average characteristics approach can be adapted to aquifers whose characteristics like hydraulic resistivity, heat diffusivity, and thickness vary moderately in

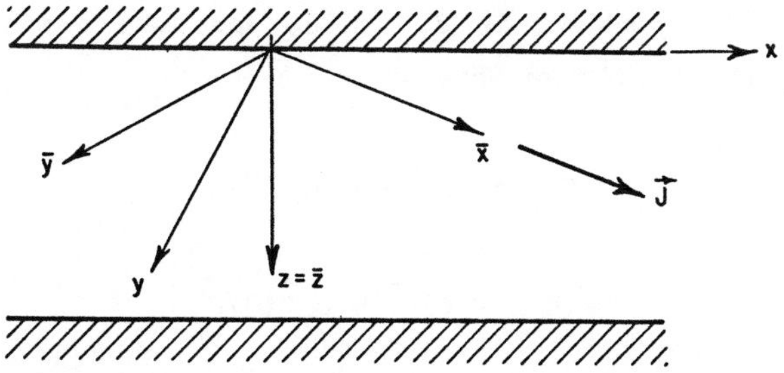

Fig. 4. Relationship between the coordinate system of reference and the principal directions of the average dispersion tensor.

the horizontal direction. Referring to Figure 1, the aquifer's characteristics may vary moderately in the horizontal directions. If in such a case the angle θ is not very small, namely, the horizontal component of the average specific flux vector is not much smaller than the vertical one, then we may assume that variations of the aquifer's local characteristics along the horizontal projection of the inclined plane are much smaller than their variations along the vertical projection of the inclined plane. Therefore in the calculation of the aquifer's average characteristics we may neglect the horizontal variation of the local characteristics. Such calculations lead to the aquifer's average characteristics which vary with the horizontal coordinates. Usage of the horizontally variable average characteristics simplifies the simulation of transport processes.

The gradient of the piezometric head may change its absolute value and direction in the aquifer. Figure 2 refers to such a case. If variations of $\vec{J}$ are moderate then by applying (37), we can calculate values of the horizontally variable principal components of the average dispersion tensor. However, due to changes in the direction of the hydraulic gradient it is possible that our reference coordinate system x, y, z, as shown in Figure 4, is different from the coordinate system, $\bar{x}$, $\bar{y}$, $\bar{z}$, representing the principal directions of the average dispersion tensor. If the absolute value of the hydraulic gradient is J, and J_x and J_y are its two components, by applying (37) we obtain

$$E_{xx} = (J_x{}^2 a + J_y{}^2 b)/J$$

$$E_{yy} = (J_y{}^2 + J_x{}^2 b)/J$$

$$E_{zz} = Jc \qquad E_{xy} = E_{yz} = (a - b)\, J_x\, J_y/J \tag{38}$$

$$E_{xz} = E_{zx} = E_{yz} = E_{zy} = 0$$

where

$$J = (J_x{}^2 + J_y{}^2)^{1/2} \qquad a = \int_0^1 (a_1/r)\, dz'$$

$$b = \int_0^1 (a_2/r)\, dz' \qquad c = 1/\int_0^1 (r/a_2)\, dz' \tag{39}$$

Considering the component of mechanical dispersion generated by the reference to average flow in the aquifer, we should note that also in this case the difference between the coordinates x, y, z and $\bar{x}$, $\bar{y}$, $\bar{z}$ should be taken into account.

If the aquifer is heterogeneous and locally anisotropic, then its local hydraulic resistivity and heat diffusivity are second-order tensors. Here the method used previously for the generation of the average solute dispersion tensor can be used for the determination of the average hydraulic resistivity and thermal diffusivity, provided that all points of the aquifer have identical directions of the local tensors and that the principal components of these tensors depend on the depth. The calculation can also incorporate moderate dependence of the principal components on the horizontal coordinates and moderate changes of the principal horizontal directions.

Summary and Conclusion

Considering total fluxes of diffused or dispersed properties through stratified aquifers, it is suggested to utilize the method of the aquifer's average characteristics. According to this method the basic equations representing transport phenomena, which depend on local characteristics of the porous medium, are replaced by basic equations considering average characteristics of the aquifer. In a locally isotropic porous medium, diffusion coefficients are scalar parameters. They generate second-order tensors representing the aquifer's average characteristics of diffusion. The principal directions of these tensors are the horizontal and vertical directions, provided that the local characteristics of the porous medium vary only with the depth.

In a stratified and locally isotropic porous medium the average solute dispersion tensor is a second-order tensor whose principal directions are the vertical direction and the two horizontal directions parallel and perpendicular to the hydraulic gradient, provided that the fluid flows horizontally and the hydraulic resistivity and dispersivities vary only with the depth.

If the local characteristics of the porous medium vary moderately in the horizontal direction, it is possible to refer to horizontally variable average characteristics of the aquifer instead of spatially variable local characteristics.

The average characteristics approach can also be used for the formulation of models referring to transport phenomena in anisotropic heterogeneous aquifers. In this case the local tensorial characteristics are replaced by other tensors representing the average characteristics of the aquifer.

Application of the aquifer's average characteristics simplifies the mathematical models and therefore leads to lower requirements for computer time and memory for the performance of the mathematical simulation.

Notation

a	parameter defined in (39).
a_1, a_2	longitudinal and transverse dispersivity, respectively.
b	parameter defined in (39).
c	parameter defined in (39).
c_s, c_w	specific heat of solid skeleton and fluid, respectively.
C	solute concentration.
C_o	solute concentration of reference.
d	aquifer's thickness.
$\overline{\overline{e}}$	local dispersion tensor.
e_{ij} $(i,j = x,y,z)$	components of the local dispersion tensor.
e_1, e_2	longitudinal and transverse dispersion coefficient, respectively.
$\overline{\overline{E}}$	average dispersion tensor.
E_{ij} $(i,j = x,y,z)$	components of the average dispersion tensor.
g	gravitational acceleration.
$\overline{\overline{I}}$	unit matrix.
$\vec{J}$, J	average gradient vector and its absolute value, respectively.
J_h, J_v	horizontal and vertical component of the average gradient.
J_i $(i = x,y,z)$	component of the average gradient.
$\vec{J}_\ell$	local gradient vector.
$\vec{k}$	unit vertical vector in the downward direction.
k	local mobility.
$\overline{\overline{k}}$	average mobility tensor.
k_h, k_v	horizontal and vertical mobility.
k_{ij} $(i,j = x,y,z)$	component of the average mobility tensor.
K	permeability.
$\vec{n}$, n	average specific flux vector and its absolute value, respectively.

n_h, n_v horizontal and vertical components of the average specific flux.

n_i $(i = x,y,z)$ components of the average specific flux.

$\vec{n}_\ell$ local specific flux vector.

N total flux.

N_h, N_v horizontal and vertical total flux.

N_i $(i = x,y,z)$ total fluxes in the directions of the coordinates.

P pressure.

$\vec{q}$, q specific discharge vector and its absolute value, respectively.

r local hydraulic resistivity.

$\overline{\overline{r}}$ average hydraulic resistivity tensor.

r_h, r_v horizontal and vertical hydraulic resistivity.

r_{ij} $(i,j = x,y,z)$ components of the average hydraulic resistivity tensor.

t time.

T temperature.

T_0 temperature of reference.

x,y,z coordinates.

$\bar{x},\bar{y},\bar{z}$ principal directions represented in Figure 4.

z' parameter defined in (13).

α coefficient of volumetric expansion.

α_s coefficient relating solute concentration with density.

γ coefficient defined in (6).

θ angle represented in Figure 1.

θ_i $(i = 1,2,3)$ angles represented in Figure 2.

κ local thermal diffusivity.

κ_s molecular solute diffusivity.

$\overline{\overline{\kappa}}$ average thermal diffusivity tensor.

Acknowledgment. The authors are indebted to Richard L. Naff from USGS Denver Federal Center, who reviewed the manuscript and provided valuable comments and suggestions.

References

Aris, R., On the dispersion of a solute in a fluid flowing through a tube, Proc. R. Soc. London, Ser. A, 235, 67-77, 1956.

Bentley, C. B., Aquifer test analyses for the Floridan aquifer in Flagler, Putnam and St. Johns counties Florida, U.S. Geol. Surv. Water Resour. Invest., 77-36, 1977.

Bermes, B. J., G. W. Leve, and G. R. Tarver, Geology and ground-water resources of Flager, Putnam and St Johns counties Florida, Fla. Geol. Surv. Rep. Invest., 32, 1963.

Gheorghitza, St. I., On the plane steady flow through inhomogeneous porous media, in Fundamentals of Transport Phenomena in Porous Media, pp. 73-85, Elsevier, New York, 1972.

Hyde, L. W., Principal aquifers in Florida, Map Ser. 16, Fla. Div. of Geol., Tallahassee, 1965.

Lagarde, A., Considerations sur le transfert de chaleur en milieu proeux, Rev. Inst. Fr. Pet., 20(2), 383-446, 1965.

Lichtler, W. F., Appraisal of water resources in the east-central Florida region, Rep. Invest. 61, Fla. Dep. of Nat. Resour., Bur. of Geol., Tallahassee, 1972.

Parker, G. F., G. E. Ferguson, and S. K. Love, Water resources of southeastern Florida, U.S. Geol. Surv. Water Supply Pap., 1255, 1955.

Pascale, C. A., Estimated yield of freshwater wells in Florida, Map Ser. 70, Fla. Bur. of Geol., Tallahassee, 1975.

Pfannkuch, H. O., Contribution a l'etude des dēplacement de fluides miscible dans un milieu poreux, Rev. Inst. Fr. Pet., 18(2), 215-270, 1963.

Puri, H. S., and R. O. Vernon, Summary of the geology of Florida and a guidebook to the classic exposures, Spec. Publ. 5, Fla. Geol. Surv., Tallahassee, 1964.

Rektory, K., Survey of Applicable Mathematics, p. 695, M.I.T. Press, Cambridge, 1969.

Rubin, H., Thermal convection in a nonhomogeneous aquifer, J. Hydrol., 50, 317-331, 1981.

Rubin, H., and B. A. Christensen, A simplified numerical simulation of mineralization processes in an aquifer, paper presented at the Summer Computer Simulation Conference, Soc. Computer Simulation, Denver, Colo., 1982a.

Rubin, H., and B. A. Christensen, Simulation of stratified flow in Floridan aquifer, paper presented at the 1982 ASCE Irrigation and Drainage Division Specialty Conference, Orlando, Fla., 1982b.

Saffman, P. E., Dispersion due to molecular diffusion and macroscopic mixing in flow through a network of capillaries, J. Fluid Mech., 6(3), 321-349, 1960.

Taylor, G. I., Dispersion of soluble matter in solvent flowing slowly through a tube, Proc. R. Soc., London, Ser. A, 219, 186-203, 1953.

Taylor, G. I., The dispersion of matter in turbulent flow through a pipe, Proc. R. Soc., London, Ser. A, 223, 446-468, 1954.

A Galerkin-Finite Element Two-Dimensional Transport Model of Groundwater Restoration for the In Situ Solution Mining of Uranium

James W. Warner and Daniel K. Sunada
Colorado State University, Fort Collins, Colorado 80523

Introduction

Contaminant transport problems in groundwater are becoming increasingly more common and increasingly more complex. Traditionally, transport problems have been concerned with contaminant migration away from waste disposal sites and with determination of the fate of various chemical constituents in the groundwater. More commonly, only conservative transport problems have been considered. Developing technologies such as in situ solution mining represent a new more complex contaminant transport problem in groundwater site restoration.

This paper describes a mathematical treatment of the transport and site restoration of contaminants subject to adsorption and desorption on the solid aquifer material. This exchange process was treated as an equilibrium-controlled reversible binary cation exchange reaction. This assumption requires that both chemical species involved in the exchange process be followed as they flow with the groundwater through the porous media. The mathematical formulation of the problem includes one equation describing groundwater flow plus two additional equations for solute transport (one equation for each of the chemical species involved in binary cation exchange). The transport equations are coupled through two additional equations that describe the cation exchange process. The partial differential equation for groundwater flow is solved first for the head distribution in the aquifer, and the two coupled partial differential equations for solute transport are then solved simultaneously for the dual changes in dissolved concentration for both chemical species. The adsorbed concentration for both exchanging solutes is then also solved for using the two equations describing the cation exchange process.

The above process for the solution of the transport equations with cation exchange was originally formulated by Rubin and James [1973] for one-dimensional groundwater flow with constant groundwater velocity. The method has received very little attention until now because of much simpler methods such as the Freundlich isotherm, which is applicable, in general, to contaminant migration problems. The method of Rubin and James is in this paper extended to two dimensions and coupled with the solution of the partial differential equation for groundwater flow. A computer model was constructed and applied to an actual field problem of groundwater restoration for a pilot scale in situ solution mining operation.

Cation Exchange

The exchange of cations adsorbed on a porous medium with cations contained in water flowing through the porous medium is generally referred to as cation exchange. This exchange process requires that a cation in solution be adsorbed onto the porous medium and simultaneously an adsorbed cation be released from the porous medium into solution. This cation exchange reaction is important in contaminant migration because it retards the movement of many contaminants which are in solution in the groundwater. Cation exchange is also important in solution mining because it causes certain contaminants to be accumulated during the mining process on the porous medium when the contaminant is in high concentrations in the groundwater. Later, during the restoration process when contaminant concentrations in the groundwater are lower, the contaminant is released back to the groundwater.

The exchange capability of a porous medium is expressed as the cation exchange capacity (CEC), which is a measure of the number of exchange sites that are available and is assumed to be a constant for a given porous medium. These exchange sites occur on the surface of the individual particles composing the porous medium. In natural groundwater systems the geologic materials that account for most of the CEC are the clay minerals. The structure of these clay minerals is such that there results a negative charge imbalance on the surface or between the lattices of the clay mineral which is neutralized by cations in the surrounding solution. Because this charge imbalance on the porous medium is negative, the anions in solution in the groundwater are mostly unaffected by this exchange process. The total number of sites available for cation exchange is constant, and these sites are always filled. Thus to remove one cation, another cation must replace it.

While the CEC of a porous medium is necessary to determine the magnitude of the cation exchange process, it is not the only factor which affects this process. The affinity for cation exchange varies considerably for different cations. In general, the affinity for adsorption onto the porous medium increases for a cation with an increase in the valence and in the atomic weight. Another important factor affecting this relationship is the ionic strength of the solution. For cations of different valences, the preference for adsorption for the cation with the higher valence decreases as the ionic strength of the solution increases. However, the ionic strength of the solution has little effect when the cations are of the same valence.

The relative proportion of the exchange sites filled by each cation in the exchange process can be expressed using an experimentally determined adsorption or exchange isotherm. The adsorption isotherm is a plot of the relative concentration in solution for a given cation, expressed as a ratio of dissolved concentration C to the total solution concentration C_0 versus the relative adsorbed concentration of that cation, expressed as a ratio of the adsorbed concentration $\bar{C}$ to the cation exchange capacity CEC. An example of an adsorption isotherm is shown on Figure 1. In experi-

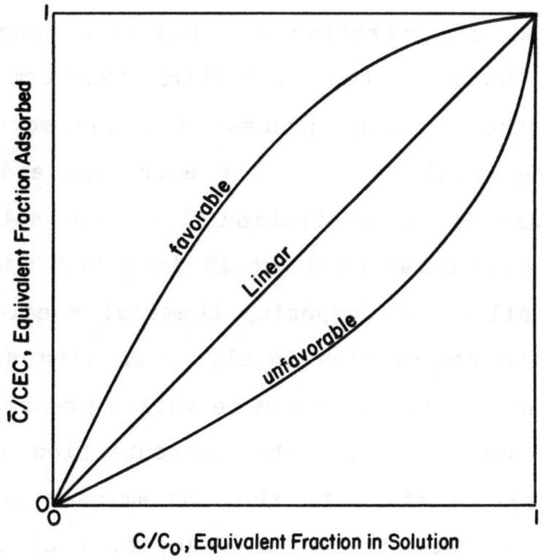

Fig. 1. Adsorption isotherm.

mentally determining an adsorption isotherm, the total solution concentration is held constant. Different adsorption isotherms are obtained for each chemical species. In addition, an infinite set of adsorption isotherms are obtained, depending on the nature of the porous medium, the total solution concentration, and the solution composition.

In contaminant migration problems the Freundlich isotherm is often used because of its simplicity to describe quantitatively the cation exchange reaction. The Freundlich isotherm is an empirical relationship given as [Freeze and Cherry, 1979].

$$\bar{C} = K_d C^{\alpha} \tag{1}$$

where

- C dissolved concentration;
- $\bar{C}$ adsorbed concentration;
- K_d constant called distribution coefficient;
- α constant.

The constants K_d and α are determined as the best fit to the experimentally determined adsorption isotherm. The Freundlich isotherm works well for most contaminant migration problems in which the total solution concentration and solution composition remain fairly constant. However, the Freundlich isotherm does not adequately describe the exchange process for the more general conditions of varying total solution concentration and solution composition. The adsorbed concentration $\bar{C}$ is not solely a function of the dissolved concentration C but is dependent upon the relative concentration of all other competing chemical species in solution. In groundwater site restoration involving in situ solution mining, the total solution concentration can be varied greatly. In essence, what is commonly done is that the concentration of a specified cation is increased relative to the contaminant concentration in order to facilitate the eluting or desorption of the contaminant from the porous medium. Under these more general conditions, the use of the Freundlich isotherm yields erroneous results.

In flowing groundwater systems the cation exchange reaction is normally viewed as being very rapid relative to the flow of the groundwater. Thus the process of adsorption and desorption is often considered as an equilibrium-controlled cation exchange reaction. For binary cation exchange (exchange between two cations) this can be expressed as

$$(Z_2\bar{C}_1)_{Ads} + (Z_1C_2)_{sol} \rightleftharpoons (Z_1\bar{C}_2)_{Ads} + (Z_2C_1)_{sol} \tag{2}$$

where C_i, $\bar{C}_i$, and Z_i represent the dissolved concentration, adsorbed concentration, and valence, respectively, for cation i. The preference for exchange of one cation for another cation can be expressed in this case in mathematical terms by the law of mass action given as [Helfferich, 1962]

$$K = \frac{[\bar{N}_2]^{Z_1}\,[C_1]^{Z_2}}{[\bar{N}_1]^{Z_2}\,[C_2]^{Z_1}} \tag{3}$$

where the square brackets denote activities, $\bar{N}_i$ denotes the absorbed concentration expressed in equivalent fractions ($\bar{C}_i$/CEC) and K is a constant called the selectivity coefficient.

The relationship between activities and concentrations for the solution phase is given as [Garrels and Christ, 1965]

$$[C_i] = \gamma_i C_i \tag{4}$$

where γ_i is the individual ion activity coefficient (dimensionless) for cation i in the solution phase.

Similarly, the relationship between activities and concentrations for the adsorbed phase is given as [Garrels and Christ, 1965]

$$[\bar{C}_i] = \bar{\lambda}_i \bar{N}_i \tag{5}$$

where $\bar{\lambda}_i$ is the rational activity coefficient (dimensionless) for cation i in the adsorbed phase.

Substitution of (4) and (5) into (3) yields

$$K = \frac{(\bar{\lambda}_2\ \bar{N}_2)^{Z_1}\ (\gamma_1\ C_1)^{Z_2}}{(\bar{\lambda}_1\ \bar{N}_1)^{Z_2}\ (\gamma_2\ C_2)^{Z_2}} \tag{6}$$

Equation (6) is then used to describe the exchange process. For the case of binary cation exchange with varying total solution concentration, (6) completely describes the exchange process. For multiple cation exchange with varying total solution concentration and composition, an equation similar to (6) is needed for each pair of competing cations to describe the exchange process.

Contaminant Transport With Adsorption

The equation describing the two-dimensional mass transport for a reacting solute subject to adsorption in flowing groundwater may be written as [Warner, 1981]

$$-\frac{\partial C}{\partial t} - \frac{\partial \bar{C}}{\partial t} = \frac{\partial}{\partial x_i}(C\ V_i) - \frac{\partial}{\partial x_i}(D_{ij}\frac{\partial C}{\partial x_j}) + \frac{wC'}{\phi b} \tag{7}$$

where

C dissolved concentration, m/l^3;

$\bar{C}$ adsorbed concentration, m/l^3;

C' dissolved concentration of the solute in the source or sink fluid, m/l^3;

V_i average interstitial velocity in the x_i direction, L/T;

D_{ij} component of the coefficient of hydrodynamic dispersion, L^2/T;

b saturated thickness, L;

ϕ effective porosity, dimensionless;

w volume flux per unit area, L/T;

t time, T.

Equation (7) is referred to as the convection-dispersion equation with adsorption and incorporates the effects of (1) convective transport in which chemical constituents are carried with the average motion of the flowing groundwater, (2) hydrodynamic dispersion, in which primarily variations in local velocity cause a spread of the chemical constituents from the average direction of groundwater flow, (3) fluid sources, in which water of a certain chemical concentration is injected into water of a different chemical concentration, and (4) adsorption reactions. Equation (7) contains two unknowns, C and $\bar{C}$. Thus at least one additional equation is needed for solution.

In contaminant migration problems the desirability of use of the Freundlich isotherm is readily apparent in the solution of (7). Equation (1) can be differentiated with respect to time and substituted directly into (7) to yield an equation in terms of only the dissolved concentration C. This equation may then be solved independently of all other equations. However, as was pointed out earlier, the Freundlich isotherm is applicable only under some rather stringent conditions, namely, that the total solution concentration and composition remain fairly constant.

A more general solution will now be given with particular application in groundwater site restoration of solution mining. For simplicity, the case of binary cation exchange will only be considered but the method can be extended to any number of exchanging chemical species. Both chemical species involved in the exchange process are followed as they flow with the groundwater through the porous media. To do this, (7) is written for both chemical species in terms of C_1 and C_2 as

$$- \frac{\partial C_1}{\partial t} - \frac{\partial \bar{C}_1}{\partial t} = \frac{\partial}{\partial x_i} (C_1 V_i) - \frac{\partial}{\partial x_i} (D_{ij} \frac{\partial C_1}{\partial x_j}) + \frac{wC'_1}{\phi b} \tag{8}$$

and

$$- \frac{\partial C_2}{\partial t} - \frac{\partial \bar{C}_2}{\partial t} = \frac{\partial}{\partial x_i} (C_2 V_i) - \frac{\partial}{\partial x_i} (D_{ij} \frac{\partial C_2}{\partial x_j}) + \frac{wC'_2}{\phi b} \tag{9}$$

Equations (8) and (9) are coupled through two additional equations which describe the cation exchange process. The first of these equations is the law of mass action given in (6). The second equation is obtained from noting that all of the exchange sites are filled and for binary cation exchange are filled with either $\bar{C}_1$ or $\bar{C}_2$. Also recall that the total exchange sites for a given porous medium is a constant and is equal to the cation exchange capacity. This can be expressed mathematically as

$$\bar{C}_1 + \bar{C}_2 = CEC \tag{10}$$

Thus (6), (8), (9), and (10) define a system of four equations with four unknowns. The direct simultaneous solution of these equations could be accomplished but is not desirable since (1) the problem would be very large (with any sort of numerical solution of value of C_1, $\bar{C}_1$, C_2, and $\bar{C}_2$ must be solved for at each of the n nodes, the dimension of the problem would therefore be 4n x 4n), and (2) the inclusion of the law of mass action results in a nonlinear set of equations which is difficult to solve. The complexity and size of the problem is reduced using the following procedure. Equation (6) is rewritten as

$$K = \frac{(\bar{\lambda}_2 \ (\bar{C}_2/CEC))^{Z_1} \ (\gamma_1 \ C_1)^{Z_2}}{(\bar{\lambda}_1 \ (\bar{C}_1/CEC))^{Z_2} \ (\gamma_2 \ C_2)^{Z_1}} \tag{11}$$

Substitution of (10) into (11) and differentiation with respect to time yields

$$K\bar{\lambda}_1^{Z_2}\,(\bar{C}_1/CEC)^{Z_2}\,\gamma_2^{Z_1}\,Z_1C_2^{Z_1-1}\,\frac{\partial C_2}{\partial t} + K\bar{\lambda}_1^{Z_2}\,\gamma_2^{Z_2}\,C_2^{Z_1}\,Z_2\,\frac{\bar{C}_1^{Z_2-1}}{CEC^{Z_2}}\,\frac{\partial \bar{C}_1}{\partial t}$$

$$-\,\bar{\lambda}_2^{Z_1}\,(1-\bar{C}_1/CEC)^{Z_1}\,\gamma_1^{Z_2}\,Z_2C_1^{Z_2-1}\,\frac{\partial C_1}{\partial t} \tag{12}$$

$$+\,\bar{\lambda}_2^{Z_1}\,\gamma_1^{Z_2}\,C_1^{Z_2}\,Z_1\,\left(1-\frac{\bar{C}_1/CEC}{CEC}\right)^{Z_1-1}\,\frac{\partial \bar{C}_1}{\partial t} = 0$$

Now let

$$g_1 = -\bar{\lambda}_2^{Z_1}\,(1-\bar{C}_1/CEC)^{Z_1}\,\gamma_1^{Z_2}\,Z_2C_1^{Z_2-1} \tag{13}$$

$$g_2 = -\,K\bar{\lambda}_1^{Z_2}\,(\bar{C}_1/CEC)^{Z_2}\,\gamma_2^{Z_1}\,Z_1C_2^{Z_1-1} \tag{14}$$

and

$$g_3 = -K\bar{\lambda}_1^{Z_2}\,\gamma_2^{Z_1}\,C_2^{Z_1}\,Z_2\,\frac{\bar{C}_1^{Z_2-1}}{CEC^{Z_2}} - \bar{\lambda}_2^{Z_1}\,\gamma_1^{Z_2}\,C_1^{Z_2}\,Z_1\,\frac{(1-\bar{C}_1/CEC)^{Z_1-1}}{CEC} \tag{15}$$

Substitution of (13), (14), and (15) into (12) yields

$$g_1\,\frac{\partial C_1}{\partial t} - g_2\,\frac{\partial C_2}{\partial t} = g_3\,\frac{\partial \bar{C}_1}{\partial t} \tag{16}$$

or

$$\frac{\partial \bar{C}_1}{\partial t} = \frac{g_1}{g_3}\,\frac{\partial \bar{C}_1}{\partial t} - \frac{g_2}{g_3}\,\frac{\partial C_2}{\partial t} \tag{17}$$

From the differentiation of (10) the following relationship is also obtained:

$$\frac{\partial \bar{C}_1}{\partial t} = -\,\frac{\partial \bar{C}_2}{\partial t} \tag{18}$$

Substitution of (18) into (17) yields

$$\frac{\partial \bar{C}_2}{\partial t} = \frac{g_2}{g_3} \frac{\partial C_2}{\partial t} - \frac{g_1}{g_3} \frac{\partial C_1}{\partial t} \tag{19}$$

Further substitution of (17) and (19) into (8) and (9), respectively, yields

$$-\left(1 + \frac{g_1}{g_3}\right) \frac{\partial C_1}{\partial t} + \left(\frac{g_2}{g_3}\right) \frac{\partial C_2}{\partial t} = \frac{\partial}{\partial x_i} (C_1 V_i) - \frac{\partial}{\partial x_i} \left(D_{ij} \frac{\partial C_1}{\partial x_j}\right) + \frac{wC_1'}{\phi b} \tag{20}$$

and

$$-\left(1 + \frac{g_2}{g_3}\right) \frac{\partial C_2}{\partial t} + \left(\frac{g_1}{g_3}\right) \frac{\partial C_1}{\partial t} = \frac{\partial}{\partial x_i} (C_2 V_i) - \frac{\partial}{\partial x_i} \left(D_{ij} \frac{\partial C_2}{\partial x_j}\right) + \frac{wC_2'}{\phi b} \tag{21}$$

The unknowns $\bar{C}_1$ and $\bar{C}_2$ do not appear explicitly in either (20) or (21) but are embedded in the variable coefficients g_1, g_2, and g_3. This allows a sequential solution technique to be used wherein (20) and (21) are solved iteratively with the partial differential equation for groundwater flow given as

$$\frac{\partial}{\partial x_i} \left(T_{ij} \frac{\partial h}{\partial x_j}\right) = S \frac{\partial h}{\partial t} + w \tag{22}$$

where h is the potentiometric head (L), T_{ij} is the transmissivity (L^2/T), S is the storage coefficient, and all other variables are as previously defined. The groundwater flow equation (22) is first solved for the head distribution in the aquifer at any specified time. From this head distribution the values of groundwater velocity V and the dispersion coefficient D are obtained. Using the initial values of C_1, $\bar{C}_1$, C_2, and $\bar{C}_2$, initial values of g_1, g_2, and g_3 are obtained from (13), (14), and (15). The parameters V, D, g_1, g_2, and g_3 are then held constant over the next time interval, and the two coupled solute transport equations (20) and (21) are then solved simultaneously for new values of C_1 and C_2. These new values of C_1 and C_2 are used in (6) and (10) describing the cation exchange reaction to calculate new values of $\bar{C}_1$ and $\bar{C}_2$. The groundwater flow equation (22) is then solved again for the head distribution in the aquifer at a new specified time.

From this new head distribution, updated values of groundwater velocity V and dispersion coefficient D are obtained. Using the new values of C_1, $\bar{C}_1$, C_2, and $\bar{C}_2$ obtained in the last iteration, updated values of g_1, g_2, and g_3 are calculated, and the iterative procedure is repeated as before.

Numerical Solution

The partial differential equation for groundwater flow (equation (22)) and the two coupled partial differential equations for solute transport (equations (20) and (21)) were solved subject to the appropriate boundary conditions by the Galerkin finite element method. In the finite element method, approximating integral equations are formed to the original partial differential equations. The integration of these integral equations is required. Because the transport equations contain the nonlinear variable coefficients g_1, g_2, and g_3, which are dependent on concentration changes, these integrations must be repeated frequently. Triangular elements and linear shape functions were used in the solution by the Galerkin finite element method. This allowed the use of some very powerful integration formulas which considerably reduced the computational effort and time required to make these integrations. The interested reader is referred to Warner [1981] for a complete description of the finite element solution to this problem or to Segerlind [1976] or Pinder and Gray [1977] on the general procedure of the finite element method and to Pinder and Frind [1972] for particular application to the groundwater flow equation or to Pinder [1973] for particular application to the solute transport equation without adsorption.

Application to a Field Problem

In situ solution mining is a relatively new technology that is being used with increased frequency throughout the uranium mining industry. The method consists of injecting through wells a lixi-

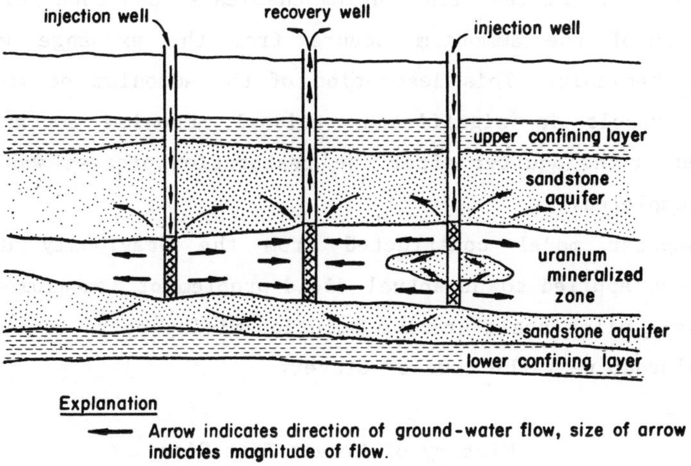

Fig. 2. Flow pattern through a vertical section of a single cell in a leach field.

viant (consisting of a leaching chemical and an oxidizing agent) into the sandstone formation which contains the uranium deposits. The uranium ore is preferentially dissolved from the host rock and the uranium-bearing groundwater is recovered through pumping wells. The method is diagrammatically shown on Figure 2. The interested reader is referred to Warner [1981] or Larson [1978] concerning the chemical reactions involved in the leaching of the uranium. A solution of ammonium bicarbonate-hydrogen peroxide is the most commonly used lixiviant in the solution mining of uranium. The ammonium is adsorbed on the clays in the aquifer during mining. Immediately after mining, the groundwater is contaminated with high levels of many chemical constituents including ammonium. Much concern has been expressed about the relatively high concentrations of ammonium found in the groundwater immediately following mining.

After mining, restoration of the contaminated aquifer is required. The normal procedure is to pump the contaminated groundwater from the aquifer which is replaced by groundwater entering the mined area from the surrounding unaffected aquifer; or alternatively the contaminated groundwater is pumped from the aquifer, purified and possibly fortified with eluting chemicals, and then reinjected. Desorption of the ammonium occurs from the exchange with the eluting chemicals. This desorption of the ammonium occurs slowly and may result in significant residual ammonium concentrations adsorbed in the aquifer after the restoration process was thought to be completed.

A computer model constructed from the previously developed theory was applied to an actual field problem of ammonium restoration for a pilot scale uranium solution mining operation in northeast Colorado near the town of Grover.

History of Grover Test Site

The Grover uranium deposit was discovered in 1970, with the major uranium mineralization occurring at a depth of about 200-250 ft (60-75 m) below land surface in the Grover sandstone member of the Laramie Formation. In 1976, an application was made to the State of Colorado by a private venture to operate a pilot scale in situ uranium solution mine at this site. The permit was granted and in June 1977, solution mining was initiated using a three five-spot pattern with a well spacing of 40 ft (12 m) (Leach Field 1, Figure 3). An ammonium bicarbonate-hydrogen peroxide solution was used as the lixiviant. In September 1977, after about 90 days of operation, mining was terminated at this leach field.

Fourteen monitoring wells were installed to detect migration of any contaminated groundwater from the mine site (Figure 3). No migration of the lixiviant was ever detected in any of the monitoring wells. After mining, core holes were drilled to determine the areal extent of groundwater contamination outside the leach field. It was determined that the ammonium contamination of the groundwater extended less than one cell distance (approximately 40 ft (12 m)) from the leach field. In May 1978, site restoration was initiated. The restoration was conducted in three separate phases. During the first phase, clean water recycling was used. This consisted of removing the contaminated groundwater by pumping, purifying the

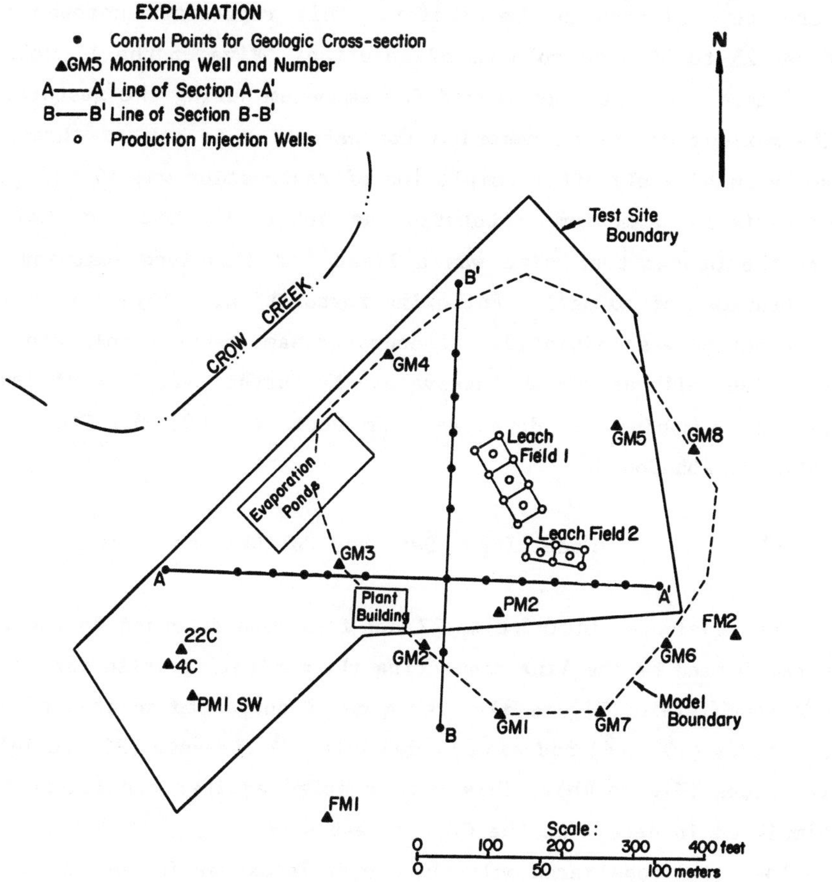

Fig. 3. Layout of the Grover test site.

groundwater by reverse osmosis, and reinjecting into the aquifer. The first phase lasted approximately 54 days. To facilitate desorption of the ammonium, a calcium chloride solution was injected during phase 2. In this second restoration phase the calcium concentration in the injection solution was increased gradually in increments of 250 mg/l to a level of about 1000 mg/l. The ammonium in solution was removed using an air stripping technique. This chemical treatment technique lasted approximately 64 days. During phase 3, clean water recycling was reimplemented. This third phase of restoration lasted approximately 118 days.

During restoration approximately 9.7 million gallons of water were recycled through the aquifer. This represents somewhere between 25 to 50 pore volumes of solution. Five of the 11 wells in the leach field were monitored for ammonium during the restoration. The maximum dissolved ammonium concentration in any of these five wells immediately after completion of restoration was 15 mg/l. The criteria for ammonium restoration as set by the State of Colorado for the Grover test site was a limit for dissolved ammonium concentrations of 50 mg/l. Following restoration, a 1-year stabilization period was initiated. Groundwater samples were collected from the five wells at 90-day intervals. No further significant desorption of ammonium was detected. In October 1980, the Grover test site was abandoned.

Model Input Data and Calibration

The model area included the leach field and extended to the perimeter formed by the line connecting the monitoring wells surrounding the leach field (Figure 3). The model encompassed an area of about 5.7 acres (.02 km^2) and was divided into 204 elements (Figure 4a) and 122 nodes (Figure 4b). This grid enabled aquifer conditions to be simulated in detail at the Grover test site.

The model considered only the Grover Sandstone in the vicinity of the test site. The average thickness of the aquifer within the model area was 70 ft (21 m). The transmissivity of the Grover Sandstone was about 103 ft^2/d (9.6 m^2/d) and the storage coefficient was about 0.3×10^{-4}. The effective porosity of the aquifer based on 21 samples was 37%. The natural regional direction of groundwater flow within the model area was to the southeast with an average gradient of 15.1 ft/mi (2.86 m/km). No field data were available on dispersivity. A value for dispersivity of 20 ft (6 m) was used in the model which ensured numerical stability. The cation exchange capacity (CEC) of the Grover Sandstone was determined from 12 samples and ranged from 4.4 to 21.0 meq/100 g of solid sample with an average of 9.3 meq/100 g. These values of CEC are typical of sandstone. During the calibration of the model the CEC of the

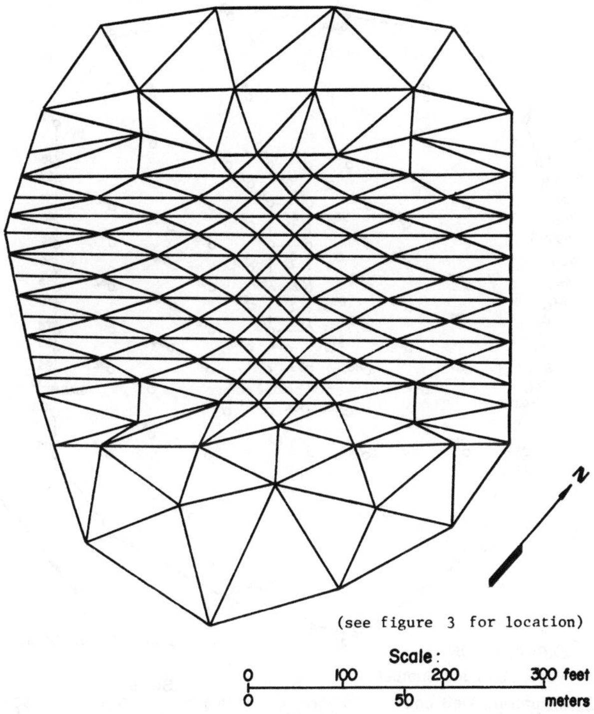

Fig. 4a. Model grid.

aquifer was chosen within the limits of the field data to achieve the best fit.

The model simulated the binary exchange between ammonium and calcium. This necessitated lumping all major cations except ammonium (i.e., calcium, magnesium, sodium, and potassium) into an equivalent concentration of calcium. The resulting premining concentration of calcium was 92 mg/l. The premining concentration of ammonium was essentially zero. In the leach field the resulting postmining concentration of calcium was 327 mg/l, and the postmining concentration of ammonium was approximately 500 mg/l. The adsorbed concentrations of calcium and ammonium were calculated using the model assuming equilibrium conditions with the solution concentrations. For the initial premining condition essentially all of the exchange sites were assumed filled by the calcium and none by the ammonium. For the postmining condition, model calculations indicated that

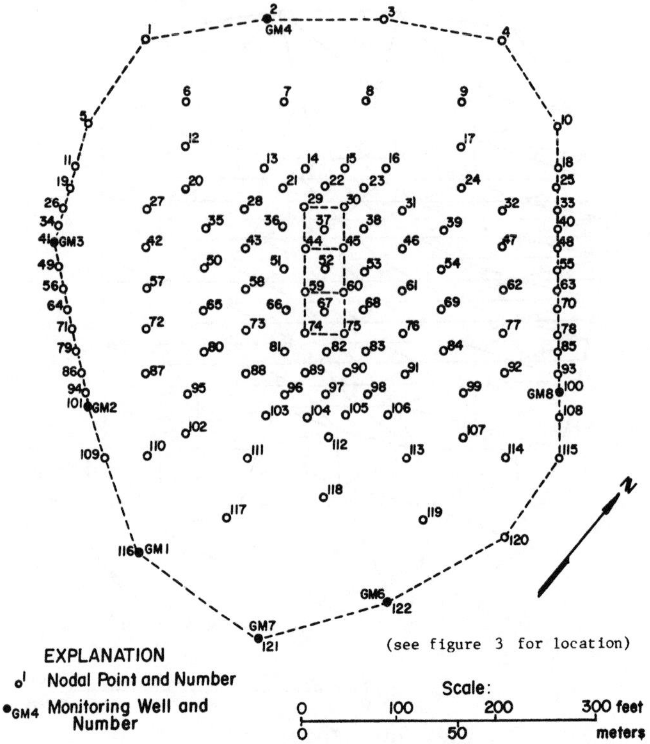

Fig. 4b. Nodal numbering system.

approximately 54% of the exchange sites were filled by the ammonium and, approximately 46% of the sites were filled by the calcium.

The adsorption isotherm for the exchange of ammonium and calcium for a montmorillonite clay defined by equation (6) was obtained from data from Laudelout et al. [1968]. The activity coefficients for ions in solution in (6) were obtained from the Debye-Huckel equation [Hem, 1961]. However, the activity coefficients for the adsorbed phase were not known. Because of this, the selectivity coefficient K_c was defined as

$$K_c = \frac{K \, \bar{\lambda}_1^{Z_2}}{\bar{\lambda}_2^{Z_1}} \tag{23}$$

where K_c is the selectivity coefficient for activity in the solution phase but uncorrected for activity in the adsorbed phase. Recall

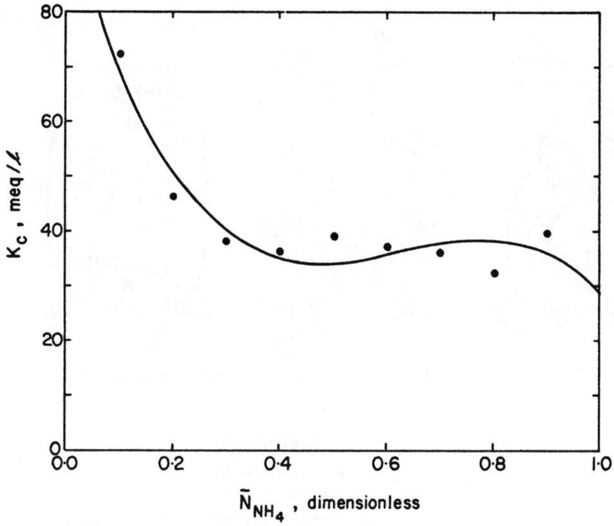

Fig. 5. K_c versus $\bar{N}_{NH_4}$.

that the selectivity coefficient K was a constant but the selectivity coefficient K_c is a variable. From the data of Laudelout et al. [1968] the plot of K_c versus $\bar{C}_{NH4}/CEC$ shown on Figure 5 was obtained. A cubic equation for K_c was fitted to these data and used in the model.

During restoration, three different pumpage and injection patterns (Figure 6) were used to ensure a fairly clean sweep of the aquifer. During phase 1 and 2 of restoration the pumpage rate was approximately 40,000 gal./d. During phase 3 the pumpage rate was reduced to about 29,000 gal./d. During all three phases, pumpage rates slightly exceeded injection rates.

Calibration of the transport model consisted of a comparison between observed or measured concentrations and model-calculated ammonium concentrations. A plot of ammonium concentrations versus cumulative water recovered during restoration is shown on Figure 7 for both observed and model-calculated values. There were some erratic fluctuations in the observed ammonium concentrations indicating noise in the data, probably due to either analytical measurement errors or unknown variations in aquifer properties. The plot

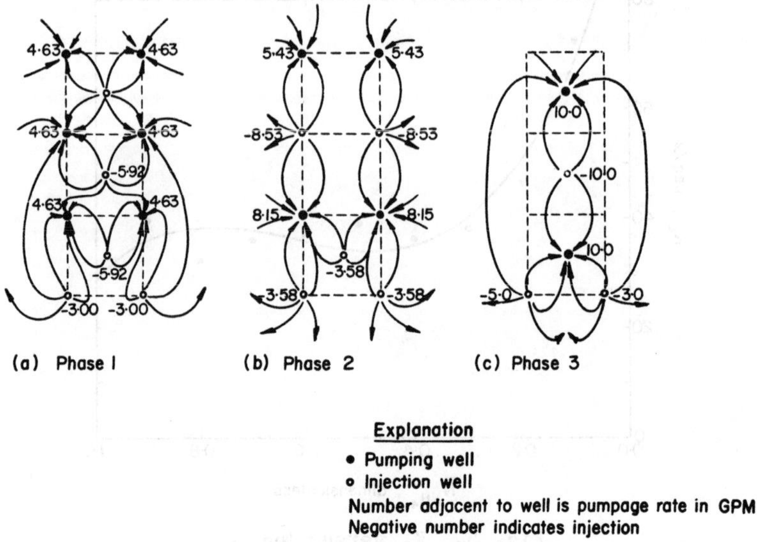

Fig. 6. Pumpage and injection rates.

was visually smoothed to eliminate this noise. For comparison, a plot of model-calculated ammonium concentrations using conservative transport is also shown on Figure 7. The model was calibrated to reduce the difference between the observed and the model-calculated total ammonium removed during restoration.

During phase 1, clean water recycling was used, and the observed ammonium concentrations dropped in the recovery water from an initial level of about 490 mg/l to an estimated 85 mg/l (Figure 7). The model calculated a similar but somewhat larger drop to about 61 mg/l (Figure 7). Most of the error between the observed and model-calculated ammonium concentrations is introduced during the initial 0.5 million gallons of water recovered. After that, the model calculations roughly parallel observed concentrations during the phase 1 restoration. The difference may be due to many factors, such as poorly defined initial concentrations of ammonium in the model or poor initial efficiency of the reverse osmosis unit used to purify the contaminated water removed from the aquifer. During phase 1 of the restoration the contaminated recovery water was processed through a single reverse osmosis unit. During later

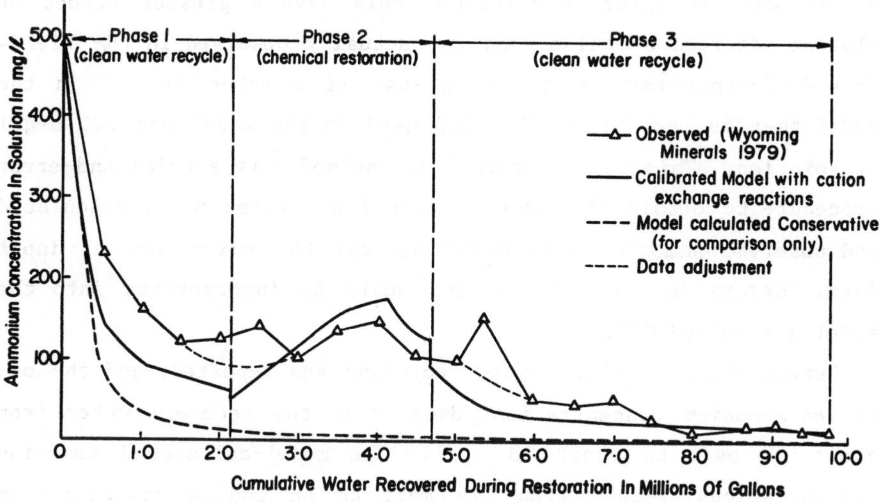

Fig. 7. Ammonium concentration versus cumulative water recovered during restoration.

restoration phases, secondary and tertiary reverse osmosis units were also used. In the model an average efficiency rate of removal of 95% for ammonium and 97.5% for calcium was used.

During phase 2 a calcium chloride solution was injected to increase the rate at which the adsorbed ammonium was desorbed from the aquifer. This resulted in observed ammonium concentrations in the recovery water increasing from an estimated 85 mg/l at the beginning of phase 2 to a peak of 163 mg/l. No measurable increase in calcium concentrations in the recovery water occurred during phase 2 until shortly before the peak in ammonium concentrations was reached (roughly at about the 4 million gallon point on Figure 7). The model-calculated ammonium concentrations agreed remarkably well with the observed concentrations during phase 2. The model-calculated ammonium concentrations increased from about 61 mg/l at the beginning of phase 2 to a peak of 172 mg/l. The model-calculated peak occurred at nearly the same time as did the peak in ammonium concentrations observed in the field (Figure 7). The model-calculated ammonium concentrations at the end of phase 2 was 124 mg/l compared to 108 mg/l measured in the field.

The model calculations indicated that during phase 2 the addition of the calcium chloride solution would have a greater effect of eluting adsorbed ammonium than was actually observed in the field. This difference may be caused by use of a value of CEC in the model that is too large. The CEC used in the model was 300 meq/l of solution. This was determined in the model by a trial-and-error procedure to obtain the overall best fit between model-calculated and observed ammonium concentrations for the given set of input data. Errors in these input data would be incorporated into the model determined CEC.

During phase 3, clean water recycling was repeated, and the observed ammonium concentrations dropped in the recovery water from about 108 mg/l to about 13 mg/l. The model-calculated ammonium concentrations dropped from 124 mg/l to an ending concentration of about 16 mg/l.

In general, the agreement is fairly close between the observed and model-calculated ammonium concentrations in the recovery water shown on Figure 7. Differences are attributed to four primary factors: (1) errors in the model input data, (2) channelizing of the flow in the aquifer in the field, (3) insufficient number of nodes used in the model, particularly in the vicinity of the leach field to provide a more detailed definition of model-calculated ammonium concentrations in the recovery water, and (4) assumptions such as two-dimensional groundwater flow and binary cation exchange upon which the model was developed may contribute also to the difference. Errors in the model input data include unknown variations in aquifer properties (e.g., permeability, effective porosity, CEC), poorly defined initial concentrations of ammonium, and errors in the reconstruction of the restoration processes used (e.g., errors in pumpage and injection rates, errors in efficiency of reverse osmosis process used to purify the recovered contaminated groundwater).

Discussion of Results

In the field, groundwater was sampled during restoration for ammonium contamination at only a few selected points, namely at five of the production and recovery wells in the leach field. The extent of groundwater contamination after restoration beyond these few points could only be surmised. One of the major advantages of computer modeling is the capability to simulate conditions in the aquifer at essentially any place any time.

The model for the Grover test site was used to calculate both dissolved and adsorbed postrestoration ammonium concentrations at all 122 nodes in the model. The model results indicate that at all but two of the nodes the dissolved ammonium concentrations were less than the 50-mg/l limit. Thus, considering the uncertainties in the model data, the model results do not contradict and, in general, support that the aquifer was restored to within the set limit.

The model results for postrestoration adsorbed ammonium concentrations indicate that potentially large quantities of adsorbed ammonium remain in the aquifer at the Grover test site. The maximum adsorbed ammonium concentrations calculated by the model was greater than 1000 mg/l. At 30 of the nodes in the model the adsorbed concentrations were greater than 250 mg/l.

In general, the restoration of the aquifer was more complete within the leach field than for the immediate surrounding area which had the highest model-calculated ammonium concentrations. Through the processes of hydrodynamic dispersion and convective transport, the ammonium was spread to the area surrounding the leach field. Slight overpumping in both the mining and restoration phases maintained regional groundwater gradients toward the leach field. During mining and during certain phases of the restoration, the outer wells in the leach field were used for injection. This caused local groundwater gradients near these wells opposite to the regional groundwater gradient and caused some of the groundwater to follow a path initially away from the leach field into the surrounding aquifer. The model results indicate that this ammonium remained more or less in place during later restoration efforts.

The mobility of ammonium in groundwater is low, and the process of cation exchange by itself would not probably result in high future dissolved ammonium concentrations in the groundwater. However, the conversion of the ammonium to nitrate would increase its mobility substantially and coupled with the process of cation exchange could possibly result in future high nitrate concentrations in the groundwater. The potential to convert ammonium to nitrate in groundwater is not known but is thought to be a reasonable possibility.

Conclusion

In situ solution mining represents a new, more complex contaminant transport problem in site restoration than traditional contaminant transport problems. The simulation of groundwater site restoration of contaminants subject to adsorption and desorption on the solid aquifer material requires a greater sophistication than previous solute transport models because of the complex chemistry involved. The solute transport model developed in this study was demonstrated to be capable of simulating binary cation exchange between two reacting solutes. The model was applied to an actual field problem of groundwater restoration for a pilot scale solution mining operation. This represents the first application known to the author of a two-dimensional transport model incorporating cation exchange reactions to a complex contaminant transport problem of site restoration for actual field data.

If in situ solution mining is to meet expectations that it will become a new major mining method, then it is important to evaluate the environmental impact of the method on the groundwater system. The model developed in this study is presented as a basic working tool to be used by regulatory agencies, mining companies, and others concerned with groundwater restoration for in situ solution mining. The model should be readily adaptable to many other field problems. The model can also be used as a predictive tool for evaluating alternative restoration strategies. The usefulness of such a predictive tool for planning purposes would include an assessment of the tradeoff benefits between differing levels of restoration efforts versus desired restoration objectives, restoration time requirements, total volumes of water required for restoration, effects of varying pumpage and injection rates, and effects of varying eluting concentrations.

Acknowledgments. The authors would like to acknowledge the support of the Environmental Protection Agency, the U.S. Geological Survey, and the Colorado State Experiment Station Project 110 for support in funding this study.

References

Freeze, R. A., and J. A. Cherry, Groundwater, 604 pp., Prentice-Hall, Englewood Cliffs, N.J., 1979.

Garrels, R. M., and C. L. Christ, Solutions, Minerals and Equilibria, 450 pp., Freeman, Cooper, San Francisco, Calif., 1965.

Helfferich, F., Ion Exchange, 624 pp., McGraw-Hill, New York, 1962.

Hem, J. D., Calculation and use of ion activity, U.S. Geol. Surv. Water Supply Pap., 1535-C, 17 pp., 1961.

Larson, W. C., Uranium in situ leach mining in the United States, Inf. Circ. 8777, U.S. Bureau of Mines, 68 pp., 1978.

Laudelout, H., et al., Thermodynamics of heterovalent cation exchange reactions in a montmorillonite clay, Trans. Faraday Soc., 64(6), 1477-1488, 1968.

Pinder, G. F., A Galerkin-finite element simulation of groundwater contamination on Long Island, New York, Water Resour. Res., 9(6), 1657-1669, 1973.

Pinder, G. F., and E. O. Frind, Application of Galerkin's procedure to aquifer analysis, Water Resour. Res., 8(1), 108-120, 1972.

Pinder, G. F., and W. G. Gray, Finite Element Simulation in Surface and Subsurface Hydrology, 295 pp., Academic, New York, 1977.

Rubin, J., and R. N. James, Dispersion-affected transport of reacting solutes in saturated porous media: Galerkin method applied to equilibrium-controlled exchange in unidirectional steady water flow, Water Resour. Res., 9(5), 1332-1356, 1973.

Segerlind, L. J., Applied Finite Element Analysis, 422 pp., John Wiley, New York, 1976.

Warner, J. W., A Galerkin-finite element two-dimensional transport model of groundwater restoration for the in situ solution mining of uranium, Ph.D. dissertation, Colo. State Univ., Ft. Collins, 1981.

A Unified Approach to Regional Groundwater Management

Robert Willis
Humboldt State University, Arcata, California 95521

Introduction

The management of groundwater resources and the evaluation of the hydrologic and environmental impacts associated with groundwater development is commonly approached using simulation or optimization models of the aquifer system. Simulation models are predictive models of the hydraulic response of the groundwater system. In simulation modeling, a set of groundwater management policies is analyzed to determine a probable response of the aquifer system. From this information, a policy is then determined which best meets the objectives of the management problem. However, in simulation the policies are inherently nonoptimal. They are nonoptimal in an operational sense in that only a limited number of alternatives can usually be analyzed. Furthermore, the trade-offs associated with the system's economic or hydrologic objectives are difficult to determine. In contrast, however, optimization modeling represents a unified approach to groundwater management. Optimization modeling identifies the optimal planning, design, and operational policies and the trade-offs in the system's objectives. Moreover, optimization modeling can also generate the set of noninferior solutions to multiobjective groundwater planning problems.

The objective of this paper is to present an optimization methodology for regional groundwater management. Specifically, it will be shown how the response equations for confined and unconfined aquifer systems can be incorporated within the framework of an optimal planning model. As a result, the hydraulic response of the aquifer system is an integral part of the optimization model. In the optimization methodology, the groundwater planning problem is formulated as a multiobjective optimization model. The methodology is applied to the Yun Lin Basin, Taiwan, to determine the optimal groundwater extraction pattern.

Response Equations

The response or transfer equations of the groundwater system are those equations relating the stated variables of the aquifer and the proposed planning or management policies. As has been discussed by Maddock [1972], Willis and Dracup [1973], and Aguado and Remson [1974], the technique transforms the partial differential equation of the groundwater system via Green's functions, finite difference or finite element methods. These resulting equations may be imbedded within the constraint region of the planning or design problem, or equivalently, the problem can be formulated as a problem in optimal control [Willis and Newman, 1977].

Confined or Leaky Aquifer System

We assume that the surface-groundwater system may be represented by the vertically averaged continuity equation for a leaky aquifer [Cooley, 1974]:

$$\nabla \cdot \underline{\underline{T}} \nabla h + \sum_{i \varepsilon \Omega} Q_i \; \delta(x-x_i) \; \delta(y-y_i) \pm S^* = S \frac{\partial h}{\partial t} \tag{1a}$$

where $\underline{\underline{T}}$ is the transmissivity tensor (L^2/T), h is the hydraulic head (L), S is the storage coefficient, and S* is a source or sink term, e.g., leakage. Ω is an index set defining the location of all wells in the basin and $\delta(\;)$ is the Dirac delta function.

The boundary conditions of the aquifer system may be expressed as

$$h(\underline{x},t) = h^* \text{ on } \mu_1 \tag{1b}$$

$$\underline{q} \cdot \underline{n} = q^* \text{ on } \mu_2 \tag{1c}$$

where μ_1 and μ_2 define the boundary of the basin, h* is the known potential, $\underline{n}$ is the outward pointing unit normal to μ_2, and q* is the given flux. Generally, these equations are time-dependent boundary conditions.

Equation (1) may be transformed into a system of ordinary differential equations with the Galerkin finite element method. The transformed equations may be written as [Pinder and Frind, 1972]

$$C\dot{h} + H\underline{h} + \underline{f} = \underline{0} \tag{2a}$$

$$\underline{h}(0) = \underline{h}^o \tag{2b}$$

where $\underline{h}$ now represents the finite element approximation to the hydraulic head; $\underline{h}^o$ are the initial conditions for the problem. The C and H coefficient matrices contain the storage coefficients and transmissivities, respectively. The f vector contains the Dirichlet and Newmann boundary conditions and importantly, the planning policies [Willis, 1976b]. Equation (2) can also be explicitly written as a system of ordinary differential equations in time as

$$\dot{\underline{h}} = A\underline{h} + \underline{g} \tag{3}$$

where $A = -C^{-1}$ H and $\underline{g} = -C^{-1}$ $\underline{f}$.

Unconfined Aquifer System

Assuming Dupuit assumptions are valid for unconfined groundwater, the vertically averaged Boussinesqu equation can be expressed as [Cooley, 1974]

$$\nabla\cdot\underline{\underline{k}}\nabla h^2+2\sum_{i\varepsilon\Omega} Q_i\ \delta(x-x_i)\delta(y-y_i)+2R=2Sy\frac{\partial h}{\partial t} \tag{4}$$

where $\underline{\underline{k}}$ the hydraulic conductivity tensor [L/T], Sy is the specific yield, and R[L/T] is the recharge occurring in the aquifer.

Equation (3) is, however, a nonlinear function of the hydraulic head. Boundary and initial conditions for the problem are again summarized in (1). Finite difference or finite element methods may be used to transform the partial differential equation into a system of nonlinear ordinary differential equations. These transformed equations may be expressed as

$$D\underline{h} + E\underline{h}^2 + \underline{r} = \underline{0} \tag{5}$$

where the coefficient matrices D and E contain the specific yield and conductivity. Planning or operational policies, the recharge, and boundary conditions are contained in the $\underline{r}$ vector. Again, $\underline{h}$ represents the vector of the hydraulic head at all nodal points in the system.

Simplifying (5), we have

$$\underline{h} = A\underline{h}^2 + \underline{g} \tag{6}$$

where now $A = -D^{-1}E$ and $\underline{g} = -D^{-1}\underline{r}$. As will be discussed, we choose to linearize these equations using quasilinearization [Bellman and Kalaba, 1965]. Assuming a trial solution to (6), $\underline{h}^k$, and expanding about the solution using a generalized Taylor series, we have

$$\underline{h}^{k+1} = A\underline{h}^{2,k} + \underline{g}^k - 2A\underline{h}^{2,k} + 2AH^k\underline{h}^{k+1} \tag{7a}$$

where H^k is a diagonal matrix containing $\underline{h}^k$; that is $H_{11}{}^k = h_1{}^k$, $H_{22} = h_2{}^k$, etc. Simplifying, we have the linear system of ordinary differential equations,

$$\dot{\underline{h}}^{k+1} = A^k\underline{h}^{k+1} + g^k \tag{7b}$$

where $A^k = 2AH^k$ and $g^k = \underline{g}^k - A\underline{h}^{2,k}$.

Solution of the Response Equations

The response equations of the groundwater system are usually solved using conventional finite difference approximations. Here, however, (3) or (7) will be solved analytically by using the matrix calculus. The general solution of these equations is [Bellman, 1960]

$$\underline{h}(t) = e^{At}\,\underline{h}_o + \int_0^t e^{A(t-\tau)}\,\underline{g}(\tau)d\tau \tag{8}$$

Assuming that the planning or management policies and the system's boundary conditions are constant over a period T,

$$\underline{h}(t) = e^{At}\underline{h}_o + (A^{-1}e^{At} - A^{-1})\underline{g} \qquad 0 \leq t \leq T \tag{9}$$

The matrix exponential e^{At} can be evaluated by $A=RQR^{-1}$. The matrix R contains the eigenvectors of A, and Q is a diagonal matrix containing the eigenvalues of A. As a result $e^{At}=e^{RQR^{-1}t}$ is simply $R\hat{Q}R^{-1}$, where $\hat{Q}$ is again a diagonal matrix; however, the elements are now $e^{\lambda_i t}$, where λ_i is the ith eigenvalue of the system. Simplifying, we have

$$h(t) = R\hat{Q}R^{-1}\underline{h}_o + A^{-1}(R\hat{Q}R^{-1} - I)(-C^{-1})\underline{f} \tag{10a}$$

$$\underline{h}(t) = A_1(t)\underline{h}_o + A_2(t)\underline{f} \tag{10b}$$

here,

$$A_1(t) = RQR^{-1} \text{ and } A_2(t) = A^{-1}(I - R\hat{Q}R^{-1})C^{-1}$$

For a series of planning periods t_1, t_2, t_m of equal length T, the equations may be expressed as

$$\underline{h}^1 = A_1(T)\underline{h}^o + A_2(T)\underline{f}_1$$

$$\underline{h}^2 = A_1(T)\underline{h}^1 + A_2(T)\underline{f}_2 \tag{10c}$$

$$\underline{h}^m = A_1(T)\underline{h}^{m-1} + A_2(T)\underline{f}_m$$

or, functionally,

$$\tilde{\underline{g}}(\underline{h}^n, \underline{h}^{n-1}, \underline{f}_{n-1}) = \underline{0}, \ \forall n \tag{10d}$$

The Planning Model

We consider a groundwater system located in an agricultural river basin. The planning problem is to determine the optimal groundwater pumping pattern to satisfy the agricultural water demands of the basin. Assuming that the planning horizon consists of m operating periods, the policy variables of the model are the groundwater extraction rates for each well site in the basin. Recognizing that the objectives of the system may reflect economic, hydrologic, and environmental considerations, the objective function of the model may be expressed as

$$\max z = \sum_{n-1}^{m} \alpha_n \sum_{p} \lambda_p f_p(\underline{n}^n, Q^n) \tag{11}$$

where f_p is the pth objective and λp is the weight or preference associated with objective p [Cohon and Marks, 1975]. Q^n is the total groundwater discharge during period n; α_n is the discount factor. The policy variables $\underline{h}^n$ and $\underline{Q}^n$ are constrained to satisfy (1) the water demand in each irrigated area ℓ, or

$$\sum_{i\varepsilon\ell} Q_\ell^n \geq D_\ell^n \qquad \forall n \tag{12}$$

(where D_ℓ^n represents the demand in irrigation system ℓ in period n demand less effective precipitation and surface water availability), (2) the balance constraints,

$$Q^n = \sum_{i\varepsilon\ell} Q_\ell^n \tag{13}$$

(3) the response equations (equations (10d)) and, possibly, lower bounds or head gradient constraints to minimize subsidence or seawater intrusion. These constraints may be written as compactly as

$$h_j^n \gtrless h_j^*, j\iota X, \forall n \tag{14}$$

where X is an index set defining the location of the control points in the basin and h_j^* are the desired bounds on the head.

We also have the well capacity restriction,

$$Q_i^n \leq Q_{i,max} \tag{15}$$

where $Q_{i,max}$ is the maximum pumping rate at well site i. Finally, the nonnegativity restrictions of the decision variables,

$$Q_i^n, h^n \geq 0 \quad, \forall i,n \tag{16}$$

The planning-optimization model has several important attributes. First, the constraint set is a convex set. This was essentially the rational for linearizing the unconfined flow equations. Second, if the objectives are separable concave (or convex if minimizing) functions of the decision variables, then globally optimal solutions will be obtained to the planning problem. Third, for the linearized unconfined flow problem, a series of optimization problems will be solved. The head distribution from one solution is then the basis for updating the response equations in the next solution of the planning model. This convergence and theoretical properties of the algorithm are presented by Rosen [1966] and Meyer [1970]. An application of the procedure to parameter estimation problems is discussed by Willis [1976a].

Model Application

Over the past 2 years, as part of an international cooperative research program, the multiobjective planning model has been applied to the water resources problems of the Yun Lin Basin, Taiwan. The overriding objectives of the research program are to develop (1) planning and operational policies allocating surface and groundwater resources to agricultural water demands within the basin, (2) to determine the trade-offs associated with additional groundwater development and agricultural water demands, and (3) to minimize the potential impacts of saltwater intrusion. We consider here,

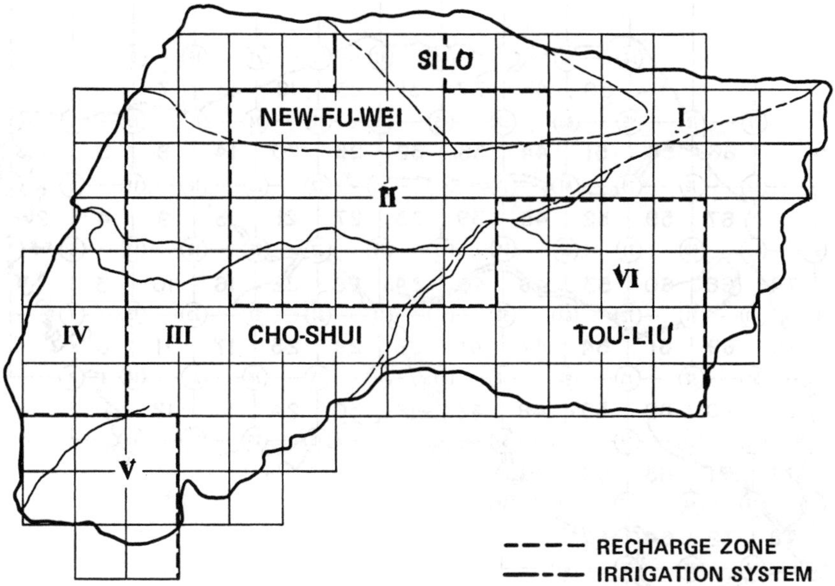

Fig. 1. The Yun Lin groundwater basin.

however, one particular application of the planning model involving the determination of the optimal pumping pattern for two different scenarios regarding groundwater development. In the first, groundwater extractions are determined assuming a well capacity restriction of 15,000 m^3/d (the current maximum). In the second case, this bound is increased to 50,000 m^3/d to reflect the potential for additional groundwater development. Other uses of the model are presented by Willis [1981] and Willis and Liu [1981].

The Yun Lin groundwater system is essentially a semiconfined aquifer. The aquifer, which is located in the Cho Shui alluvial fan, is composed primarily of unconsolidated sand and gravel materials. The aquifer depth ranges from 40 m in the eastern portion of the basin to more than 1000 m in the Peikang area. Approximately 76% of the total groundwater recharge occurs via infiltration of precipitation and seepage from the numerous streams in the basin [Water Resources Planning Commission (WRPC), 1976]. The Cho Shui River, which forms the northern boundary of the study area, is the principal recharge boundary of the system. The Peikang River in the

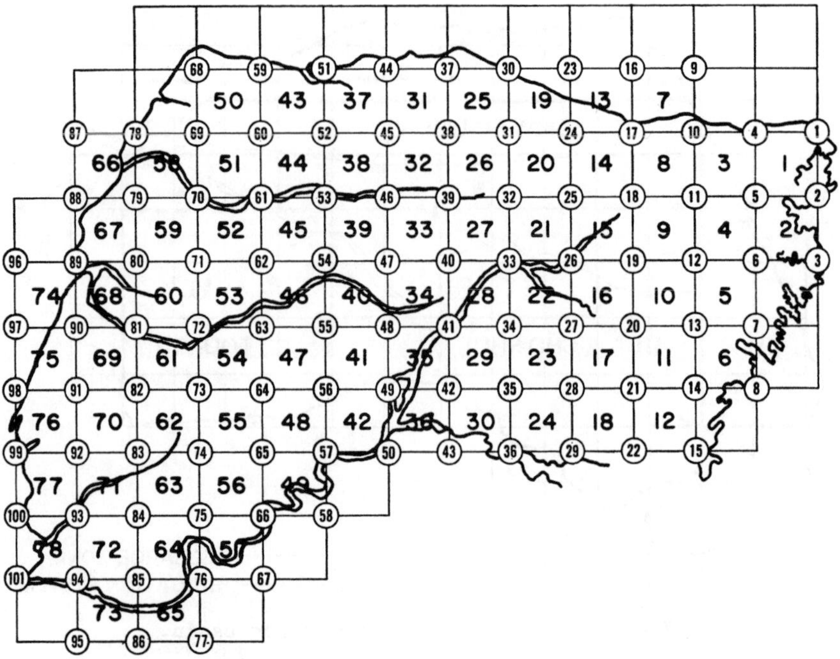

Fig. 2. Finite element grid: Yun Lin groundwater basin.

south, does not however interact with the Yun Lin aquifer system (Figure 1).

Water resources in the basin are distributed via four irrigation systems: the Cho Shui, Fu Wei, Si Lo, and Tou Liu systems. Each irrigation district is administered by the Yun Lin Irrigation Association. The association controls the allocation of surface water, originating from the Cho Shui River, and groundwater from the 500 association wells in the basin. Currently, the total irrigated area in the basin is approximately 43,260 ha.

The hydrology of the basin is characterized by distinct rainy and dry seasons. The rainy period, which extends from May through October, is dominated by typhoon-producing thunderstorms. Seventy six percent of the total rainfall occurs during this period [Water Resources Planning Commission, 1980]. During the dry season, northeast monsoons produce the majority of the precipitation. However, streamflow in the dry season is insufficient to supply the agri-

TABLE 1. Recharge Parameters for the Yun Lin Basin

Recharge Zone	Recharge Rates, x 10^4 m/d					
	November	December	January	February	March	April
1	8.52	8.69	15.42	17.75	12.76	18.50
2	3.47	4.94	10.23	13.83	11.11	17.17
3	2.03	3.07	3.81	5.10	5.80	6.79
4	0.29	0.62	0.59	3.32	4.17	10.32
5	5.48	3.55	3.24	3.23	3.97	3.57
6	1.41	2.57	2.95	1.78	1.74	2.64

cultural water demands of the basin. Typically, groundwater extractions account for more than 90% of the total water usage during the dry season. As a result, it is during the dry season that the groundwater system is most highly stressed. For this reason, the groundwater pumping pattern is determined during this period.

A Galerkin finite element simulation model was developed to predict the hydraulic head distribution in the Yun Lin system and to generate the response equations for the optimization analysis [Tsao et al., 1980]. The system was discretized into 78 (4 by 4 km) linear quadrilateral elements; the system has 101 nodal points. The finite element grid for the basin is detailed in Figure 2.

The validation and calibration of the model is discussed by Willis [1981], and Tsao et al. [1980]. The model's groundwater and hydrologic parameters are summarized in Tables 1, 2, and 3. The

TABLE 2. Mean Dry Season Hydrology

Irrigation System	Mean Precipitation, mm	Mean Surface Water,* m^3/dry season	Water Target,* m^3/dry season
Cho Shui	194.	1.28	9.13
Si Lo	232.	1.34	7.03
Fu Wei	212.	0.16	1.81
Tou Liu	355.	1.20	11.95

*X10^8.

TABLE 3. Hydraulic Parameters of the Yun Lin Basin

Material Zone	Transmissivity, m^2/d	Storage Coefficient
1	57600.	0.0254
2	28800.	0.0254
3	14400.	0.0005
4	7200.	0.0254
5	7200.	0.0005
6	4320.	0.0254
7	4320.	0.0008
8	4320.	0.0005
9	2880.	0.0254
10	2880.	0.0008
11	2880.	0.0005
12	1584.	0.0254
13	2880.	0.0014
14	2880.	0.0008
15	2880.	0.0001
16	1080.	0.0014
17	1080.	0.0008
18	1080.	0.0001
19	720.	0.0008

demand data, which represent a variety of cropping patterns in the Yun Lin Basin, was obtained from the Yun Lin Irrigation Association [Ko, personal communication, 1981].

Model Preliminaries

Initially, the dynamic response equations of the aquifer system are generated using a series of Matrix Eigensystem Routines [1976]. The response equations analyzed the hydraulic response of the aquifer system during the November through April dry season. The response equations incorporated the time-dependent boundary conditions; these conditions were expressed as piecewise linear functions of time over the 180-day planning period.

Two objectives were considered in the analysis: (1) maximize the sum of the hydraulic heads over all the planning period and (2) minimize the total water deficit for all irrigation systems. The first objective is a linear surrogate for minimizing the ground-

TABLE 4. Pumping Rates

Well Site (Node Number)	Nov.-Dec.	Jan.-Feb.	March-April	Constant Pumping
22	15000	15000	15000	15000
50	12270.	15000	15000	12270.
58	1834.	2821.	7630.	1834.
66	5594.	6837.	15000	5594.
84	3383.	2804.	14842.	3383.
92	15000	14739	15000	15000.
95	716.	869.	3333.	716.

In m^3/d; $Q \leq 15,000\ m^3/d$.

water extraction costs; the latter objective reflects the losses from decreased agricultural production. In this preliminary analysis the objective weights were both set to one, indicating equal preference for the objectives. The hydraulic head was also bounded at -20 m to reflect current groundwater conditions.

Agricultural Production

The resulting linear optimization model has 225 constraints and 438 decision variables (not including upper and lower bounds on the head values and pumping rates). The APEX-III large-scale optimization package was used to solve the model [Control Data Corporation, 1980]. Typical solution times averaged 800 CPU seconds; central memory requirements are approximately 200K (octal).

TABLE 5. Pumping Rates

Well Site (Node Number)	Nov.-Dec.	Jan.-Feb.	March-April	Constant Pumping
22	50000.	50000.	50000.	50000.
50	2982.	1986	1830.	2986.
58	0.	0.	2555.	2124.
66	9612.	8411.	7663.	6334.
84	5244.	4761.	11716.	3716.
92	16327.	14268.	18688.	15683.
95	0.	0.	3415.	765.

In m^3/d; $Q \leq 50,000\ m^3/d$.

TABLE 6. Irrigation Deficits

	Constant Pumping	Nov.-Dec.	Jan.-Feb.	March-April
Cho Shui	3761255.	3761255.	3723352.	3648026.
Si Lo	2943389.	2943389.	2943389.	2943389.
Fu Wei	739575.	789575.	789575.	789575.
Tou Liu	5414156.	5414156.	5411427.	5411427.
Total	12908375.	12908375.	12872743.	12792417.

In m^3/d.
$Q \leq 15{,}000\ m^3/d$.

Model Results

The results of the optimization analyses for the two different combinations of pumping upper bounds are summarized, for selected wells and sites in Tables 4 and 5. Several things are apparent from the tables. First, given the opportunity to pump more, the model increased pumpage in those regions which are more highly permeable. As a result, extractions are increased in certain areas, while they are reduced in the less permeable regions of the aquifer. For example, consider node 92. The pumping rate has been increased in the first and third periods with a minimal change in the pumping occurring during the second planning period. This is with the identical lower bound restriction on the head values.

Second, the ability to shut off the pumps to allow recovery of the head levels, also is effective in increasing the yield of the aquifer. This, in conjunction with increased pumping from the more permeable regions of the aquifer, has the effect of increasing the groundwater yield without violating the minimum head restrictions in the basin.

Third, in comparison with a constant dry season pumping pattern, the groundwater yield can be significantly increased. For example, Tables 4 and 5 show the optimal constant pumping schedule [Willis and Liu, 1981]. The corresponding water deficits, for all possible cropping patterns, are represented in Tables 6 and 7. In comparison

TABLE 7. Irrigation Deficits

	Constant Pumping	Nov.-Dec.	Jan.-Feb.	March-April
Cho Shui	3691368.	3699342.	3621014.	3495427.
Si Lo	2663389.	2663389.	2663389.	2663389.
Fu Wei	613735.	607227.	582389.	579575.
Tou Liu	4994501.	4989748.	4994958.	4999813.
Total	11962993.	11959756.	11861750.	11738204.

In m^3/d.
$Q \leq 50{,}000\ m^3/d$.

with the constant pumping pattern, the transient schedule reduces the overall deficit in the second and third operational periods by 36,000 and 116,000 m^3/d. The situation is more dramatic when the pumping upper bound is increased to 50,000 m^3/d. The water deficit is reduced in each operational period. In the first period, the deficit decreases by 3200 m^3/d (Tou Liu and Fe Wei regions). This is balanced by an increase in the deficit in the Cho Shui area. The potential deficit in the second period, however, is reduced by over 100,000 m^3/d. Pumping has increased in the Cho Shui and Fe Wei irrigation districts finally during the third period. The deficit has been decreased by 224,000 m^3/d, again primarily from increased extractions in Cho Shui and Fe Wei. The significant result is the increased yield does not degrade the aquifer below the current groundwater conditions, even with the increased well capacity of the system.

Conclusions

This paper has presented a unified approach to groundwater management using an optimization methodology. The optimal planning models are predicated on the response equations of the aquifer system. These same equations, which normally would be used in a simulation approach, can be incorporated directly within the framework of optimization modeling. In contrast to simulation modeling, the optimization approach identifies the optimal planning or operational policies. In conjunction with multiobjective programming techniques, the system trade-offs and the set of noninferior solutions can also be identified. The methodology has been applied to Yun Lin Basin, Taiwan. Groundwater extraction rates were determined for two scenarios, reflecting alternative groundwater development scenarios. The results demonstrate the utility of optimization modeling in identifying the potential safe yield of regional groundwater systems.

References

Aguado, E., and I. Remson, Ground-water hydraulics in aquifer management, J. Hydraul. Div. Am. Soc. Civ. Eng., 100(HY1), 103-118, 1974.

Bellman, R., Introduction to Matrix Analysis, McGraw-Hill, New York, 1960.

Bellman, R., and R. Kalaba, Quasilinearization, Elsevier, New York, 1975.

Cohon, J., and D. Marks, A review and evaluation of multi-objective programming techniques, Water Resour. Res., 11(2), 208-220, 1975.

Control Data Corporation, APEX III - optimization package, Minneapolis, Minn., 1980.

Cooley, R., Finite element solutions for the equations of groundwater flow, Tech. Rep. Ser. H-W Publ. 18, Desert Res. Inst., Reno, 1974.

Maddock, T., Algebraic technological function from a simulation model, Water Resour. Res., 8(1), 129-134, 1972.

Matrix Eigensystem Routines: EISPAK, Springer-Velag, New York, 1976.

Meyer, R., The validity of a family of optimization methods, J. SIAM Control, 8(1) 41-54, 1970.

Pinder, G. F., and E. Frind, Application of Galerkin's procedure to aquifer analysis, Water Resour. Res., 8(1), 108-120, 1972.

Rosen, J. B., Iteractive solution of non-linear optimal control problems, J. SIAM Control, 4(1), 223-244, 1966.

Tsao, Y.-S., et al., Finite element modeling of the Yun Lin groundwater basin, report prepared for the Provincial Water Conservancy Bur., Tai Chung, Taiwan, 1980.

Water Resources Planning Commission, Hydrologic Features of Taiwan, Republic of China, Taipei, Taiwan, 1980.

Willis, R., Optimal management of the subsurface environment: Parameter identification, technical completion report, Office of Water Resour. and Technol., Cornell Univ. Water Resour. Cent., Ithaca, N.Y., 1976a.

Willis, R., Optimal groundwater quality management: Well injection of waste waters, Water Resour. Res., 12(1), 47-53, 1976b.

Willis, R., A conjunctive surface-groundwater planning model, in Proceedings of ASCE Conference, Water Forum '81, San Francisco, 1981.

Willis, R., and J. A. Dracup, Optimization of the assimilative waste capacity of the unsaturated and saturated zones of an unconfined aquifer system, Eng. Rep. 7394, Univ. of Calif., Los Angeles, 1973.

Willis, R., and P. Liu, Optimization model for groundwater planning, J. Water Resour. Plann. Manage. Div. Am. Soc. Civ. Eng., in press, 1981.

Willis, R., and B. Newman, A management model for groundwater development, J. Water Resour. Plann. Manage. Div. Am. Soc. Civ. Eng., 103(WR1), 159-171, 1977.